MATERIALS AND TECHNOLOGY FOR SPORTSWEAR AND PERFORMANCE APPAREL

MATERIALS AND TECHNOLOGY FOR SPORTSWEAR AND PERFORMANCE APPAREL

Edited by
Steven George Hayes
Praburaj Venkatraman

CRC Press
Taylor & Francis Group
Boca Raton London New York

CRC Press is an imprint of the
Taylor & Francis Group, an **informa** business

CRC Press
Taylor & Francis Group
6000 Broken Sound Parkway NW, Suite 300
Boca Raton, FL 33487-2742

© 2016 by Taylor & Francis Group, LLC
CRC Press is an imprint of Taylor & Francis Group, an Informa business

No claim to original U.S. Government works

Printed on acid-free paper
Version Date: 20150616

International Standard Book Number-13: 978-1-4822-2050-6 (Hardback)

Library of Congress Cataloging-in-Publication Data

Materials and technology for sportswear and performance apparel / editors, Steven
   George Hayes and Praburaj Venkatraman.
      pages cm
   Includes bibliographical references and index.
      ISBN 978-1-4822-2050-6 (acid-free paper) 1. Sport clothes. I. Hayes, Steven G., 1970-
   II. Venkatraman, Praburaj.

TT649.M375 2016
646.4'7--dc23   2015022175

Visit the Taylor & Francis Web site at
http://www.taylorandfrancis.com

and the CRC Press Web site at
http://www.crcpress.com

Contents

Preface...vii
About the Editors..ix
Contributors..xi

1. An Overview of the Sportswear Market 1
   Sam Dhanapala

2. Fibres for Sportswear...23
   Praburaj Venkatraman

3. Fabric Properties and Their Characteristics..53
   Praburaj Venkatraman

4. Fabrics for Performance Clothing 87
   Tasneem Sabir and Jane Wood

5. Composite Fabrics for Functional Clothing...103
   Jane Ledbury and Emma Jenkins

6. Smart Materials for Sportswear 153
   Jane Wood

7. Applications of Compression Sportswear..171
   Praburaj Venkatraman and David Tyler

8. Impact-Resistant Materials and Their Potential.................................205
   Praburaj Venkatraman and David Tyler

9. Seamless Knitting and Its Application..231
   Kathryn Brownbridge

10. Garment Fit and Consumer Perception of Sportswear......................245
    Simeon Gill and Jennifer Prendergast

11. Evaluating the Performance of Fabrics for Sportswear......................261
    Praburaj Venkatraman

12. Application of Pressure Sensors in Monitoring Pressure 289
    David Tyler

13. Body Scanning and Its Influence on Garment Development............ 311
    Simeon Gill

14. Eco-Considerations for Sportswear Design.. 327
    Jennifer Prendergast and Lisa Trencher

Index..343

Preface

Sportswear and other performance apparel arguably fill a specific niche market and at the same time permeate all aspects of the fashion business. It is easy to view it from the purely technical perspective of material specification and performance, but that would fail to appreciate the importance of the aesthetic in all its incarnations. To accommodate this it is usual to speak in terms of performance or leisure sportswear, functional or fashionable, technical or consumer. But again, these distinctions miss the point: The aesthetic appeal of the track athlete's apparel, or that of the firefighter, has a bearing on how they feel about wearing it, their performance and the way they are perceived by those around them. In this book, we bring together aspects of materials, performance, technology, design and marketing that we hope will enable the reader to bridge this gap in appreciation. To do this effectively, the reader should place the contents of this book within the context of a user-centred design and concurrent product development process, with an eye on design for manufacture. Several models exist to describe these methods; choose the one that speaks to you most. Crucially, this approach allows you to be focused on the user needs (be that an individual, a team, a service group or a demographic) whilst avoiding the inherent pitfalls of a sequential, opaque development process which has little consideration for the creation of the garment as well as its conception.

A chronological list of chapters here would be superfluous, but it is worth emphasising the scope of this book and the breadth of knowledge and ideas brought to you. Within the text we move from the marketing scenario to the eco-friendly dimensions of sportswear and performance apparel via consideration of fibres, fabrics and fit with the inclusion of specific information on smart materials, impact-resistant fabrics and pressure sensing. The focus on fit incorporates research and practice into the use of 3-D body scanning and its influence on pattern engineering for apparel product development. The ability to evaluate materials for use in sportswear and performance apparel is key, and as such permeates the whole text and is given specific treatment in the latter section of the book. The chapters serve a purpose individually. But as a whole we feel they consolidate all the important facets of materials and technology for sportswear and performance apparel.

This book represents the efforts of many people from within the Department of Apparel (Manchester Metropolitan University) – past and present – without whom it could not have been realised: a heartfelt thanks to all. However, praise must go to Dr. Praburaj Venkatraman for conceiving and driving the creation of this book. His inspiration and diligence have been pivotal in its completion. Both he and I hope you find the book informative, interesting and of use in whichever aspect of sportswear and performance apparel you are involved.

Dr. Steve Hayes

About the Editors

Steven George Hayes, BSc, PhD, CText FTI, FHEA, is a senior lecturer in fashion technology management in the School of Materials at The University of Manchester, where he is engaged in all aspects of learning and teaching, research and enterprise within the field of fashion technology and management.

Dr. Hayes initially worked as a maintenance and then production engineer in the garment manufacturing industry in the UK and Morocco. He was both a student and a lecturer at UMIST and Manchester Polytechnic/Manchester Metropolitan University. In 1997, he took up the post of senior lecturer (manufacturing technology) within the Department of Clothing Design and Technology at Manchester Metropolitan University and later became a principal lecturer for technology where he was responsible for the management and development of their technology provision.

His original research was concerned with the dynamics of lockstitch formation, the on-line monitoring of stitch formation and the effect of machine adjustments on thread consumption. Since then, his interests have diversified and now include the study of ergonomic clothing comfort (functional clothing) from both a subjective and objective perspective, technology absorption in newly industrialised countries, design for manufacture in the clothing industry and the employment and economic implications of off-shore manufacturing. He is the editor-in-chief of the *Journal of Fashion Marketing and Management* and has co-authored, edited and contributed to several books exploring the fields of fashion technology and management.

Praburaj Venkatraman (Prabu), PhD, is a textile technologist and a chartered member of the Textile Institute (CText ATI), a senior member at AATCC (American Association for Textile Colorists and Chemists) and a professional member of ITAA (International Textile Apparel Association) US, who has a keen interest in the area of technical textiles with specific focus on functional and performance apparel. He did his BTech (textile technology) from KCT (Kumaraguru College of Technology), Coimbatore, India, and MSc in technical textiles (2001) and his PhD in medical textiles (2005) both from the University of Bolton. He is also a fellow of the Higher Education Academy, UK. He worked in the apparel industry (India) for 4 years, which involved technical textiles, and he specialised in nonwoven fabrics ranging from material testing and product development to garment manufacturing.

Dr. Venkatraman has been a senior lecturer in textile technology at the Department of Apparel, Manchester Metropolitan University (MMU), UK, since 2010 and he has led the BSc (Hons) Fashion Materials and Technology Programme since 2012. He has experience in the design and development of protective garments for sportswear (rugby) using protective pads and generating design principles of developing compression garments for sportswear and to enhance well-being. His other areas of interest are in the development of sustainable methods in denim manufacturing and finishing process and development of smart wearable textiles for monitoring health. He maintains research collaboration with academic institutes in Australia, the United States and India. He has two PhD completions and currently supervises two PhD students and an MSc research student. He also serves as external examiner for PhD exams overseas and as a regular reviewer of journal articles and monographs. He supervises undergraduate projects, MPhil students and MSc by research students. He widely disseminates his research work in various journals, workshops and conferences. He teaches textile materials, technical textiles – performance apparel, textile testing, garment quality control and management – and product development to a wide range of undergraduate and postgraduate students at MMU. He also provides technical support to industry in developing innovative textile materials.

Contributors

Kathryn Brownbridge
Manchester Metropolitan University
Manchester, UK

Sam Dhanapala
Manchester Metropolitan University
Manchester, UK

Simeon Gill
University of Manchester
Manchester, UK

Emma Jenkins
Manchester Metropolitan University
Manchester, UK

Jane Ledbury
Manchester Metropolitan University
Manchester, UK

Jennifer Prendergast
Manchester Metropolitan University
Manchester, UK

Tasneem Sabir
Manchester Metropolitan University
Manchester, UK

Lisa Trencher
Manchester Metropolitan University
Manchester, UK

David Tyler
Manchester Metropolitan University
Manchester, UK

Praburaj Venkatraman
Manchester Metropolitan University
Manchester, UK

Jane Wood
Manchester Metropolitan University
Manchester, UK

1

An Overview of the Sportswear Market

Sam Dhanapala

CONTENTS

1.1   Introduction 2
1.2   Definition of the Sportswear Market 2
1.3   Sportswear Clothing Market 3
1.4   Competitive Position 4
1.5   Consumers and Sportswear 6
      1.5.1   Women and Sportswear 8
1.6   Branding Strategies 9
1.7   Product Life Cycle 10
1.8   Key Markets 12
1.9   Channels to Market 12
      1.9.1   Wholesale 12
      1.9.2   In-Store Formats 13
      1.9.3   Flagship Stores 13
      1.9.4   Factory Outlets 13
      1.9.5   Online 13
1.10 Market Drivers and Emerging Trends 14
      1.10.1 Economy-Based Issues 14
           1.10.1.1 Aging Population 14
           1.10.1.2 Growth of China, Russia, India and Brazil 15
           1.10.1.3 Major Sporting Events 15
           1.10.1.4 Sports Participation 16
      1.10.2 Celebrity Endorsement and Sports Sponsorship 16
      1.10.3 Technology 19
      1.10.4 Fashion versus Function 19
      1.10.5 Mass Customisation 20
1.11 Conclusion 20
References 21

1.1 Introduction

The sportswear market is fundamentally changing because of globalisation and the popularity of sportswear as part of mainstream fashion. This has opened up many opportunities for sports and fashion brands alike. This chapter considers some of the core aspects of sportswear from its definition to the market driver, considering aspects such as celebrity endorsement, mass customisation and emerging economies.

1.2 Definition of the Sportswear Market

Sportswear or activewear by definition is about functionality, comfort and safety with the specification developed and designed to deliver a product that fits in with the performance needs of the sportsman and sportswoman. In competitive sports, the implication of the performance of sportswear can often be the difference between winning and second place.

However, sportswear today is far more than this; with the growth of sportswear as a fashion, the term has expanded to include clothing and footwear worn for leisure as well as clothing and footwear worn by people watching sports events. Hence, sportswear needs to deliver fashion and, in the process, the marketplace has seen a convergence between performance, functionality and fashion, providing further complexities and opportunities for companies operating in the market.

Size has delivered competitive advantage to the major players in the market, such as Nike, Adidas and Puma, with the financial capability to innovate through technology and fashion. Top athletes are often the face of these brands, in a symbiotic relationship for mutual benefit. The status of sportsmen and -women as celebrities has grown; this, together with sportswear's inclusion into mainstream fashion, has allowed companies such as Nike to capitalise on the trend, with sales growing from $3.9 billion to $27.8 billion from 1993 to 2014 (Nikeinc.com, 2014).

Inspiring mainstream fashion, today's sportswear design inspires aspects of clothing from work wear to evening wear. An example of this is the recent history of leggings. The introduction of Lycra® made leggings the staple basic for aerobic fans of the 1970s and 1980s. Since then, leggings have become a fashion staple and appear in all different guises, varying in length, material, pattern, colour and texture. This chapter examines the sportswear market, the companies operating in it and the consumers they serve, looking at the drivers and issues in the marketplace.

Sportswear can be defined as performance-driven functional clothing and footwear designed for and worn when playing sports or undertaking recreational pursuits. This is expanded to include sports clothing and footwear design, inspired by performance and function, reflected in clothing worn for fashion and leisure.

In examining the sportswear market, the focus is on both functional and fashion sportswear that includes both clothing and footwear designed, manufactured or distributed by sports brands as well as clothing chains' production of their own label sportswear brands. As the two areas of function and fashion merge, it is increasingly difficult to distinguish one from the other, with performance clothing and footwear being fashion oriented as well.

1.3 Sportswear Clothing Market

Sportswear is a global phenomenon, according to Euromonitor; as reported in *Retail Week*, the sportswear market represented $282 billion worldwide in 2014, a 7% increase on the previous year and outpacing the growth of broader apparel, which stands at 5.8% of the market (Bearne, 2014; Thomasson and Bryan, 2014). At this rate, the sportswear market by the end of 2015 could be worth over $300 billion globally.

Even though there is strong growth in China, India and Russia, by far the largest market is the United States, which accounts for over 35% of sales globally (Kondej, 2013). It is predicted that growth in sports-inspired clothing will be partially evident as function and fashion further converge.

The overall sportswear market can be broken down into sectors by sporting pursuit and leisure wear:

- *Outdoor pursuits* include such things as cycling, hiking, mountaineering, snow sports and sailing, including a wide variety of leisure wear for both function and fashion.
- *Leisure wear excluding outdoor wear* includes items like T-shirts and polo shirts, sold under the sportswear brand or sportswear category.
- *Team sportswear* is items worn for football and rugby by the players as well as the exponential number worn by the many supporters are included in this category.
- In *running* there has been a surge in the running sportswear market due to innovation in running gear. The growth in use amongst women has been particularly noticeable as it is seen as a simple, fast, accessible way to fitness and good health without the need of a partner, special equipment, a gym membership or even much time.
- For *aerobics and indoor fitness*, weight loss is still the key driver that instigated gym membership, together with a proliferation of aerobic type exercise classes. The rapid increase of obesity rates in most

of the Western world has resulted in government intervention and the aerobic industry targeting this sector. Even though this is not a professional sport, the technology behind aerobic clothing, together with its fashion, is an important driver.

- *Swimwear* includes performance swimwear, high-fashion bathing suits and everything else in between. More so than most other sportswear segments, aesthetics and fashion play an important role, with the majority of sales being for women's swimwear.

- In *racket sports*, stars such as Serena and Venus Williams have revolutionised tennis sportswear, where the unconventional has become normal, with a deviation from the classic white suit. Venus has designed her own collection, Eleven, and Serena has a range developed by Nike.

- In *golf*, the clothing market, dominated by Nike Golf, consists of a range developed for the golfer that features performance, comfort and fashion.

1.4 Competitive Position

There appears to have been a polarisation in the major sportswear brands with Nike and Adidas being the largest globally, as seen in Figure 1.1. Nike's market share not only is the largest, but is also growing, with sales for the year ending 31 May 2014 at $27.8 billion representing a 52% increase over the previous 5 years.

The strong market position of Nike and Adidas points toward an oligopolistic market, where a few players control a large proportion of the market. This has come about through high barriers to entry, where the investment required for performance materials and technology and functional yet fashionable design is high. Included in the barriers to entry are the extremely high marketing budgets required for brand building, sponsorship and the requirement to compete on a global level. To compete effectively investment is required in most of these areas, not just one.

However, it is important to recognise that there are a number of relatively smaller players in the top 20 companies – some operating worldwide, others focusing on markets such as North America. VF Corporation, Puma and Asics, although relatively smaller, are large competitors in the market and represent a dominant force. It is interesting to note that Ante and Li Ning, Chinese companies supplying predominantly the Chinese market, are included in the list of top 20 sportswear companies.

Mainstream retailers have also recognised the potential opportunities in sportswear, with H&M having developed a collection with the help of Swedish Olympians for the Summer Olympics and Paralympics in Rio de

Company	Major brands	Sales revenue 2013/2014 in $ billion	Headquarters	Type of company
Nike Inc	Nike Converse Hurley Jordan	$27.80	USA	Public
Adidas AG	Adidas Reebok Rockport	$18.63	Germany	Public
VF Corp	Timberland The North Face Nautica Vans	$6.10	USA	Public
Puma SE	Puma	$3.84	Germany	Public
Asics Corp	Asics	$3.02	Japan	Public
Amer Sports Corp	Amer	$2.74	Finland	Public
Jarden Corp	Jarden	$2.72	USA	Public
New Balance Athletic Shoes Inc	New Balance Warrior PF Flyers	$2.39	USA	Private Private
Under Armour Inc	Under Armour	$2.33	USA	Public
Skechers USA Inc	Skechers	$1.85	USA	Public
Quicksilver Inc	Quicksilver	$1.81	USA	Public
Columbia Sportswear Co	Columbia Sorel Mountain Hardware	$1.69	USA	Public
Mizuno Corp	Mizuno	$1.68	Japan	Public
Hanes Brands Inc	Champion	$1.30	USA	Public
Anta Sports Products Ltd	Anta Fila*	$1.19	Hong Kong	Public
Billabong Int Ltd	Billabong	$1.00		
Li Ning Co Ltd	Li Ning	$0.95	Hong Kong	Public
Fila Korea Ltd	Fila	$0.75	North or South Korea (or both)	Private

* Mainland China, Hong Kong and Macau.

FIGURE 1.1
Top sportswear companies, their brands and sales revenue. (Individually sourced from company annual reports.)

Janeiro 2016. H&M has agreed to create the outfits for both the opening and closing ceremonies, as well as a wardrobe for the sportsmen and -women to wear around the Olympic village. The focus will be more about fashion than functionality. Topshop has also seen the potential and gone into a 50/50 joint venture with Beyoncé to launch Parkwood Topshop Athletic Ltd, formed to produce a global athletic street-wear brand to launch in winter 2015 (Arcadia

Group, 2014). Whilst unlikely to pose an imminently serious challenge, a holistic attack by mainstream retailers through price advantage has the potential of chipping away at the sales of casual and fitness wear of major sportswear brands such as Nike Inc. and Adidas AG.

Nike, Adidas and Puma have invested heavily in technical fabrics that deliver improved sports performance. For example, H&M's fitness tights, although fashionable at half the price, can't compare on performance with the equivalent Adidas running tights made out of the German firm's patented quick-drying Climalite material (Adidas, 2015).

1.5 Consumers and Sportswear

When examining the sportswear consumer, there are a number of different aspects to consider; this includes the different perception of fashion and sportswear across the world. In particular the Asian markets are significant, with some Asian economies the fastest growing in the world. China has become a particularly important market for retailers; with its economy growing at approximately 7% per year, it is projected to become the largest world economy by 2021 (*Economist*, 2014). Also, it is important to consider the drivers and use of sportswear, as well as the profile of the consumers who purchase sportswear.

As people take up sports as part of an active lifestyle, supporting health and well-being, wearing sportswear represents an outward sign that you care about your well-being and are striving to gain a better quality of life. Maslow refers to clothing as one of the very basic elements in his 'Hierarchy of Needs' (Maslow, 1954), as it is a basic need for survival and keeping warm as shown in Figure 1.2. Fashion, on the other hand, is more to do with self-esteem, where fashion represents image and personality. Sportswear to a broad degree fits into self-esteem; however, brands such as Nike, using straplines such as 'Just do it' and 'Game on World' reaches out and inspires the type of consumer looking for self-actualisation.

This model views an individual as autonomous and free to make decisions based in the main on personal desires and wants, perhaps a Western perspective. However, this model does not always hold true when considering Asian cultures and peoples (Figure 1.3) (Schütte and Ciarlante, 1998). Whilst the two lower levels remain the same, it is suggested that the three highest levels emphasise the importance of social needs. Once an Asian individual is accepted by a group, the 'affiliation needs' must be satisfied. The next level, 'admiration needs', are fulfilled through actions in a group that derives respect. At the top of the pyramid is status within society as a whole. Luxury and branded sportswear displays a symbol of wealth and capability that others will notice. This, to an extent, explains the growth of sportswear

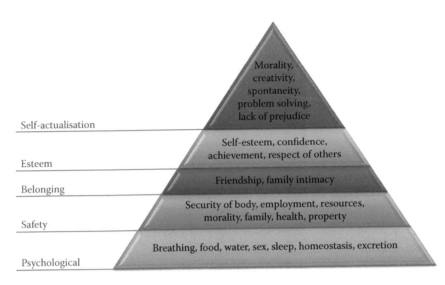

FIGURE 1.2
Maslow's hierarchy of needs. (From Maslow, A. 1954. *Motivation and Personality*. New York: Harper.)

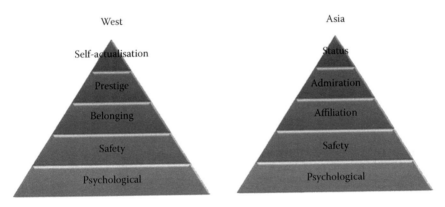

FIGURE 1.3
Maslow's hierarchy of needs and the Asian equivalent. (From Schütte, H. and Ciarlante, D. 1998. *Consumer Behavior in Asia*. Washington Square, NY: New York University Press.)

and luxury goods in Asian countries. Whilst the growth in prosperity of Asian economies correlates with the growth of sportswear and luxury in these economies, there is evidence to suggest that a higher proportion of income is spent on these items than in Western society.

The drivers to purchase sportswear can be grouped into the specific purposes: firstly to actively participate in sports, secondly to wear sportswear for fashion and thirdly to wear it for leisure and comfort. Each item of sportswear in a consumer's wardrobe may have a different function and their functions may not be mutually exclusive, as can be seen in Figure 1.4.

FIGURE 1.4
Function of sportswear.

In order to understand who actually wears sportswear, it is possible to use commercial segmentation systems such as Mosaic from Experian, which is available in most of the major countries in the world (Experian, 2015). Experian has gone one step further in the UK to devise 'fashion segments'. Using Experian's fashion segments, it is possible to identify the profile of the UK sportswear consumer. The profiling system classifies all 44 million UK adults, split by gender into 20 female profiles and 15 male profiles categorised by their attitudes and behaviour when purchasing clothing.

Using fashion segments it is clear that the male consumer has a younger profile but is not necessarily sports active. Interestingly, most males under 35 years shop in sportswear stores and wear sportswear; the direct target market represents about 30% of the UK male population. Women's wear clothing is by far the largest clothing sector. Fashion segments identify 20 female profiles, from which sportswear straddles across all profiles of women under 45 years, worn for casual leisure wear, fashion and sports exercise (Experian, 2009). This is represented by the strategy adopted by both Adidas and Nike, for whom women have become a key opportunity for growth and gaining market share, with not only products but also campaigns and sub-brands created for this market.

1.5.1 Women and Sportswear

The sporting arena has been dominated by men, especially in team sports such as football, where the women's teams are almost unknown. In fact, in Brazil women's football was banned between 1965 and 1982 by the ruling military government. However, prominent sportswomen in individual sports include Serena Williams and the IndyCar racer Danica Patrick. In addition, there is a close affiliation between women and sports apparel, with Jessop (2013) recently reporting women's college sports apparel sales seeing triple-digit growth. The consumer apparel market is gender specific with apparel brands overhauling their women's apparel to create designs that fit women's bodies and appeal to their tastes not only aesthetically but also functionally.

1.6 Branding Strategies

The dominant market position gained by Nike and Adidas has not taken place by accident; it has come about through careful consideration of customer needs and wants. In order to do this they have developed a portfolio of brands, put together in a coordinated and complementary way so that the customer perception of each brand is clear and distinctive. Sub-brands have also been developed to take account of different consumer needs in terms of performance, fashion and function.

However, the challenge is to manage a portfolio of brands in a way that the cannibalisation of one's own customers is minimised. Cannibalisation means that marketing efforts are duplicated as well as confusing to the customer. As the sportswear market has developed the needs and wants of each customer, it has become more specific; sportswear is not just categorised between men and women and footwear and clothing, but can also be segmented by type of sport played, the level of technology and performance as well as differing needs of performance against function and fashion.

Nike's brand architecture reveals a distinctive portfolio. The Nike core brand is broken down into sport and sportswear. *Nike Sport* splits products by type of sport and training undertaken. The proposition appears to be that the functional performance sportswear focus is on both clothing and footwear. The *Nike Sportswear* range, on the other hand, focuses on fashion-driven casual clothing. The Jordan range is a premium collection of sportswear inspired by Michael Jordan – but still very much part of Nike, with the distinctive swish appearing alongside the Jordan logo. Converse, Hurley and Nike Golf also have very distinctive propositions, differentiated from the company's core brand. In 2012 and 2013, Nike disposed of both Umbro and Cole Haan brands. Umbro, even though very much soccer inspired, did show a conflict with the Nike brands, so the disposal was understandable.

In comparison, brand architecture adopted by Adidas has a slightly different orientation. Adidas Sports Performance focuses on the five key areas of football, basketball, running, training and the outdoors with the aim of making athletes 'better by making them faster' (Figure 1.5). Adidas Originals is based on iconic and authentic sportswear, incorporating style and functionality. Sub-brands such as Neo and Y-3 focused on style and fashion rather than function. Collaboration with Stella McCartney to create the premium range 'Adidas by Stella McCartney' fits within the performance range and

Brand architecture portfolio strategy

Brand differentiation

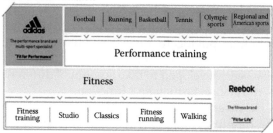

FIGURE 1.5
Adidas brand architecture portfolio strategy. (From Adidas Group. 2012. Annual Report. [online] Available: http://www.adidas-group.com/media/filer_public/2013/07/31/gb_2012_en .pdf [accessed 17 February 2015].)

is seen as a synergy of performance with style, whereas Y-3, the successful collaboration designed by Yohji Yamamoto which has been going on for over 10 years, is more to do with sports-inspired fashion. Reebok fits within the brand portfolio as the brand focusing on fitness. Looking at the overall brand architecture it is clear that the company is reaching out to a broad audience. The broad reach of Adidas and its sub-brands is set up to reach almost anyone and positions Adidas well in its challenge to Nike to be the leading sportswear brand in the world.

1.7 Product Life Cycle

Traditionally, products follow a life cycle. Kotler and Armstrong (2004) identify that through the life cycle a product goes from introduction to growth, maturity and, finally, decline, and the timescale can vary; this can be seen in Figure 1.6. Clothing is no exception, but due to the fast and fickle nature, fashion products have short life spans and hence a shorter time for a business to reap the rewards of a fashion trend. On the other hand, functional

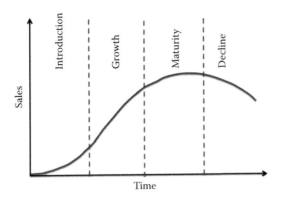

FIGURE 1.6
Product life cycle. (From Kotler, P. and Armstrong, G. 2004. *Principles of Marketing*. Upper Saddle River, NJ: Pearson.)

sportswear appears to have a longer product life cycle; historically, sportswear items are purchased as a replacement or a spare item. However, with the influence of fashion and the different reasons for wearing sportswear, the life cycle is shortening and now follows fashion life cycles more closely. This could provide a rationale for growth in the market as spending per head on sportswear increases.

The highly competitive nature of the market suggests that one brand will innovate a product and then others will imitate and copy. It is suggested that this process results in an increase in the life span of a product (Figure 1.7) (Bass, 1969).

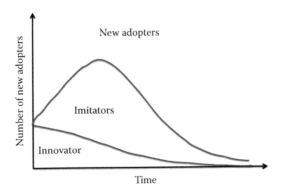

FIGURE 1.7
Innovation and imitation life cycle model. (From Bass, F. M., Trichy, V. K. and Jain, D. C. 1994. *Marketing Science* 13 (2): 203–223.)

1.8 Key Markets

The largest market for sportswear is the United States, which accounts for over a third of global sales; China represents the second largest market, with 10% share of sales and Japan is third with 7% (Kondej, 2013). The economic growth of the BRIC countries (Brazil, Russia, India and China) is well documented, with these countries expected to be profoundly influential for sportswear.

As well as being the largest market, growth in the sportswear market has steadily continued in North America. In 2010, the Centers for Disease Control and Prevention (CDC) in the United States reported 35.7% of American adults and 17% of American children as obese; this had been the culmination of an upward trend since the 1960s (Ogden and Carroll, 2011). This has instigated a drive toward health and well-being in America, which has accelerated the growth of the sportswear market.

The Chinese market provides an enormous opportunity for sportswear brands; Adidas recently commenced a mass media campaign targeted at women, seeing this as a key area for sportswear growth in China. Penetration into this market is seen as a major opportunity by sportswear brands. Domestic brands such as Li Ning and Anta operate mainly in China, yet they are in the top 20 in the world by sales. Japan, on the other hand, spends more per capita on sportswear than most other nations – in fact six times more than China (Kondej, 2013).

1.9 Channels to Market

1.9.1 Wholesale

In order to gain global reach, wholesale is a financially safe and fast option to market. The major sportswear retailers have taken full advantage of this approach, allowing them to cover nearly every country in the world through independent distributors and licensees. Major wholesale customers include Decathlon, Intersport and Footlocker.

Traditionally, the primary channel to market for sportswear brands has been via wholesale. Proportionally, 81% of Nike's revenue was via wholesale in 2013 (including Nike direct to consumer via wholesale), but even though it remains the major avenue, year after year there has been a fall in the proportion of sales via this channel (Nikeinc.com, 2014). On the other hand, wholesale for Adidas accounted for 65% of sales in 2014, even though sales from wholesale has grown year on year as a proportion of overall sales, there is a downward trend over the last 5 years.

1.9.2 In-Store Formats

The growth of online sales and the confidence of sportswear retailers such as Nike and Adidas have resulted in the brand going straight to the consumer. Adidas's retail sales account for only 26% of all sales; clearly, it is a growing channel even though there is still a heavy reliance on wholesale.

1.9.3 Flagship Stores

These stores provide the opportunity to showcase the brand and have been an important strategy in internationalisation, where the brand experience can be gained by visiting the store. Flagship stores in key cities have been a highly successful strategy for luxury brands focused on image. Likewise, sports brands are able to use flagship stores to transcend the image the brand wants to portray. Flagship stores allow control over brand portrayal and have a key role to play in the brand strategy. The flagship is a showcase with the main objective being to draw attention and build awareness and reinforce the brand identity. In addition, it provides an experience to consumers so they can get wrapped up in the true essence of a brand. It is identified by its prestigious location, and by being large and opulent; it provides a shopping experience and, in terms of visual merchandising, is usually an engaging feast for the eye. Adidas is expanding the concept as part of its expansion across the UK. The company has opened a flagship store in Leeds, UK, with the full range of Adidas products; this is a key moment in Adidas's expansion drive with the vision to open stores across all major cities in the UK. There is also a drive to make the stores more women friendly, with plans to launch running clubs and yoga classes.

Similarly, the Nike flagship store 'Niketown', strategically placed in key cities around the world, has the same impact in terms of developing brand image. The location of these stores is critical: Having mainline brands in close proximity in terms of location aligns with the theory of association.

1.9.4 Factory Outlets

Such stores have also become a major strategy for sportswear brands. Similarly to that adapted by luxury retailers, the outlet is usually the main channel for clearance and excess stock and return, and it forms an important growth strategy for both Nike and Adidas. Factory outlet shopping has become a major shopping habit of North American consumers; having vast amount of land to develop these outlets has also been a key driver. During 2011/2012, Adidas had 45% of its direct sales (excluding wholesale) from factory outlets, with 47% coming from concept stores and only 5% from online (Adidas Group, 2013).

1.9.5 Online

Online shopping has become the norm and part of everyday life. Shoppers have the bargaining power where they are able to use the Internet to seek out the best prices and offers without national or international barriers. In addition, the size of a business does not necessarily limit the opportunity to trade online.

Nike's first website was created as far back as 1996; sales from the website form a tiny proportion of the company's sales online (as with Adidas), accounting for only 1.5% of total sales. Putting this into perspective, by looking at retail sales overall in the United States across all sectors, it accounts for nearly 6%, and in the UK online sales account for over 10% (Nikeinc.com, 2014). Even though sportswear brands have not taken full advantage of online opportunities, there is a major opportunity for sportswear brands to develop their online presence and increase sales through the web. The website development seen by all five major sportswear brands has represented a revolution. NikeID and MiAdidas take buying online to a new level, allowing the customer to customise items on the website and have them delivered. Companies such has Burberry are referred to as being omnichannel, seamlessly integrating all their channels; however, sportswear brands have not been able to develop their offering to quite the same level yet, but this will change over the next 5 years.

1.10 Market Drivers and Emerging Trends

1.10.1 Economy-Based Issues

The global recession continues to leave its footprint on the world economy, with much of the Western world still laden by mounting debt. The implications have been far reaching: Still relatively high unemployment in the Western hemisphere together with suppressed consumer confidence and consumer spending has resulted in challenging times for fashion retailing. In the UK, the major victim on the High Street was JJB Sports. However, interestingly, despite the obligatory drop in sales in 2009 after the impact of the global banking crisis, at a local level both in the United States and other key markets, there has been a growth in sportswear sales. The apparel market has been affected by the recession, as have most sectors; however, after the initial dip in 2008 and 2009, there has been steady growth in both apparel and sportswear (Kondej, 2013). Hence, whilst the market may not be immune from recessionary conditions, the brands and retailers that have ridden the wave show its strength and endurance.

1.10.1.1 Aging Population

Overall, the world's population is still increasing and within this, the aging population (people aged 60 years or over) is set to increase from 11.7% of the world's population in 2013 to 21.1% by 2050 (United Nations Department of Economic and Social Affairs, 2013). Traditionally, there has been a negative correlation between age and propensity to purchase clothing. In addition, sportswear is purchased mainly as a replacement rather than an addition to the wardrobe by the older market. The impact could be limited growth globally and, importantly, in developed growth nations.

1.10.1.2 Growth of China, Russia, India and Brazil

Growth will be significant in the developing nations. It is widely expected that, by 2020, China will overtake the United States as the world's largest economy. The world population is over seven billion, of which nearly 20% are in China (Kondej, 2013). This, together with the growing economic wealth of the country, will create a wealthier Chinese population and open up new opportunities for sportswear brands. Brazil, India and Russia represent important growth markets. Nike's sponsorship of the Indian cricket team in the Bleed Blue campaign represents the passion felt by the team and the hearts and minds of the millions of its fans. The passion for cricket in India, a country with a population of over a billion people, is at all levels and hence such advertising appeals to a mass audience. This provides Nike the opportunity to develop mass appeal for its products (Figure 1.1).

1.10.1.3 Major Sporting Events

Consumers in the UK are 12% more likely to buy sportswear when a big event is taking place (Clifford, 2011). The 2012 London Olympic Games was a significant market driver and contributed to the growth of the sportswear market globally, especially as London 2012 positioned fashion at the epicentre of the games with a shift from costume to fashion, with functionality and practicality replaced by performance and design. The British partnership with Stella McCartney positioned the UK as a market leader of fashion-focused sportswear. The US team sported a Ralph Lauren look, and Giorgio Armani was responsible for the Italian uniform (Figure 1.8) (Ukman, 2012).

Sales of branded products increased 250% from the Beijing Games (Adidas Group, 2012); in addition there was a 14% increase of demand for sports style products as a result of the Olympics, indicating that the market still offers opportunities for growth (Adidas Group, 2012). According to Simon Fowler, managing director, John Lewis, Oxford Street:

> London 2012 gave John Lewis the opportunity to tell our unique British story for the first time to a global audience whilst being part of the greatest show on Earth. While customer interest in 2012 was about official merchandise, we are now seeing a great uplift in sportswear sales triggered by the 'Olympic effect'. (Somerville, 2013)

Brand	Olympic Team Kit Sponsorship
Adidas	Olympic Teams of Australia, France, Germany, Great Britain and South Africa
Anta Sports	China Olympic Team
Asics	Olympic Teams of Ireland, Japan and the Netherlands
Cedella Marley for Puma	Jamaica Olympic Team
Dirk Bikembergs 4 Adidas SLVR	Ceremonial Outfits, France
Fila	Hong Kong Olympic Team; Hats, South Korea Olympic Team
Giorgio Armani	Italian Olympic Team
Hudson's Bay	Ceremonial Outfits, Canadian Olympic Team
Hermès	French Olympic Equestrian Team
Li Ning	Ceremonial Outfits, Sweden, Chinese Olympic Team; Basketball Teams: Argentina and Spain; Badminton Teams: Australia, New Zealand and Singapore, Track and Field: Eritrea and Zimbabwe
Next	Ceremonial Outfits, Great Britain
Nike	Brazil, Canada, China, Estonia, Germany, Russia, Qatar, Ukraine, US Olympic Teams; US Men's Basketball Team; US and Russia Track and Field Teams
Prada	Italian National Sailing Team
Ralph Lauren	Team USA
Stella McCartney for Adidas	Team Great Britain

FIGURE 1.8
Olympics, London 2012 sponsorship of team kits. (Adapted from Ukman, L. 2012. The Fashion Games. [online] Available: http://www.sponsorship.com/About-IEG/Sponsorship-Blogs/Lesa-Ukman/August-2012/The-Fashion-Games.aspx [accessed 17 February 2015].)

1.10.1.4 Sports Participation

Sports participation influences sportswear sales. Statistics from the US Sports and Fitness Industry Association (2013) show that running, followed by basketball and then swimming, have the largest level of participation. In terms of growth, running and jogging, shows a growth of 4.7% over the 5 years ending 2012 (Sports and Fitness Industry Association, 2013). The picture in the UK is different: The Olympic effect results in a growth in sports participation. According to Sport England (2014), in 2013/2014, 15.6 million adults played a sport at least once a week; this is 1.7 million more than in 2005/2006.

1.10.2 Celebrity Endorsement and Sports Sponsorship

Celebrity endorsement is a form of marketing communications where the celebrity is the brand ambassador and represents the brand and its personality in his or her actions, words and images. Celebrity endorsement is a way of harnessing the specialism, popularity and personality of the celebrity into the brand. Today celebrity endorsement represents far more than just being a spokesperson for the brand; the celebrity and the brand have to be intrinsically connected. For global brands such as Nike, Adidas and Puma, this has transcended nationalities and cultures. At a national level sports people's and athletes' link to a brand can be very powerful for a brand, providing the ability to enter a market and gain dominance.

Celebrity endorsement has become synonymous with major sports brands, used as the main brand-building tool. In theory, sportswear is in synergy with sports; in the same way there is a strong association with sportswear and the athlete. By their very nature driven to perform and ultimately to win, athletes have become celebrities in their own right. Icons in sports such as David Beckham and Michael Jordan have resilience and longevity – so much so that Beckham and Jordan have become brands in themselves. The association of sports brands with such icons of sport can transcend an image of a brand that is difficult to capture in any other way.

For a brand to successfully use the sponsorship of a sporting celebrity, the celebrity first must have the capacity to win and succeed in the chosen sport in order to have a positive impact on the brand, but this is not enough in today's world. The celebrity must be attractive or have enduring qualities, have intellectual capability as well as a lifestyle that reflects and represents the brand and should also be credible with the target audience. This in itself should suffice; however, for the celebrity to truly represent the brand there should be a meaningful transfer between the brand and the celebrity, thereby representing compatibility so that the brand is eventually synonymous with the sports person – almost a part of his or her DNA in terms of identity, personality and positioning. It is also important that the sports person connects with the target audience in order for the celebrity endorsement to work effectively.

In the three-way relationship as shown in Figure 1.9, the celebrity endorsing the product should have credibility, have expertise in his or her field, be trustworthy, attractive and successful. The celebrity should also be familiar

FIGURE 1.9
Three-way relationship between brand, personality and consumer.

to the audience as well as have an affinity to the brand. The aim is to create the appropriate consumer attitude toward the celebrity and hence the brand.

To find the perfect match between the celebrity and the brand is not simple; however, it is possible, even when there is not a perfect match, to focus on the elements that do work. Wayne Rooney's sponsorship with Nike has been challenging at times; however, Rooney has emerged as a highly successful brand ambassador, with passion for his sport, as reflected in the 2006 World Cup advertising campaign. The controversial advertisement showed Rooney drenched in warrior style red paint. It was a powerful statement representing the raw desire to succeed, which had a strong resonance with Rooney's personality.

The budget for a celebrity endorsement runs into millions. Usain Bolt's sponsorship with Puma is worth $8.6 million yearly (Weir, 2013). Puma not only sponsored Bolt but also enlisted Cedella Marley to design the 2012 Olympic kit for the Jamaican team.

Once they are on top of their game sports stars are capable of fusing the world of fashion and sport. Sometimes the relationship goes further and they engage with fashion design as well. Serena Williams and David Beckham are such examples.

More than occasionally, the sports personality does not live up to expectations and, even worse, negative issues in his or her life can influence the appeal of the brand. Lance Armstrong, Oscar Pistorius and Tiger Woods have all had a negative influence and damaged Nike's brand; in such cases the brand dissolves the sponsorship and distances itself from the sporting celebrity.

Yet sponsorship remains an important tool for sports brands. Sponsorship without words implicitly links the brand with heroes of the hour as they are seen wearing the logo. The International Events Group (IEG) $14.35 billion was spent in 2014 on sporting sponsorship in North America; accounting for 70% of the total spent overall for sponsorship in the region. On a global level IEG estimates the sponsorship industry is worth over $55 billion in 2014 (Sponsorship.org, 2015). Clearly, sports sponsorship is important because it is seen ultimately as a driver of sales for the respective brands. However, there appears to be a swing in the type of sponsorship offered by sports brands. Rather than sponsoring individual athletes, there appears to be a shift toward sponsorship of teams and events.

Nike's estimated $2.7 billion annual marketing budget is being spent (Nikeinc.com, 2013) on supplying the NFL league with uniforms for all its 32 teams, and it is the official soccer ball sponsor for the English Premier League, the FA Cup and the national football teams of Brazil, England, the United States, the Netherlands, France and Portugal, to name but a few. In addition, it sponsors major soccer teams such as Manchester United, Paris St. Germain, Porto and Inter Milan. The company's 'Find Your Greatness' campaign during the London Games focused on everyday athletes. There is a growing belief that sponsorship of a team rather than an individual will bring about the required revenue. Nike is aggressively pursuing some of Adidas's team sponsorships as well, recently announcing that, beginning in 2017, it will be the sponsor for the national UK Athletics team. The deal makes it a bit less conflicting for British athletes such as Mo Farah, who wore Nike shoes but was clothed in Adidas-branded team gear in his winning Olympic 2012 races.

1.10.3 Technology

There has been an intrinsic change in sportswear with a parallel evolution in the technology used. Sportswear is getting lighter, more breathable and improving in wicking, flexibility, fit and strength, to name but a few technological developments. Added to this there is a continuous drive toward performance-driven design. This design is further enhanced to take account of style and fashion – so much so that the crossover between sportswear and fashion is increasingly seen in mainstream fashion. Rather than evolutionary, this could be a revolutionary change in the way sportswear is worn and used and what it represents.

1.10.4 Fashion versus Function

As the purpose of sportswear evolves and diffuses, fashion and sport have an even closer association. Fashion influences are seen all the way from performance sportswear to casual sportswear; similarly, the look and feel of sportswear is present in many collections from couture to ready-to-wear to mass market.

It can also be argued that sports clothing and footwear are now predominantly purchased with limited consideration of the products' intended active purpose and instead have been popularised for everyday use, further blurring distinction between fashion and sportswear. The drive for this collaboration of fashion and function had been customer and brand driven. From a brand perspective, it presents an opportunity to expand product usage and frequency of purchase as well as appealing to new market segments – hence, the opportunity for growth. The consumer, on the other hand, has seen a relationship between sportswear and the image it represents when worn. It represents modernity, health and well-being, all linked to self-esteem and fashion.

Bringing fashion into performance sportswear is about making clothing worn for sports fashionable. The reasons are many: for example, to increase the frequency of purchase so that sportswear is bought not just as a replacement but also as an addition to the wardrobe. It also inspires individuals to take up sport who might not have done so otherwise – thus growing the market and hence sales and the profile of the brand in terms of standing out from the competition. Whether seen being worn in the gym or running or in taking part in a competitive sport, this clothing more and more needs to represent the individual's fashion sense. Celebrity culture and the need to look good no matter what the occasion is have, to a degree, fuelled this drive.

There is also a change in lifestyle where someone may need to go straight from work to the gym. Yoga pants can be worn with boots to work and then to the gym with trainers; it is this dual functionality that has come into sportswear and hence a drive to be fashionable. On the other end, ideas from sportswear performance design and technology are inspiring fashion. Yohji Yamamoto's decade-old collaboration with Adidas on the Y-3 collection is a marriage of the avant-garde design of Yohji Yamamoto and innovation and technology behind Adidas.

1.10.5 Mass Customisation

Both Nike and Adidas have been offering mass customisation for over a decade. However, the real opportunity for growth from mass customisation has come through advances in functionality of the websites, allowing consumers to customise products online and see graphically what they are producing. Traditionally, Nike and Adidas have had a vertically integrated footwear strategy. Therefore, the structure was in place to facilitate this. However, the ability for consumers to customise their own footwear on the website makes it convenient and flexible. In addition, customers can play around with the item to get the look they want. This closer contact between the end consumer and the brand provides an engagement with the brand and the end consumer, which is an important direction for both Nike and Adidas as they strategically build their direct business in a drive to get closer to the consumer – an opportunity that is difficult to create through wholesale channels. From the customer perspective the issue is that there is no return unless the item itself is faulty and the lead times on average are about 3 to 4 weeks. In addition, there is no sales advisor to guide and advise them. Nevertheless, this has been a strategic move by the larger players to engage with customers directly and build their direct business.

1.11 Conclusion

It is clear that the potential of sportswear marketing is just being realised and that, as competitors scrabble for market dominance, innovative ideas and practices have resulted in sportswear that is both fashionable and functional. This chapter has considered some of the current issues and future directions that are emerging.

Responding to increased competition from High Street retailers, sports brands are focusing far greater emphasis on direct consumer sales through innovative in-store approaches and online developments that deliver an enhanced customer experience. In addition, sports brands are slick in their capability to forge strong symbiotic links with key individuals through celebrity sponsorships and designer collaborations.

Whilst the size and dominance of Adidas and Nike have shadowed many of the other players in the market, globalisation and the growth of the Chinese markets have seen the emergence of Asian brands such as Ante and Li Ning. Is there an opportunity for these brands to capture a larger share of the rapidly growing Asian market or even to enter Western markets with greater impetus?

References

Adidas. (2015). Textiiliteknologiat. Available: http://www.aditeam.fi/textiiliteknologiat (last accessed 10 February 2015).

Adidas Group. (2012). Q3 Nine months report January–September 2012. Available: http://www.adidas-group.com/media/filer_public/2013/07/31/q3_2012_en.pdf (last accessed 18 February 2015).

Adidas Group. (2013). Annual Report 2013. Available: http://www.adidas-group.com/media/filer_public/2014/03/05/adidas-group_gb_2013_en.pdf (last accessed 23 February 2015).

Arcadia Group. (2014). Beyoncé and Topshop create joint venture company. New line in stores fall 2015. Available: https://www.arcadiagroup.co.uk/press-relations/press-releases-1/beyonce-and-topshop-create-joint-venture-company (last accessed 16 February 2015).

Bass, F. (1969). A new product growth for model consumer durables. *Management Science* 15 (5): 215–222.

Bass, F. M., Trichy, V. K. and Jain, D. C. 1994. Why the Bass model fits without decision variables. *Marketing Science* 13 (2): 203–223.

Bearne, A. (2014). The battle of the sportswear market. Available: http://www.retail-week.com/blog-the-battle-of-the-sportswear-market/5057892.article (last accessed 2 January 2015).

Clifford, E. (2011). Sports Clothing and Footwear - UK - August 2011. Mintel Group Ltd.

Economist. (2014). Catching the eagle. Available: http://www.economist.com/blogs/graphicdetail/2014/08/chinese-and-american-gdp-forecasts (last accessed 23 February 2015).

Experian. (2009). Fashion segments. Available: http://www.experian.co.uk/business-strategies/fashion-segments.html (last accessed 18 February 2015).

Experian. (2015). Experian mosaic. Available: http://www.experian.co.uk/marketing-services/products/mosaic-uk.html (last accessed 23 February 2015).

Jessop, A. (2013). Women's college sports apparel sales see triple-digit growth. Available: http://www.forbes.com/sites/aliciajessop/2013/09/07/womens-college-sports-apparel-sales-see-triple-digit-growth/ (last accessed 23 February 2015).

Kondej, M. (2013). The sportswear revolution: Global market trends and future growth outlook. Available: http://go.euromonitor.com/rs/euromonitorinternational/images/The-Sportswear-Revolution-Global-Market-Trends-Euromonitor.pdf (last accessed 2 January 2015).

Kotler, P. and Armstrong, G. (2004). *Principles of marketing*, 10th ed. Upper Saddle River, NJ: Pearson Education.

Maslow, A. (1954). *Motivation and personality*. New York: Harper.

Nikeinc.com. (2014). 2014 annual report. Available: http://investors.nike.com /investors/news-events-and-reports/default.aspx?toggle=reports (last accessed 20 December 2014).

Ogden, C. and Carroll, M. (2011). Prevalence of overweight, obesity, and extreme obesity among adults: United States, trends 1960–1962 through 2007–2008. Available: http://www.cdc.gov/nchs/data/hestat/obesity_adult_07_08/obesity _adult_07_08.htm (last accessed 18 February 2015).

Schütte, H. and Ciarlante, D. (1998). *Consumer behavior in Asia*. Washington Square, NY: New York University Press.

Somerville, M. (2013). Brits get fit as sportswear sales rise. Available: http://www .retailgazette.co.uk/articles/02042-brits-embrace-the-fitness-as-sportswear -sales-rise (last accessed 23 February 2015).

Sponsorship.org. (2015). Sponsorship Spending Report. Available: http://www .sponsorship.com/IEG/files/4e/4e525456-b2b1-4049-bd51-03d9c35ac507.pdf (last accessed 7 September 2015).

Sport England. (2014). National picture. Available: https://www.sportengland.org /research/who-plays-sport/national-picture/ (last accessed 23 February 2015).

Sports and Fitness Industry Association. (2013). 2012 Sports, fitness and leisure activities topline participation report. Available: https://www.sfia.org/reports/all/ (last accessed 18 February 2015).

Thomasson, E. and Bryan, V. (2014). H&M lines up to compete in booming sportswear market. Available: http://www.reuters.com/article/2014/01/09/olympics -hm-idUSL6N0JY3CI20140109 (last accessed 2 January 2015).

Ukman, L. (2012). The fashion games. Available: http://www.sponsorship.com /About-IEG/Sponsorship-Blogs/Lesa-Ukman/August-2012/The-Fashion -Games.aspx (last accessed 20 February 2015).

United Nations Department of Economic and Social Affairs. (2013). World population ageing 2013. Available: http://www.un.org/en/development/desa/population /publications/pdf/ageing/WorldPopulationAgeing2013.pdf (last accessed 18 February 2015).

Weir, K. (2013). Bolt signs $10 million deal to stay with Puma. Available: http:// uk.reuters.com/article/2013/09/24/uk-athletics-bolt-puma-idUKBRE98N0AH 20130924 (last accessed 20 February 2015).

2

Fibres for Sportswear

Praburaj Venkatraman

CONTENTS

2.1 Introduction 24
2.2 Fibre Properties and Modifications 26
2.3 Terms Used for Assessing Fibre Performance 26
2.4 Physiological Parameters 29
    2.4.1 Mechanism of Body Perspiration and Temperature
        Regulation 30
    2.4.2 Stretch and Recovery 31
2.5 Fibres for Sportswear 33
    2.5.1 Microfibres 34
    2.5.2 Hollow Fibres 34
    2.5.3 Bicomponent Fibres 35
    2.5.4 Thermoregulation Fibres 36
        2.5.4.1 Cellulose Blends 40
    2.5.5 New Developments in the Fibre Industry 40
2.6 Market Trend and Overview 41
    2.6.1 Market Drivers 41
2.7 Widely Used Fibre Types for Sportswear 41
2.8 Fibre Types and Blends Affecting the Performance of Garments 42
2.9 Moisture Management 43
    2.9.1 Moisture Management Fibres 45
    2.9.2 Wicking in Activewear Products 46
    2.9.3 Maintaining Body Temperature 47
2.10 Discussion and Summary 47
References 49
Useful Resources 51

2.1 Introduction

Fibres have a wavy undulating structure and contribute many characteristics to the fabrics that are significant for the performance of functional clothing and sportswear. Sportswear depicts a multitude of attributes; for instance, it provides functional support, enhances performance, protects athlete from strain/injury, promotes sporting activity, communicates fashion and style and, more importantly, offers the wearer comfort. The most vital factor that fibres/filaments contribute toward wearer comfort is moisture and thermal balance leading to a suitable microclimate next to the skin. O'Mahony and Braddock (2002) highlighted that in the UK, 'sportswear' often refers to active, performance clothing designed and manufactured for sports-related activities. However, in the United States it includes casual leisure wear. In this chapter, sportswear refers to those garments intended for professional sports.

Fibres influence the overall comfort of the wearer, mainly in providing a balance between heat loss and body perspiration. In recent years, there has been a tremendous increase in the development of new fibres to cater to the fast growing sportswear and functional clothing market. The demand for performance sportswear drives innovation in fibres and fabrics (Rigby, 1998).

Nylon was the first synthetic fibre (made in 1935 by E. I. DuPont de Nemours, Wilmington, Delaware), and since then a number of fibres have been produced. Nylon was originally used in toothbrush bristles, and the first recognised textile product made of nylon was nylon stockings replacing silk stockings. Commercial nylon stockings appeared in the market in 1938 (Humphries, 2009). Polyester had been used in sportswear since the 1970s because of its dynamic properties resulting in fabrics that are resilient, dimensionally stable, easy care, durable and sunlight and abrasion resistant. These properties make it ideal for an array of sportswear applications (Kadolph, 2007). Synthetic fibres have been widely preferred for activewear due to the multitude of performance enhancements they offer compared to natural fibres (Kirkwood, 2013). Natural fibres are often blended with synthetic fibres to achieve an optimised performance. For example, cotton is used in apparel because it absorbs perspiration, but it saturates quickly, causing discomfort due to fabric cling. Hence, cotton and polyester are often blended to gain comfort without cling.

Today, highly functional fibres (Hongu and Phillips, 1997), microfibres (Purane and Panigrahi, 2007), nanofibres (Brown and Stevens, 2007) and smart fibres (Tao, 2001) have been used in functional clothing. Many claims have been reported, particularly moisture management, thermoregulation and performance-monitoring attributes. Fibres used in sportswear and functional clothing are multidimensional and require a number of characteristics apart from possessing a length-to-width ratio for making a yarn. In the fibre industry, the parameters that influence performance are fibre fineness, fibre shape, molecular structure and adding finishes (Hongu and Phillips, 1997;

Kadolph, 2014). Typical properties of fibres for sportswear include durability, absorbance, high moisture regain, lightweight, extensible, colourfast, dimensionally stable and washable. For instance, phase change materials are quite popular among athletes for thermal regulation, as discussed in this chapter. Hence, textile fibres contribute toward moisture control, thermal regulation and breathability, cooling effects, softness, stretch and UV protection. Synthetic fibres are found in countless applications in apparel and functional clothing due to mechanical and chemical properties (Ravandi and Valizadeh, 2011) compared to natural fibres.

There is evidence that innovation in sportswear and performance products is limited in volume and that use of leisure wear and sports-related fashion clothing is on the rise, which drives the mass market and increases the consumption of fibres. This could be arguable in the sense of the mass market reaching a broad population and whether mass customisation is either a follower or driver in sportswear. Rigby (1998) described a volume versus performance pyramid, which indicated how market volumes change as fabric/garment performance increases and hence the price of garments. However, the volume of garments under production remains higher for low-performance sports-related street-wear clothing, whilst volume decreases for high-performance sportswear but the price increases. Innovation is often linked with product branding and is not related to volume of production. Lenzing, a leading cellulose fibre manufacturer, reported that during 2012, 84 million tonnes of fibre were produced. According to CIRFS (the European Man-Made Fibres Association), a review of the world production of cotton, wool and man-made fibres (1991–2012) showed that the production of cotton and wool decreased. Cotton production decreased from 46% in 1992 to 31% in 2012, whilst wool production decreased from 5% in 1992 to 1% in 2012. However, there were positive trends for production of man-made fibres, which increased from 49% in 1992 (CIRFS, 2014). This indicates that the consumption of man-made fibres is increasing globally.

In this chapter, widely used fibre types are highlighted and their characteristics are critically appraised. Various technical terms used to ascertain the performance of fibres are also explained. During intense activity, human bodies generate heat and sweat, and it becomes essential to understand these physiological changes in the context of sports activity. Sportswear requirements differ from those of fashion apparel and fibres are often blended to utilise the combined effect of two or three fibres (a typical example would be wool, polyester and elastomeric fibres). The effect of blending fibre types on the performance of fabrics is discussed. New developments in the area of fibres, such as Trinomax AQ®, Outlast®, Nilit® Breeze, Trevira® and TENCEL®, are reviewed. The market trend for new and smart fibres is ever increasing and is highlighted in the context of sportswear and functional clothing. The fibres used in the area of moisture management, wicking and thermal regulation are discussed and evidence from a wide range of resources is also presented in the context of those garments worn next to the skin, such as base layer garments, compression vests and trousers, thermal underwear and stretch tights. The information provided here should be regarded as essential for effective design and development of performance clothing, particularly sportswear.

2.2 Fibre Properties and Modifications

Fibres possess a wavy undulating physical nature called *crimp* that can affect the **warmth** and **resiliency** of fabrics. Gupta and Kothari (1997) describe a textile fibre as a long, thin material which has a high degree of fineness and outstanding flexibility. In addition, they add that fibres should have good dimensional and thermal stability, acceptable strength and extensibility. Fibres are broadly classified into natural, regenerated and man-made fibres (see Denton and Daniels, 2002). Wool fibres possess a natural three-dimensional crimp. In the case of synthetic fibres, a *texturing process* can impart crimp to manufactured fibres or yarns. Fabrics made from crimped yarn or fibres tend to be more resilient and have increased bulk, cohesiveness and warmth. **Cohesiveness** is the ability of fibres to cling together and is important in making yarns. **Fibre resiliency** is the capability of the material to spring back to shape after being creased, twisted and distorted. **Thick fibres** possess greater resiliency because there is more mass to absorb the strain; in addition, the fibre shape affects fibre resiliency.

US-based Invista promoted its Coolmax® fabric which is a popular brand among athletes for its ability to wick moisture from the skin, to absorb and spread moisture to enhance drying rates and to keep the wearer cool and dry. It is made of lightweight hydrophilic polyester, which is **channelled**. The perspiration wicks away from the skin due to capillary action of channelled fibres (tetra and hexa channels).

2.3 Terms Used for Assessing Fibre Performance

In order to understand the characteristics and performance of fibres, it is essential to be familiar with the following technical terms (Collier, Bide and Tortora, 2009; Denton and Daniels, 2002):

- **Absorbency:** ability to take in moisture (expressed as the moisture regain), which is the amount of water a dry fibre absorbs from the air under standard conditions of 21°C and 65% relative humidity
  - Table 2.1 presents some widely used fibres and their moisture regain

TABLE 2.1

Moisture Regain of Fibre

Fibre Type	Moisture Regain (%)
Acetate	6.5
Acrylic	1.5
Cotton	7.0
Nylon	4.5
Polyester	0.4
Rayon	11.0
Rubber	0
Spandex	1.3
Wool	13.6

Source: ASTM D 1909-04.

- **Absorption:** a process whereby the liquid is fully absorbed by the fibre through its structure that is quite common among hydrophilic fibres (Figure 2.1)
- **Adsorption:** a process where liquid is taken up between the fibres on their surface rather than being held within the fibre
- **Wicking:** the ability of fabric to transfer moisture adsorbed on fibres from one section to another. Wicking can also occur if liquid is absorbed within the fibre. A smooth surface reduces wicking action. Cotton (hydrophilic) and olefin (hydrophobic) fibres both possess a good wicking action. Fine fibres with channels are capable of transporting liquids (e.g. four-channelled INVISTA Coolmax fabric)
- **Moisture regain:** the ratio of the mass of moisture in a material to the oven dry mass, usually expressed in percentage. A dry fibre placed in a humid atmosphere will absorb moisture
- **Hydrophilic:** a fibre that has high regain (that absorbs water); also referred to as 'water loving'

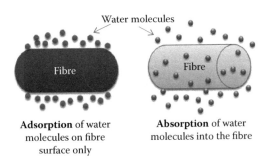

Adsorption of water
molecules on fibre
surface only

Absorption of water
molecules into the fibre

FIGURE 2.1
Mechanism of water adsorption and absorption.

- **Hydrophobic:** a fibre that does not absorb water; referred to as 'water hating'. Regarding the dimensional stability in water, hydrophobic fibres shrink less when washed than hydrophilic fibres. With regard to stain removal, it is easier to remove stains from hydrophilic fibres because water and detergent are absorbed into the fibre

Fibre absorption affects the overall comfort of the wearer, and some of the relevant terms are discussed below:

- **Skin discomfort:** little absorption or wicking and movement of perspiration, resulting in a fabric clammy feeling
- **Static buildup:** fabric clinging to the body or sparks occurring with fabric made of hydrophobic fibres because of no moisture content to help dissipate the built-up charge on fibre surface (dirt also clings onto the fibre surface)
- **Water repellence:** fibre that resists penetration of water or wetting but that is not waterproof. Nonabsorbent fibres (polyester, acrylic) help repel water. Water-repellent finishes (use of fluorochemicals) on fabrics are popular
- **Wrinkle recovery:** better wrinkle recovery by hydrophobic fibres when laundered as they do not swell or absorb water
- **Hygroscopic:** fibres that absorb moisture without feeling wet
- **Thermophysiological:** related to optimum balance between the heat loss and heat generated by the body. This depends on the body metabolism, thermal regulation mechanism and individual health condition. One of the important properties of fibres used for functional and activewear clothing is the elasticity and flexibility
- **Fibre elasticity:** the ability to increase in length under tension (elongation) and return to original length when tension is released. Stretch and recovery of fibre/fabric makes a comfortable garment and causes less seam stress. Complete recovery prevents bagginess in fabric (knees, elbows), which can be noticed during a stretch and when wearing close-fitting garments. Fibres that elongate 100% are *elastomeric fibres* and typical examples are spandex, elasterell-p, lastol and rubber
- **Fibre flexibility:** the ability of fibre to bend easily and repeatedly without breaking. Acetate fibre is a flexible fibre used in garments; glass fibre is a stiff fibre widely used in technical textiles

Hence, in order to understand the performance of fibres, their characteristics and properties – particularly in the design and development of functional clothing and sportswear – it is necessary to be familiar with various technical terms used in the industry as well as to utilise them during communication.

2.4 Physiological Parameters

In this section, three important factors that are vital for the designing of sportswear are discussed – namely, **body sweat patterns, thermal heating patterns** and **stretch and recovery** requirements of an athlete. Smith and Havenith (2012) reported the body mapping of sweating patterns in athletes: the regional sweat rates were compared between 13 aerobically trained females and 9 aerobically trained males. For female participants the regional sweat rate (RSR) showed highest at the central upper back, heels and foot and between the breasts; the lowest RSRs were observed over the breast and middle and lower back. These researchers further added that despite some differences in distribution, both sexes showed highest RSR on the central upper back and the lowest toward the extremities. Hence, when designing garments with various fabric components, it is essential to consider the sweat patterns of the body. For instance, in the case of a sports bra, it is necessary to have a fabric made of filaments/fibres that wick away the body perspiration – particularly in the lower central back – and facilitate breathability whilst preventing fabric sticking to the body and inhibiting free movement.

Because designing functional clothing (particularly activewear) requires knowledge of body heat patterns, designers can use the body mapping technology which W. L. Gore Associates have developed. The body mapping provides information on heat and moisture formation on various zones of the body (Performance Apparel Markets, 2012). According to the body mapping technology in the case of men (front side/anterior part), sweat zones are higher in the central torso and shoulders, followed by limbs and arms. On the back side, the central back, lower back, arms and shoulders are more prone to sweat formation. In men, the heat zones are neck, shoulders, chest, ribs, lower limbs and thighs and, in the back part, the upper back, shoulders, lower back, back thighs and limbs. In the case of women, the heat zones in the front include shoulders, neck, waist, lower limbs and arms. In the case of the back side/posterior part, shoulders, central back, thighs and lower limbs are affected. Hence, whilst designing garments for athletes, care should be taken to ensure that the different needs of men and women are factored in and, at the same time, that thermophysiological comfort is balanced by using appropriate fibres/fabrics in clothing.

Swerev (2003) stated that comfort is not a subjective feeling as it is a physiological process which the body attempts to balance between heat loss and production. However, most researchers have defined comfort of clothing differently; for instance, thermophysiological comfort is the way the cloth helps to maintain heat balance during activity and skin sensational wear comfort – mechanical contact of fabric with the skin (Saville, 1999). Ravandi and Valizadeh (2011) added that comfort of clothing can also be affected by constituents such as physical and chemical properties of fibres, filaments, yarns and fabrics.

2.4.1 Mechanism of Body Perspiration and Temperature Regulation

The human body perspires in two forms: insensible (in vapour form) and sensible (in liquid form) perspiration, and, to be in a comfortable state, the sports clothing worn next to the skin should allow both types of perspiration to transmit from the skin to the outer surface (Das et al., 2009). Gleeson (1998) reported that during intensive activity, body heat production exceeds 1000 W. Some of this heat is dissipated and the remaining heat raises the body's core temperature. This rise in temperature is sensed by the skin's thermoreceptors, which produce sweat to cool the body. Sweat evaporation and increased skin blood flow are two of the body's mechanisms to dissipate heat (Figure 2.2). Inappropriate clothing may affect the ability to lose heat from the body.

Swerev (2003) reported that in order to feel comfortable thermophysiologically, there must be an equilibrium between production and heat loss or else the body's core temperature changes. Approximately 10% of human heat loss is due to breathing; the majority of heat loss is through the skin. This can happen in the form of 'dry heat transfer' – through radiation and conduction and by 'moist heat flow' due to evaporation of perspiration. In both of these methods, heat is lost via clothing. Swerev further added that how heat is dissipated depends on fibre composition, yarn and fabric construction and cut of the garment. In addition, the microclimate between the fabric and skin can be circulated (convection) and exchanged with the surrounding air (ventilation). Outlast Technologies analysed the skin temperature in various parts of the body and reported that the body's core temperature is 37°C and skin temperature is between 31°C and 35°C, including abdominal area (35°C), head (34.4°C), shoulders and upper thighs (34.3°C), wrists (31.4°C), hands (31.6°C), ankles (30.8°C) and feet (30.8°C–31.6°C). A small fluctuation in temperature will lead to discomfort and affects performance (Swantko, 2002). Hence, it can be inferred that designing functional garments with effective heat and moisture management depends on awareness of body heat and sweat pattern.

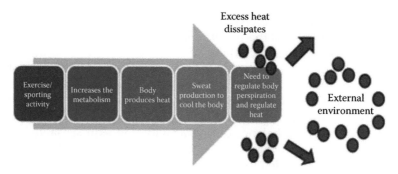

FIGURE 2.2
Mechanism of body perspiration and temperature regulation.

Gavin et al. (2001) investigated whether temperature regulation improved during exercise (running and walking) in moderate heat when wearing clothing (evaporative polyester fabric) which was supposed to promote sweat evaporation compared to conventional cotton garments. No differences were observed during and after exercise. It was concluded that clothing did not affect the thermoregulation during exercise. However, recently an Israel-based producer of nylon 6.6 yarns reported on its Nilit Breeze (http://www.nilit.com), which promoted its products, maintaining the thermal balance. The filaments are made of a flat cross section that quickly transfers body heat; the polymer contains inorganic particles that increase surface area and promote cooling. Textured filament offers bulkiness for added ventilation and breathability. The producer further added that the Nilit Breeze sleeve lowered the body temperature by 1°C compared to a standard sleeve when tested on a thermal sweated mannequin. Hence, there is mixed reaction with regard to regulation of clothing addressing body temperature.

2.4.2 Stretch and Recovery

Stretch fabrics have become a staple in sportswear apparel due to the increased comfort and fit the highly extensible fibre can offer. Voyce, Dafniotis and Towlson (2005) explained how a person's skin stretches considerably, with some key areas of stretch including 35% to 45% at knees and elbows; with sporting activities increasing such numbers, stretch of sportswear apparel is key for comfort. Although elastomeric fabrics are used to improve comfort, tighter compression garments are becoming more desirable in the market. Normal body movement expands the skin by 10% to 50% and strenuous movements in sports will require least resistance from a garment and instant recovery. Stretch and recovery of garments depends on the fibre constituents (textured, elastomeric, microfibre), yarn formation (ply yarn or core yarn), fabric structure (woven or knitted) and any fabric finishes. Elastane yarns are characterised by their ability to recover from stretch and they are often used in stretch fabrics. Figure 2.3 illustrates various body postures that undergo stretch and require support.

BISFA (International Bureau for the Standardization of Man-Made Fibres) has described elastane as 'a fibre composed of at least 85% by mass of a segmented polyurethane which, if stretched to three times its un-stretched length, rapidly reverts substantially to the original length when the tension is removed' (BISFA, 2014). Widely used elastane Lycra® was developed by INVISTA. In a recent report INVISTA claimed that Lycra SPORT fabric was specifically created to improve athletic performance and comfort. It added further that sportswear characteristics, including compression, freedom of movement and comfort, are essential for athletes of all levels and in most sports, from the fastest sprinters and cyclists in the world to swimmers, gymnasts or rugby players. INVISTA developed a three tiered

FIGURE 2.3
Different body stretch movements: (a) men and (b) women. (Courtesy of Shutterstock.)

end use performance standard for athletes: 'Lycra sport fabric' for active performance; 'Lycra sport beauty fabric', combining performance with beauty and style to assist in 'looking in shape while getting in shape™'; and 'Lycra sport energy fabric', innovated for compression fabrics used in high-intensity and high-energy-sports. Lycra sport energy provides 100% stretch in both directions, to maintain freedom of movement as compression garments are worn tight. There are minimum recovery powers at 40% and 65% of fabric stretch and normal hysteresis for consistency in stretch and recovery (Lycra, 2012).

Elastane yarns are often covered with another fibre or filament and are not used on their own. This provides more bulk and improves abrasion resistance. The main end uses for the yarns are garments and other products where comfort and fit are important. Typical examples are sports and leisure wear, swimwear, elastic fabrics and stockings (BISFA, 2014). Generally, casual garments are designed with 2%–5% elastane for added comfort and stretch; however, sportswear such as tights have up to 10% elastane. However, compression garments that apply mechanical pressure use up to 40% elastane. Hence, based on the end uses, desired stretch can be designed by varying the composition of elastane fibres.

Senthilkumar and Anbumani (2011) recently reported dynamic elastic recovery (DER) of elastic knitted fabrics intended for sportswear at different extension levels by determining the stress/strain of fabrics. They studied two types of fabrics with different types of yarn: spandex core cotton spun (SCCS) yarn and spandex back-plated cotton (SBPC) yarn with identical fabric geometry (wales per centimetre, stitch density, loop length, etc.) and evaluated the DER behaviour. They found that at 20%–30% extension, the fabric loop deformation can take place with no change in the residual energy of elastane; however, at 40%–50% extension, fabric undergoes stretch, which may cause yarn slip in the structure. They concluded that SBPC fabric had higher DER value (higher stress) than SCCS fabric, claiming that SBPC fabric had good elastic recovery, which enhances the wearer's performance and supports in muscle recovery. It can be inferred that clothing intended for sportswear with strenuous activity requires stretch and recovery, particularly those that are worn skin tight (e.g. compression garment) compared to leisure wear or basic sportswear. Careful planning and factoring these parameters results in a garment that is fit for purpose.

2.5 Fibres for Sportswear

This section discusses the importance of microfibres, hollow fibres and bicomponent fibres used in functional garments, highlighting their characteristics and typical uses.

2.5.1 Microfibres

Fine-diameter fibres less than 1 denier are often termed microfibres and have valuable properties. They are soft, durable and drapeable, possess high absorbency and are used for high-performance end uses, especially sportswear. They are produced through melt spinning with strict process control, resulting in a uniform and high-quality polymer. Fibres with <0.5 denier cannot withstand tensile forces of melt spinning. Commercially, nylon, polyester, acrylic and rayon are available in the market. Ultrafine fibres are less than 0.3 denier per filament. Microfibres are manufactured by a bicomponent process using two different polymers that do not mix (Collier et al., 2009; Purane and Panigrahi, 2007). A typical example is by producing a bicomponent fibre in islands-in-the-sea formation and then dissolution of the sea part of the fibre leaves the tiny microfibres.

2.5.2 Hollow Fibres

Hollow fibres were introduced in the 1980s; their cross section is hollow and available in round, trilobal or square shapes. The hollow fibres are resilient, have better recovery, are bulky and provide better thermal insulation by trapping air. Hollow polypropylene fibres are lightweight and soft and offer good thermal insulation used in thermal underwear (Ravandi and Valizadeh, 2011). Microfibres are finer than delicate silk and offer excellent draping, luxurious handle, resistance to shrinkage, superabsorbency and strength (Purane and Panigrahi, 2007). These properties enable them to be used in a wide range of applications.

Hollow fibres are made of a sheath of fibre material with one or more hollow spaces at its centre (Figure 2.4). It is produced using *C-shaped spinneret holes*; molten fibres relax after extrusion; open 'c-form' structure closes to produce a hollow fibre. There is also a spinneret hole with a *solid core*

FIGURE 2.4
Hollow fibres. (Source: IBWCh.)

around which polymer flows and produces hollow fibre. Hollow fibres offer increased absorbency and can act as filters for kidney dialysis, and as carriers of carbon particles for safety clothing in contact with toxic fumes (Collier et al., 2009).

2.5.3 Bicomponent Fibres

Bicomponent fibres have been in use as technical textiles for quite some time. Fibres consist of two polymers which are chemically or physically different or both. Bicomponent fibres can be produced with two variants of the same generic fibre, two types of nylon, two types of acrylic or two generically different fibres of polyester or nylon, nylon and elastane. Based in Belgium, Centexbel has an extrusion plant that manufactures bicomponent fibres. Different types of bicomponent fibres are produced based on the end use (see Table 2.2). Bicomponent fibres are made of two components distributed over the entire length of the fibre (Centexbel, 2014). They are available

TABLE 2.2

Different Types of Bicomponent Fibres

Name	Bicomponent Fibre	Characteristics and Uses
Concentric sheath core		Used in melt fibres with sheath made from polymers with a low melting point around a core with a high melting point. During heating the sheath will melt; consequent cooling will bind the structure. This is used with fibres of different melting point, for instance, sheath is made from polymers with a low melting point around a core which has a high melting point.
Eccentric sheath core		As in the preceding description, two polymers are used; however, the core is off centre. Due to different shrinking ratios of polymers the fibre will curl when heated in a relaxed state. It is possible to add crimp and volume.
Side by side		Both polymers share an equal part of fibre surface. Fibre can develop more crimp than the eccentric sheath/core.
Pie wedges		This is made of 16 adjoining pie wedges. Each pie wedge of polymer is flanked by another polymer. Microfibres (0.1 to 0.2 denier) are produced by splitting them by mechanical action. It is possible to provide a hole in the middle of the pie wedge to split the filaments more easily.
Islands/sea		In this type, one polymer (gray) represents the island and the other polymer (black) represents sea. This structure allows producing fine microfibres by dissolving the latter, which is easier than extruding fine fibres directly.

Source: Centexbel © 2014. (Images reproduced with permission.)

in staple, filament and microfibre formats. By extruding two polymers in one single fibre, the different properties of both polymers can be extracted (Figure 2.5). Bicomponent fibres are used in a wide range of functional products including sportswear.

Puma recently developed sportswear, including T-shirts, shorts and sweatshirts that incorporate patented Celliant® technology. Celliant is a brand name for a bicomponent fibre made from polyester which contains a blend of minerals and proprietary ingredients embedded in the core of the fibre (Performance Apparel Markets, 2013). Celliant technology absorbs and stores the electromagnetic energy emitted from the body (in the form of infrared light) and releases it to be reabsorbed into the skin and tissue.

2.5.4 Thermoregulation Fibres

ADVANSA Thermo°Cool® is a combination of fibre shapes – channelled fibres and hollow fibres – that creates additional spaces within the fibres that allow better circulation of air, thus improving significantly the fabric's evaporation capability (Giebel and Lamberts-Steffes, 2013). Outlast Technologies' PCMs (phase change materials) are located in the fibre. The fibres are spun into yarns and are intended for fabrics worn next to the skin. The Outlast technology uses the PCMs, which absorb, store and release the heat for optimal thermal

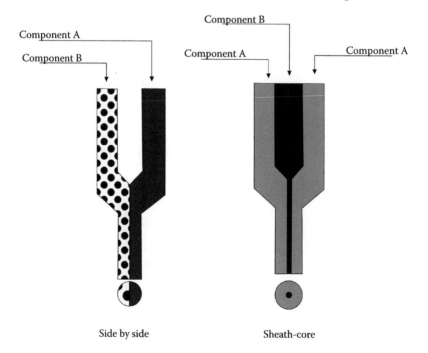

Side by side   Sheath-core

FIGURE 2.5
Bicomponent filament extrusion.

comfort. Outlast technology has the ability to continually regulate the skin's microclimate. As the skin gets hot, the heat is absorbed, and as it cools, that heat is released.

As illustrated in Figure 2.6, Outlast's technology works on the principle where excess heat generated by the body due to harsh external environment or intense activity is absorbed by the Outlast microthermal fibres. The stored heat is released back to the body as needed, maintaining the temperature. The company claims that the fabric made of Outlast fibres offers a constant microclimate next to the skin. Figure 2.7 shows the thermoregulation process pictorially, highlighting three key stages.

Recently Jockey®, a US-based apparel manufacturer, incorporated PCMs developed by Outlast into its underwear line to maintain good thermal comfort and keep the wearer comfortable by balancing the heat produced by the body. The PCMs melt when the surrounding heat rises and then store surplus energy. When the physical activity decreases, the body cools down and the PCM's Thermocules™ solidify and emit the heat which was stored. This results in improving the comfort of the wearer. Jockey has introduced its product in men's and women's underwear (Performance Apparel Markets, 2011a). Figure 2.8 illustrates the cross section of polyester fibre with Thermocules; Figure 2.9 illustrate Outlast technologies – acrylic filament with thermocules, whilst Figure 2.10 illustrate viscose filament with thermocules; they enhance durability and can be laundered many times without losing performance (Swantko, 2002).

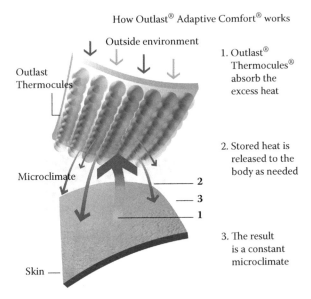

How Outlast® Adaptive Comfort® works

Outside environment

Outlast Thermocules

1. Outlast® Thermocules® absorb the excess heat

Microclimate

2. Stored heat is released to the body as needed

— 2
— 3
— 1

Skin —

3. The result is a constant microclimate

FIGURE 2.6
Adaptive comfort – Outlast Technologies. (Courtesy of Outlast Technologies LLC.)

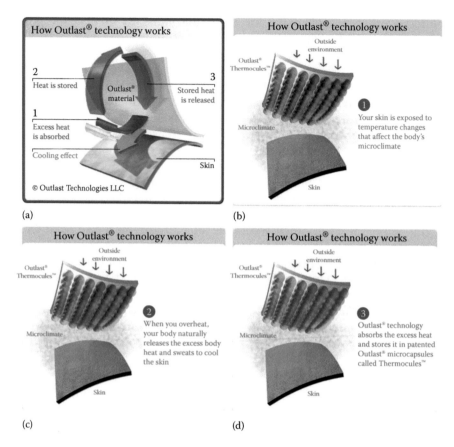

(a)   (b)

(c)   (d)

FIGURE 2.7
Outlast technology utilizes phase change materials (PCMs) that absorb, store and release heat for optimal thermal comfort. This is illustrated in panel a where it provides a cooling effect to the skin. Outlast technology has the ability to continually regulate the skin's microclimate. As the skin gets hot, the heat is absorbed, and as it cools, that heat is released. When the skin is exposed to temperature changes, it affects the body's microclimate (panel b). When the body is overheated due to intense activity or exposure to environment, the body naturally releases the excess body heat and perspires to cool the skin (panel c); Outlast technologies absorb the excess heat and store in the outlast micro-capsules called Thermocules (panel d). Outlast technology proactively manages heat while controlling the production of moisture before it begins. (Courtesy of Outlast Technologies LLC.)

Celliant is a patented technology that absorbs and stores the electromagnetic energy emissions from the human body and re-emits them back to the body, where they are reabsorbed into the muscle tissue (Future Materials, 2013) through the use of fibres, converts the body's natural energy (heat) into infrared (IR) light and emits it into the body's tissue and muscles. The result is a responsive textile clinically proven to benefit the human body, utilising a blend of minerals and proprietary ingredients embedded into the fibre's core. Celliant technology is a blend of proprietary ingredients

FIGURE 2.8
Outlast Technologies' polyester fibre cross section with thermocules. (Courtesy of Outlast Technologies LLC.)

FIGURE 2.9
Outlast Technologies' acrylic filament with thermocules. (Courtesy of Outlast Technologies LLC.)

that recycles vibrant energy (heat) leaving the body into infrared light (IR) and sends it back to the body where it is absorbed by the tissue and muscles. IR waves are beneficial as they penetrate deep into the tissue. This energy causes the body to increase circulation and oxygen wherever it is applied. Celliant is a blend of 13 thermoreactive minerals, including titanium dioxide, silicon dioxide and aluminium oxide. Additional proprietary ingredients are blended with polyester fibre to create a variety of staple fibres, spun and filament yarns and fabric blends (Celliant, 2014). According to Hologenix, a US-based company, Celliant technology has been shown to increase oxygenation in the body tissue and reduce body aches and pains. The technology uses electromagnetic energy emissions in the form of infrared light produced by the human body. Celliant technology harnesses the human body's natural energy and is clinically proven (clinical trials) to enhance tissue oxygen levels (Performance Apparel Markets, 2013).

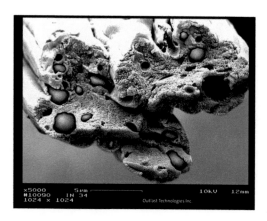

FIGURE 2.10
Outlast Technologies' acrylic fibre showing embedded thermocules. (Courtesy of Outlast Technologies LLC.)

2.5.4.1 Cellulose Blends

Generally, cellulose fabrics tend to absorb water within their fibre structure and then become heavy and take a longer time to dry. This leads to stretching of fabric and its clinging to the skin. When the intense activity ceases, the fabric leaves the wearer cold. However TENCEL® from Lenzing is a synthetic fibre, which is solvent spun from cellulose widely used in moisture management with improved aesthetics and is used in sportswear. Firgo, Suchomel and Burrow (2006) reported that TENCEL can be used effectively in performance sportswear if the fabric is carefully designed using double-layer fabrics that have better moisture absorption, moisture spreading, a quick drying rate, reduced wet-cling behaviour, better balance of water vapour permeability and thermal comfort and, more importantly, less synthetic appearance and handle. The two-layer fabric strategy which was tested is based on the fact that the fabric is made of a hydrophobic inner layer in contact with the skin where the sweat is pulled through the fabric by the hydrophilic layer, which is the outer layer (made of TENCEL). This outer layer is ideal for spreading (wicking) the moisture and evaporating it to the environment. Firgo et al. also reported that, after having researched a number of blends with polyester, 30% TENCEL and 70% polyester blended yarn fabric performed better in terms of absorbency, moisture spreading, drying rate, a good wet-cling index and intermediate water vapour permeability.

2.5.5 New Developments in the Fibre Industry

Trinomax AQ® was recently developed by a joint collaboration between Nilit, an Israel-based manufacturer of nylon 6.6, and LincSpun yarns, an Australia-based developer of intelligent yarns and filaments. Trinomax AQ is made by twisting together three types of fibres – Merino wool, textured Nilit nylon 6.6 and

Nilit Aquarius – using a proprietary LincSpun™ technology. The resulting yarn is lightweight, durable and soft, has the ability to regulate body temperature and wicks moisture. The company claims that performance is durable and lasts through repeated laundering, and it is intended for a range of products such as activewear, socks and sportswear (Performance Apparel Markets, 2011b).

2.6 Market Trend and Overview

According to the Research and Markets Report, the UK sportswear market grew 5.4% in 2012 with apparel accounting for 71% of the market value and footwear making up 29%. Among the retail stores, Pentland Group and Sports Direct have a majority of the market share after JJB Sports entered administration in 2012. Among sportswear brands Nike and Adidas lead the market (*WSA*, 2012b). Euromonitor International reported that global sales of sportswear are set to rise from US $245 billion in 2012 to US $300 billion by 2017 (Kondej, 2013).

2.6.1 Market Drivers

Major sporting events are drawing attention to various sportswear brands including sports apparel and footwear companies who want to have a market share in the following events.

- Winter Olympics, Sochi, Russia, 7–23 February, 2014
- FIFA Soccer World Cup, Brazil, 12 June–13 July, 2014
- Summer Olympics in Rio de Janeiro, Brazil, 5–21 August, 2016

These events also promote activewear and replica products among fans, supporters of sporting heroes and sports enthusiasts. In addition, the market would be driven by increasing demand in Brazil, China, the United States and India (Kondej, 2013).

2.7 Widely Used Fibre Types for Sportswear

In the past, cotton was widely preferred for a wide range of garments until the advent of synthetic fibres such as polyester, nylon, polypropylene, acrylic, etc. During the 1970s use of polyester fibres in sportswear predominated, continuing in the 1980s and 1990s. **Polyester** is preferred in sportswear because it is lightweight, cheap to produce, dye-fast, durable, easy to care for, quick drying, hydrophobic in nature and has wicking ability. In addition, the polyester filament fabrics can be given hydrophilic coating. Hence, polyester fibre-based fabrics with their hydrophobic cores and hydrophilic coatings can wick moisture away from its contact with the skin to outer surface to environment. Polyester is often blended with other natural fibres, mainly to extract its benefits to maintain moisture management and durability. CIRFS reported that among man-made fibres, polyester is a dominant fibre (*WSA*, 2012a). Many sportswear brands are progressively moving toward recycled polyester. Adidas was one of the companies that carried out a life cycle analysis by conducting research on the environmental impact of polyester. Mechanically recycled polyester has a better environmental profile than chemically recycled fibres; however, chemically recycled fibres have a wide range of applications in the industry.

Elastane is another synthetic fibre widely preferred for its elasticity (mainly for stretch and recovery). The elastic nature of filaments is used in sportswear to compress muscles, offer stretch for body movements and support in recovering from muscle soreness. A wide range of sportswear products, such as foundation garments, swimwear, base layer products, compression tights, etc. are made of elastane.

Merino wool is widely used in sportswear; for instance, superfine Merino wool possesses superior water vapour permeability and quick-drying properties. ADVANSA's Thermo°Cool can be blended with Merino wool (50%/50% or 70%/30%) for better thermoregulation and comfort (Giebel and Lamberts-Steffes, 2013). Merino fibre can absorb up to 35% of its dry weight in moisture vapour. During strenuous exercise or hot conditions, a Merino wool garment closer to the skin actively transfers moisture vapour away from the body. This causes the microclimate above the skin to become less saturated with vapour, thereby making the wearer feel less clammy and it is less likely for the vapour to form sweat droplets on the skin's surface. Recently, Pearl Izumi, a Japanese cycling and sports apparel company, promoted its cycling jersey with Merino performance technology developed by Australian Wool Innovation using 19.5 µm wool that promotes comfort and warmth (Pearl Izumi, 2010).

2.8 Fibre Types and Blends Affecting the Performance of Garments

The performance of fabrics is dependent on chemical and physical properties of constituent fibres, yarns and finishes used. Blending different types of fibres enables utilisation of the advantages of each fibre, which in turn facilitates counteracting the disadvantages of other fibres, thus enhancing the appearance, durability, comfort, maintenance, cost and overall performance.

Typical blending includes polyester/cotton, cotton/Lycra, polyester/viscose, nylon/Lycra, wool blends, Merino wool/acrylic, etc.

Polyester and viscose are quite often blended in apparel applications, since polyester has excellent durability and wrinkle resistance and is easy to care for and maintain. Viscose fibres have good absorbency and poor durability and wrinkle resistance. Das et al. (2009), who investigated the polyester and viscose fabric blends, demonstrated that blending of fibres has an important role in moisture-related comfort properties of clothing. The water vapour permeability and absorbency of fabric increases with the increase in hydrophilicity of material. However, the wicking decreases with the increase in viscose proportion. The higher hydrophilic material (viscose) provides better absorption; however, it reduces spreading, resulting in moisture accumulation and a sticky feeling. Hence, if an athlete produces a lot of perspiration, a higher proportion of polyester fibres and smaller percentage of viscose will act as quick absorption of sweat from the skin, and polyester filaments will help to spread the moisture to the outer surface.

Sampath et al. (2011) investigated the effect of optimum level on comfort properties of knitted fabrics made of 150-denier polyester filament containing 34, 48, 108, 144 and 288 filaments. The fabrics were finished with a moisture management finish and were assessed for wetting, wicking and moisture vapour transmission. The researchers reported that when filament fineness increases, wicking rate increases to a certain level. The yarn made of 108 filaments had higher wicking. The moisture vapour transmission was higher for finer fabrics than for fabrics made of coarser filaments. This study highlights the fact that the number of filaments in a yarn and filament fineness should be at optimum levels to promote moisture transmission. Filament fineness and number of filaments in a yarn play a vital role in determining the comfort characteristics of microdenier polyester knitted fabrics.

2.9 Moisture Management

One of the main requirements of sportswear is moisture management, and an optimum microclimate (including temperature and humidity) between the skin and clothing is necessary for an athlete to focus on the sport. To facilitate the moisture management, the breathability and body temperature should be regulated. Breathability is the ability of the fabric/garment to transport the moisture or perspiration from the skin to the environment. However, intense exercise or sporting activity increases body heat and sweat is produced to cool the skin's surface.

Figure 2.11 illustrates the conventional materials that react to body heat generation; during intense activity, sweat is produced and transported, accumulated in the garment and results in discomfort. However, the Outlast

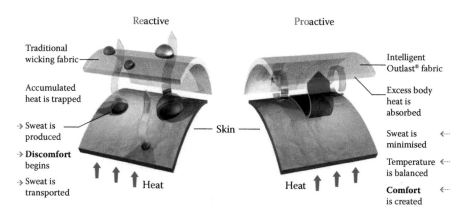

FIGURE 2.11

Moisture management between reactive and proactive materials. (Courtesy of Outlast Technologies LLC.)

technology is proactive, absorbing the excess heat generated by the body; sweat is minimised and temperature is balanced, resulting in better comfort. If excess body heat is not removed from the body, the athlete's performance decreases; in extreme cases it may result in fatigue, heat stress disorders or fainting. On the other hand, if the excess body perspiration is not removed from the body, then the fabric clings to the body, and clothing saturated with perspiration may rub the skin, resulting in blisters and rashes. The next sections highlight the importance of breathability and the effect of wicking in sportswear.

Breathability is measured by determining **resistance to evaporative heat transfer** (RET); the lower the RET value is, the higher is the fabric's breathability. Breathability can also be evaluated by determining the **moisture vapour transfer rate** (MVTR), which is the rate at which a fabric allows the moisture vapour through the fabric – that is, determining the amount of moisture vapour in grams to pass through 1 m² of fabric over a 24-hour period. The Hohenstein Institute has produced results for RET values and physiological human comfort as ratings of wear comfort (Table 2.3).

TABLE 2.3

Resistance to Evaporative Heat Transfer

RET Value	Breathability	Wearer Comfort
≤60	Highly breathable	Very good
>60 but ≤130	Highly breathable	Good
>130 but ≤200	Breathable	Satisfactory
>200 but ≤300	Slightly breathable	Unsatisfactory
>300	Not breathable	Unsatisfactory

Source: Hohenstein Textile Testing Institute GmbH & Co. KG Hohenstein Institute, Germany. 2014.

The RET value for a nude person is zero and the value increases as the clothing layers are increased, thus decreasing the overall breathability of a fabric. It is necessary to know the breathability of materials whilst designing layering of garments often used as outdoor sportswear.

2.9.1 Moisture Management Fibres

Trevira, a German-based fibre company, has developed Trevira Perform Moisture Control, which has excellent moisture management properties. Fabrics made of this fibre are intended for work wear and sportswear. The fibre has a dual channel system (Figure 2.12) that accelerates transport of condensation molecules given off from the skin, from inner to outer layer. Moisture diffuses quickly to the fabric surface, where it spreads and evaporates rapidly. The inner side of a fabric made of 100% Trevira Perform Moisture Control in contact with the skin stays drier and prevents a chill effect. The following are some of the properties that the company claims for its product:

- Very good moisture management
- Rapid dispersion and evaporation of moisture
- Quick drying
- Good pilling characteristics
- Easy care
- Washable, with good fastness standards
- Pleasant handle
- Good electrostatic properties
- Resistant to UV light and chlorine

Trevira Perform Moisture Control – dumbbell shaped with its dual channel system – accelerates the transfer of perspiration from the skin to the fabric surface, where it evaporates rapidly; the inner surface of the fabric remains dry (Trevira, 2014). The graph in Figure 2.13 illustrates the drying time for woven fabric (at 22°C and 56% relative humidity) made of Trevira Perform

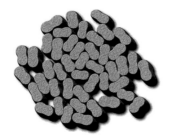

FIGURE 2.12
Trevira Perform Moisture control fibres. (Courtesy of Trevira GmbH.)

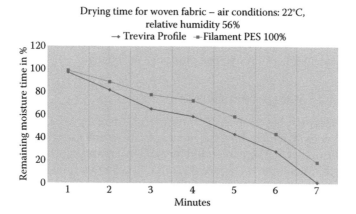

FIGURE 2.13
Drying time for Trevira Profile. (Courtesy of Trevira GmbH.)

Moisture control fibres and 100% polyester filament yarns. The graph shows that the drying time of fabric made of Trevira Perform Moisture control was better than the garment made of 100% polyester filament yarns. Moisture management using biomimetic pine-cone effect was reported recently, where the fibres become porous when they absorb moisture, and in dry conditions, the structure opens like a pine cone reducing the air permeability and increasing thermal insulation property (Inotektextiles, 2014).

Teijin Limited recently reported its development of polyester fibres with enhanced moisture absorption and quick-drying capabilities to prevent postexercise chilling and stickiness from body sweat. The product had been recommended for sportswear (Teijin Limited, 2014). The fabric made of Teijin fibres is composed of a three-layer structure with a hydrophobic contact layer that has moisture-repellent polyester fibres, a moisture-absorbent middle layer and an outer moisture diffusion layer.

2.9.2 Wicking in Activewear Products

Wicking is the transmission of liquid through a textile by capillary action (Collier et al., 2009). Wicking or liquid transport properties depend on fibre type, yarn construction, fabric structure, density and structure of yarns, finishing treatment, viscosity, relative humidity of the atmosphere and ambient temperature (Nyoni and Brook, 2010). The capillary channels in the interfibre spaces and size of the pores greatly determine the wicking. Various stages involved during the wicking process include (Nyoni, 2003):

1. Uptake of moisture from the skin surface
2. Removal of moisture from the skin and transport through the fabric surface

3. Spreading of moisture within the fabric structure

4. Absorption of moisture within suitable fibres: 'dynamic' fabrics usually

5. Containment of an outer layer of hydrophilic fibres to absorb and store sweat away from the skin's surface

6. Evaporation of moisture from the fabric surface

During the last decade or so, there were developments in fibre size and shape, particularly introducing channels in filaments, such as hexa channels and tetra channels (e.g. Coolmax), where the capillary action enhances wicking action.

2.9.3 Maintaining Body Temperature

Ultimate performance of sportswear lies in providing the optimum microclimate next to the skin, and this applies to those garments worn closer to the body, such as base layer products: sports vests, shorts, underwear, thermal wear, socks, etc. Beringer (2014) from the Hohenstein Institute, an independent laboratory in Germany, recently reported the innovative technology from CoolCore LLC (http://www.coolcore.com), a company in Portsmouth, New Hampshire, in providing cooling effects of textiles. Unlike the phase change material or latent heat storage, CoolCore Technologies uses the body's own sweat to achieve cooling and is free from chemicals. Finishes or chemicals deteriorate over prolonged usage in conventional materials; however, CoolCore uses the body's own heat to achieve cooling.

CoolCore reported temperature-regulating fibres, which are engineered to exploit the body's own sweat or added moisture to achieve a cooling effect. The CoolCore technology, which lasts the lifetime of the garment, works on three principles: wicking (moves sweat from the body), moisture transportation (to avoid saturation and accelerate drying) and regulated evaporation from the garment resulting in a greater cooling effect than that found in conventional materials. In addition, the company claims that the garment does not cling to the body due to quick wicking, making it suitable for intensive sports such as cycling, running, etc.

2.10 Discussion and Summary

This chapter highlighted the importance of fibres in development of products designed for functional clothing, particularly sportswear. Fibres form the basic constituent of a garment and garment performance relies on choosing the right type of fibre composition by keeping up to date with modern developments in this field. During the last decade or so, the world production of man-made fibres increased from 49% in 1992 to 68% in 2012 and this trend is bound to continue as the demand for performance clothing increases. This is due to the fact that, compared to natural fibres, man-made fibres are engineered to meet specific requirements of the end user. The properties and characteristics of man-made fibres differ based on the fibre-forming substance and it is essential to know the various technical terms used in the industry, which have been briefly presented with examples. It could be argued that of the various developments in the fibre industry, fibre shape and fineness have played a pivotal role in varying the performance of fabrics – for instance, affecting moisture management, resiliency, bulkiness, warmth and overall comfort to the wearer.

During the selection of fibre types for sportswear, it is essential to factor the physiological parameters – particularly, sweat patterns, thermal regulation and stretch and recovery for parts of the body. This has been highlighted by critically reviewing evidence pertaining to this area and offering new insight into development of sportswear. For instance, awareness of body sweat patterns and/or body mapping technology will enable designers to choose the most appropriate garment design. The design will elicit the usage of mesh fabric panels in the central back of garments, where the athlete sweats the most compared to other body zones, thus facilitating quick wicking and evaporation of body perspiration. In addition, it could be argued that multidisciplinary collaboration among sports science practitioners, physiologists, textile engineers and garment designers will enable development of products fit for this purpose.

The mechanism of body perspiration and regulation of body temperature have been illustrated. The most heat is lost via clothing. The human body aims to balance between heat production and heat loss, and any imbalance results in discomfort. Hence, careful consideration has to be given in regulation of body temperature for sportswear. A rise in body temperature induces the body to produce sweat to cool the body. The garment should be able to allow moisture breathability and quickly transfer the moisture from the skin to the environment.

Athletes require stretch and recovery during training and sporting activity. The stretch enables the athletes to move freely without restriction and offers support to the muscles and joints and prevents strain on the tissue. The elastane fibres play an important role in providing 100% stretch, apply compression to the muscle and prevent soft tissue injury. Fine microfibres, hollow fibres and bicomponent fibres offer numerous possibilities to blend different characteristics of two different polymers and vary shape, fineness and size to cater to demands of the end user.

Latent heat storage (phase change materials) products using Thermocules in filaments to provide optimum thermoregulation have been a breakthrough in technology in absorbing body heat and releasing it when needed; they have offered possibilities for designers to tailor products to the specific needs of athletes. Newly developed cellulose fibre TENCEL continues to provide new technologies, particularly in moisture management and appearance and fabric handle. Widely used fibre types such as polyester, elastane and wool blends will continue to be used in sports performance clothing, mainly due to their properties. Blending of fibres and filaments has been a practice in meeting the growing requirements of performance clothing. The blending of fibres enables enhancement of performance, durability and maintenance and reduction of the overall cost of the product; this trend in sportswear will continue to grow.

Fibres that have dual channels to wick perspiration, avoid condensation, accelerate drying and maintain body temperature using body heat will be implemented in various athletic wear such as vests, underwear, base layers, tights, etc. Hence, it becomes vital for apparel designers, technologists and various stakeholders in the field of sportswear to carefully select different fibre types based on the end use and requirements. Innovations in fibre technology are fuelled by the ever growing thrust among athletes to outperform in their respective fields. Recent technological developments to utilise the energy produced by the body during an activity have been tested and resulted in production of smart products with lifelong performance. It is anticipated that textiles for sportswear will witness advances and smart developments in the years ahead that will encourage athletes to wear smart garments that meet their specific requirements and enhance their performance and appearance.

References

ASTM D 1909. (2004). Standard table of commercial moisture regains for textile fibers. American Society of Textile Materials, ASTM.

Beringer, J. (2013). Innovative Technology – Cooling Power. Quality Label for Coolcore, Hohenstein Institute, Schloss Hohenstein, 74357 Bönnigheim, Germany.

BISFA. (2014). International Bureau for the Standardization of Man-Made Fibres.

Brown, P. and Stevens, K. (2007). *Nanofibres and nanotechnology in textiles*. Cambridge, UK: Woodhead Publishing.

Celliant. (2014). Celliant Technology (http://www.celliant.com). Online resource accessed on 6 June 2014.

Centexbel. (2014). Pilot line for research and prototyping, Bicomponent fibres. http://www.centexbel.be/bicomponent-fibres.

CIRFS. (2014). Worldwide production of cotton, wool and man-made fibres, 1992–2012. European Man-Made Fibres Association, Brussels.

Collier, B. J., Bide, M. and Tortora, P. G. (2009). *Understanding textiles*, 7th ed. Upper Saddle River, NJ: Pearson Education.

Das, B., Das, A., Kothari, V. et al. (2009). Moisture flow through blended fabrics – Effect of hydrophilicity. *Journal of Engineered Fibres and Fabrics* 4 (4): 20–28.

Denton, J. M. and Daniels, P. N. (2002). *Textile terms and definitions*, 11th ed. Manchester, UK: Textile Institute.

Firgo, H., Suchomel, F. and Burrow, T. (2006). TENCEL high performance sportswear. *Lenzinger Berichte* 85: 44–50.

Future Materials. (2013). Responsive textile technology. *Celliant, Future Materials,* issue 1, p. 20.

Gavin, T. P., Babington, J. P., Harms, C. A., Ardelt, M. E., Tanner, D. A. and Stager, J. M. (2001). Clothing fabric does not affect thermoregulation during exercise in moderate heat. *Medicine and Science in Sports and Exercise* 33 (12): 2124–2130.

Giebel, G. and Lamberts-Steffes, E. (2013). Merino goes technical with ADVANSA Thermo°Cool. A paper presented in Performance Days, Functional Fabric Fair, Munich, Germany, 15–16 May 2013.

Gleeson, M. (1998). Temperature regulation during exercise. *International Journal of Sports Medicine* 19 (Suppl 2): S96–S99.

Gupta, V. B. and Kothari, V. K. (1997). *Manufactured fibre technology.* London: Chapman & Hall.

Hohenstein Institute. (2014). Body mapping technology. Hohenstein Textile Testing Institute GmbH & Co. KG Hohenstein Institute, Germany.

Hongu, T. and Phillips, G. O. (1997). *New fibres,* 2nd ed. Cambridge, UK: Woodhead Publishing Ltd.

Humphries, M. (2009). Fabric reference, Pearson Prentice Hall.

Inotektextiles. (2014). Biomimetics, Innotektextiles.com/technology.

Kadolph, S. J. (2007). Quality assurance for textiles and apparel, 2nd ed, Fairchild Publications, New York.

Kadolph, S. J. (2014). *Textiles,* 11th ed. Upper Saddle River, NJ: Pearson.

Kirkwood, B. (2013). Taking the lead. Sportech, Future Materials, official publisher of *Techtextil News,* issue 6, pp. 8–9. World Textile Information Network Ltd. (WTiN).

Kondej, M. (2013). The sportswear revolution – Global market trends and future growth outlook. Webinar, Euromonitor International, London, UK.

Lycra. (2012). Lycra fibre revolutionizes sportswear, helping the fastest athletes in the world, http://www.invista.com (accessed on 6 April 2014).

Nyoni, A. B. (2003). Liquid transport in nylon 6.6 yarns and woven fabrics used for outdoor performance clothing, PhD Thesis, University of Leeds, UK.

Nyoni, A. B. and Brook, D. (2010). The effect of cyclic loading on the wicking performance of nylon 6.6 yarns and woven fabrics used for outdoor performance clothing. *Textile Research Journal* 80 (8): 720–725.

O'Mahony, M. and Braddock, S. E. (2002). *SportsTech: Revolutionary fabrics, fashion and design.* Thames and Hudson Ltd., London, Outdoor performance fabrics, PhD Thesis, University of Leeds, UK.

Pearl Izumi. (2010). Pearl Izumi includes Merino in its new cycling apparel range. *Asian Textile Journal* 19 (4): 13.

Performance Apparel Markets. (2011a). Product development and innovations, 2nd quarter. Textiles Intelligence Limited, Wilmslow, UK, pp. 11–31.

Performance Apparel Markets. (2011b). Product development and innovations, 2nd quarter. Performance Apparel Markets, Textiles Intelligence Limited, Wilmslow, UK, p. 23.

Performance Apparel Markets. (2012). Summer sportswear: Providing cool comfort, Issue 41, 2nd quarter. Textiles Intelligence Limited, Wilmslow, UK, pp. 40–41.

Performance Apparel Markets. (2013). Product developments and innovations, 1st quarter. Textiles Intelligence Limited, Wilmslow, UK, pp. 11–26.

Purane, S. V. and Panigrahi, N. R. (2007). Microfibres, microfilaments and their applications. *AUTEX Research Journal* 7 (3): 148–158.

Ravandi, S. A. and Valizadeh, M. (2011). Properties of fibres and fabrics that contribute to human comfort. In *Improving comfort in clothing*, ed. Song, G. Cambridge,
UK: Woodhead Publishing Ltd.

Research and Markets. (2013). Sports Clothing and Footwear Market Report 2013.
http://www.researchandmarkets.com/research/ntskz6/sports_clothing.

Rigby, D. (1998). Development of performance fibres and fabrics – A proactive
approach. *World Sports Activewear (WSA)*, pp. 13–17.

Sampath, M. B., Mani, S. and Nalankilli, G. (2011). Effect of filament fineness on comfort characteristics of moisture management finished polyester knitted fabrics.
Journal of Industrial Textiles, Sage Publications. doi: 10.1177/1528083711400774.

Saville, B. P. (1999). *Physical testing of textiles*. Cambridge, UK: Woodhead Publishing

Senthilkumar, M. and Anbumani, N. (2011). Dynamics of elastic knitted fabrics for
sportswear. *Journal of Industrial Textiles*, http://jit.sagepub.com/content/early
/2010/10/26/1528083710387175.

Smith, C. J. and Havenith, G. (2012). Body mapping of sweating patterns in athletes:
A sex comparison. *Journal of American College of Sports Medicine*, pp. 2350–2361,
http://www.acsm-msse.org.

Swantko, K. (2002). Adaptive comfort. *Knit Americas*, Fall, pp. 20–21.

Swerev, M. (2003). What dermatologists should know about textiles? In *Textiles and the
Skin*, eds. Elsner, P., Hatch, K. and Wigger-Alberti, W. Karger, Basel, Switzerland,
vol. 31, pp. 1–23.

Tao, X. M. (2001). *Smart fibres, fabrics and clothing*, Cambridge, UK: Woodhead Publishing.

Teijin Limited. (2014). http://www.teijin.com/products/advanced_fibers/online
resource (accessed on 6 June 2014).

Trevira. (2014). The high performance brand with climate effect. Brochure, Trevira
Perform Moisture Control.

Voyce, J., Dafniotis, P. and Towlson, S. (2005). Elastic textiles. In *Textiles in sport*, ed.
Shishoo, R., 205. Cambridge, UK: Woodhead Publishing.

WSA. (2012a). Casting a new line in polyester. Performance and sports materials.
World Sports Activewear 18 (6). Textile Trades Publishing, Liverpool, UK.

WSA. (2012b). WSA, Performance and Sports Materials. Sep/Oct 2012, p. 4.

Useful Resources

BISFA, an international association of man-made fibre producers (http://www.bisfa
.org/).

CIRFS, European Man-Made Fibres Association, is the representative body for the
European man-made fibres industry (http://www.cirfs.org/).

CoolCore, http://www.coolcore.com.

CSIRO, Fibre Science Research Program, Australia (http://www.csiro.au/Organisa
tion-Structure/Divisions/CMSE/Fibre-Science.aspx).

Fiber Source, American Fiber Manufacturers Association (http://www.fibersource
.com/fiber.html).

Hohenstein Institute, Hohenstein Textile Testing Institute GmbH & Co. KG Hohenstein
    Institute, Germany (http://www.hohenstein.com/en/home/home.xhtml).
IWTA, International Wool Textile Organisation (http://www.iwto.org/wool/).
Nilit Breeze, http://www.nilit.com.
Outlast Technologies, Outlast Europe GmbH, Germany.

3

Fabric Properties and Their Characteristics

Praburaj Venkatraman

CONTENTS

3.1 Introduction 53
3.2 Fabric Properties and Their Influence on Product Performance 54
   3.2.1 Essential and Desirable Properties of Fabrics............................57
3.3 Factors That Influence Fabric Behaviour..64
   3.3.1 Internal Factors Influencing Fabric Performance.......................64
   3.3.2 External Factors Influencing Fabric Behaviour 67
      3.3.2.1 Dry State Measurements...69
      3.3.2.2 Wet State Measurements..69
      3.3.2.3 Commercial Examples – Fabrics for Outer Wear 70
   3.3.3 Fabric and Human Body Interaction..71
3.4 Fabric Structure and Characteristics..73
   3.4.1 Structural Influence and Effect on Performance.......................74
   3.4.2 Special Multilayer Fabrics for Protection 74
3.5 Fabric Composition and Its Effect on Sportswear Performance 76
   3.5.1 Importance of Fabric Composition in Sportswear 77
   3.5.2 Natural Fibres and Their Effects on Fabric Performance...........78
   3.5.3 Synthetic and Smart Fibres Used in Sportswear.......................79
3.6 Discussion 80
3.7 Summary and Conclusions 83
References...84

3.1 Introduction

A confident understanding of fabric behaviour and characteristics is vital in the design and development of a functional garment – for instance, a warp knit mesh fabric made of 100% polyester designed to wick moisture away from the skin, with the quick-dry ability, making it ideal for everyday wear and preferred in extreme performance requirements. On the other hand, georgette is a balanced, plain-woven fabric generally made of 100% polyester with high-twist yarns giving the fabric a less smooth appearance used in fashion apparel. Textile materials have evolved in recent times and fabrics play a significant role in the development of the sportswear industry. In fact, the materials reflects the quality of a brand and its identity. The primary focus of this chapter is to present the essential and the desirable properties suitable for performance apparel, especially for sportswear. Various sportswear applications are discussed to enable the reader to understand the rationale for each parameter. Generally, garments intended for fashion apparel will have to fulfil the following characteristics: durability, strength, colourfastness, aesthetics and so forth. These properties are mandatory for everyday use and maintenance for fashion apparel. However, in the case of performance apparel, the requirements are functional and application-specific properties, such as moisture transmission, thermal resistance, wicking, waterproofing and flame resistance. The reason for requirements of such properties is because functional apparel is subjected to a wide range of end uses such that a garment will be affected by internal (fibres, yarn fineness, warp/weft movement, fabric density, thickness, fabric count) and external (exposure to sunlight, wind, rain, cold weather conditions and during use) factors. These factors affect the performance and behaviour of functional apparel discussed in the sections with examples.

In addition, the interaction between the human body and garments is significant; this is true for those close-fit garments such as base layer garments, where thermoregulation plays a vital role in the performance of an athlete. Fabrics for sportswear are either woven or knitted and often blended with synthetic and natural fibres with varying linear density to provide an optimum performance. The heavyweight fabrics for outerwear are multilayered (coated or laminated) and their properties differ from those of lightweight fabrics. Parameters of these fabrics due to structural difference are highlighted here with specific focus on functional apparel. Fabric behaviour will be affected by its composition and this is presented with examples. This chapter will empower the reader to understand the properties of fabrics for various performance applications and how these parameters will affect the overall performance of the garment.

3.2 Fabric Properties and Their Influence on Product Performance

A number of fabrics are used in performance and sportswear apparel including smart fabrics, which have an intelligent approach to high body or ambient temperature; the warmer the material gets, the faster the moisture management system functions – Burlington's smart fabric temperature management. The smart fabric technology uses micro-encapsulated phase change materials to absorb and release heat to enhance comfort.

The section below lists the various type of fabrics suitable for sportswear and performance apparel.

- Lightweight, stretchable and soft waterproof or breathable fabrics
- Fabrics made of fine microfibres with breathability
- Soft shell or three-layer fabrics, which are bonded as well as laminated, and made of tricot warp knits or woven fabric for wind insulation or waterproofing
- Knits with synthetic or natural fibre blends and up to 30% elastane fibres for stretch and recovery
- Fleece and brushed knits made of synthetic fibres that have a natural feel and stretchy and smooth surface
- Woven shirts with varying fabric weight (160 to 400 g/m^2)
- Eco-friendly fabrics including recycled polyester
- Fancy fabrics with patterns, designs and finishes
- Laser or etched burnt out, 3D knits
- Honeycomb patterns and work wear and protective-wear fabrics including durable rugged finish, cut resistant, flame retardant, resistance to abrasion, reflective facings, etc.

Figure 3.1 illustrates the types of fabrics used for fashion apparel, which include woven, knitted fabrics; braids; interlining; bonded fabrics and felt. It should be noted that woven and knitted fabrics will perform differently due to their structural variation, which is summarised in Table 3.1. Functional apparel has a combination of fabrics made of woven and knitted fabrics to suit various applications.

In order to understand the fabric properties of garments designed for activewear it is necessary to explore the sports trends: types of sports preferred, frequency of activities and garments frequently purchased. In the UK,

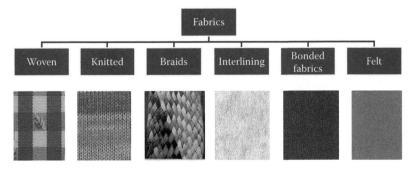

FIGURE 3.1
Common fabric types for apparel end use.

TABLE 3.1

Fabric Characteristics

Knitted Fabrics	Woven Fabrics
• Series of interconnected loops made with one or more sets of yarns.	• Two or more sets of yarns interlaced to form the fabric structure. Yarns interlace at right angles.
• Can be ravelled from top to bottom. Warp knits cannot ravel.	• Can be ravelled from any cut edge.
• Fabric can snag and run, bowed or skewed.	• May be bowed or skewed.
• Usually heavier because more yarn is used.	• Usually lighter in weight because less yarn is used.
• Possess stretch and elasticity, adapts to body movement.	• Possess limited stretch and adaptability to body movement.
• Good recovery from wrinkles; air permeable.	• Bulkiness and recovery from wrinkle depend on weave structure.
• Possess open spaces between yarns and bulky.	• Stable to stress, less air permeable, especially with dense fabric.
• Porous and less opaque.	• Provide maximum hiding power and cover.
• Less stable in use and care.	• More stable in use and care.
• Higher shrinkage unless heat-set.	• May shrink less than 2%.

a number of sporting activities are preferred; amongst many, swimming, running, cycling, tennis, golf and aerobics are widely preferred (Mintel Group, 2011). Populations involved in sports are generally younger groups whose ages range from 16 to 24 and this trend can be noted across various sports. However, populations in the age ranges of 25 to 34 and 35 to 44 are also active, particularly in swimming, using a gym, running, cycling and tennis.

There has been a gradual and consistent increase in the number of people involved in sporting activities over the last 8 years (Figure 3.2). This includes

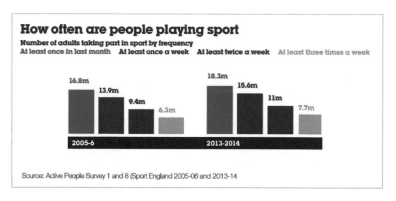

FIGURE 3.2
Number of adults taking part in sports, 2005–2014. (Source: Sport England.)

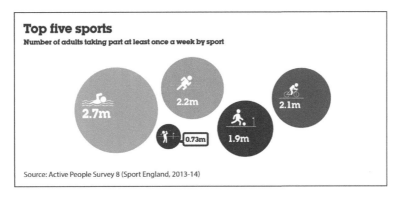

FIGURE 3.3
Top five sports played.

adults who take part once a month, once a week or two and three times a week.

A Sport England (http://www.sportengland.org) survey on the most frequently preferred sports in the UK reported that swimming was mostly preferred to stay fit and as a sporting activity (Figure 3.3), followed by athletics, football, cycling and golf. In this chapter, specific focus will be given to those fabrics used in the manufacture of garments for these sport activities.

It can be noted that most of sport loving populations purchase garments for casual end uses or as fashion apparel rather than for sporting activities. This trend was observed across all forms of garments, from trainers to fleece. It should be noted that the survey was based on Internet users ($n = 2000$) who were above 16 years of age; among these consumers, trainers, jogging trousers, T-shirts, shorts, football shirts, sweatshirts and sports jackets were popular. In addition, at the lower end of the segment, replica rugby shirts, waterproof jackets, fleece, leggings and vests were also popular (Figure 3.4). A recent report from the Mintel Group (2011) stated that 33% of consumers purchased sportswear as comfortable leisure wear, 21% of consumers stated they would prefer branded sportswear and 46% of consumers purchased sportswear clothing to stay physically active and to enhance their performance. This finding is interesting in that the sportswear market has a good base, particularly among younger age groups and most prefer garments to stay fit and support their performance. It is also necessary to note that adults active in sports prefer functional garments for leisure activities.

3.2.1 Essential and Desirable Properties of Fabrics

A fabric property is a characteristic of a material which it should possess for it to be used in a desired application satisfactorily. In other words, it can also be termed as the requirement of a textile material for a certain purpose. In this section, various fabric properties mandatory for performance

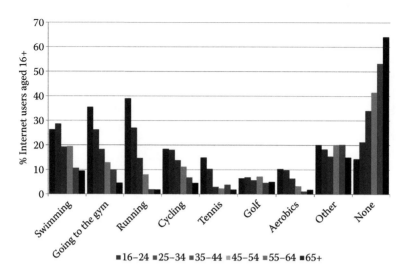

FIGURE 3.4
Sports clothing purchased during 2010/2011 in the UK. (Source: Mintel, 2011.)

and sportswear are highlighted. In order to identify the desirable and essential property it is necessary to know the requirements of a specific sport (Table 3.2). Essential properties are those that are necessary for a particular sport either due to regulations or user requirement. On the other hand, desirable properties are those preferred by users for aesthetics and appearance. Let us explore essential and desirable properties for some of the widely preferred sports. It should be noted that each and every sport has different requirements based on the nature and intensity at which it is played. However, it is assumed these are at professional levels and include those fabrics that are used for casual, fitness and sporting activities.

Some other properties of fabrics/garments required in performance apparel include:

- Absorb moisture readily
- Appearance
- Attractive
- Bend repeatedly without breaking
- Conceal or protect
- Dimensionally stable
- Easy to dispose
- Easy to maintain
- Fabric hangs freely
- Good surface texture

TABLE 3.2

Major Sports and Fabric Parameters Required

Type of Sport and Garments Used	Essential Property of Fabric	Desirable Property of Fabric
Football		
Typical garments include tops, trousers, base layer tights, socks, compression tops and shorts, soft-shell jacket and knee support	• Moisture (sweat) management • Breathable • Anti-cling • Anti-static • Lightweight fabric • Anti-odour • Durable • Washable • Colourfast	• Aesthetics • Sensory comfort • Soft next to the skin
Golf		
Shirts, trousers, jackets, waterproof jackets and socks	• Comfort • Moisture management • Thermal insulation • Durable	• Colourfast • Smooth to skin • Aesthetics – crease recovery and stiffness • Soil resistant
Cycling		
Bib shorts, cycling shoes, short-sleeved jersey, base layer vests, fingerless gloves, socks and cap	• Stretch and recovery • Sweat absorption • Wicking • Compression • Breathable • Windproof • Anti-odour • UV protection	• Long-lasting fit • Durable (good bursting strength) • Fabric stability
Swimming		
Swimsuit, board shorts, jammers, racer-back suits, swim briefs, bodysuits and soft-shell jackets	• Chlorine-resistant fabric • Low moisture absorption rate • Colour fast • Quick drying • Improved elasticity • Drag resistant • Four-way stretch	• Soft feel to the skin • Shape retention • Improved comfort • Support in garment fit • Anti-bacterial • UV protection
Athletics		
Bodysuits, tops, T-shirts, shorts, tracksuits, leotards, sports bras	• Lightweight • Keep cool • Sweat management (wicking) • Breathable • Sustainable (natural fibres) • Waterproof (jackets) • Thermal insulation (fleece) • Compression (base layer)	• Snag resistance • Aesthetics • Colourfastness • Water repellent

- Impact protection
- Lustre
- Nonabsorbent
- Protect the body
- Resilient
- Resistant to insect damage
- Resistant to mildew
- Resistant to shrinkage
- Soft next to the skin
- Stretch without breaking
- Transfer electric charges
- Transfer or maintain heat

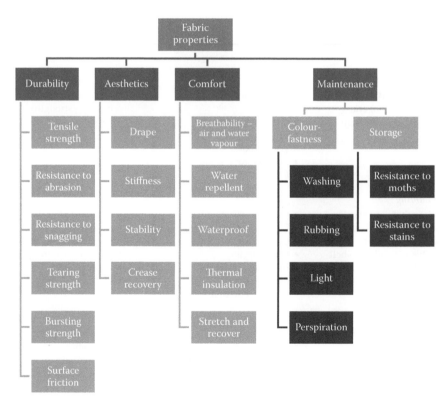

FIGURE 3.5
Fabric properties.

- Wick moisture readily
- Withstand degradation from sunlight
- Withstand pulling force
- Withstand use

Figure 3.5 illustrates the properties of fabrics classified into four sections: durability, aesthetics, comfort and maintenance of a garment on a day-to-day basis. The terminology of various fabric properties is presented in Table 3.3 with specific application.

TABLE 3.3

Fabric Properties and Their Applications

Category	Textile Parameter	Definition	Suitability
Durability	Breaking strength	The force required to break a fabric when it is under tension.	Woven fabric
	Tearing strength	The force required to continue a rip already started in the fabric.	Woven fabric
	Bursting strength	The amount of pressure required to rupture a fabric.	Knitted fabric, felts, nonwoven fabric, lace
	Abrasion resistance	Resistance to wear away of any part of material when rubbed against another material.	All fabric and applications
	Pilling resistance	Formation of pill or fuzz on the surface of the fabric.	Hydrophobic fibres and fabrics with inferior yarn quality
	Surface friction	The ability of the fabric to offer resistance to rubbing force or sliding action.	Fabrics with low yarns per inch
Aesthetics	Fabric drape	The ability of the fabric to drape or to hang on its own weight to follow the body contours and graceful folds/curves.	Woven and knitted fabrics
	Stiffness (fabric handle)	It determines the bending length of the fabric or ability of the fabric to bend under its own weight at a specified angle.	Woven, nonwoven fabrics in both directions
	Crease recovery	The ability of the fabric to resist creasing.	Wool/silk – high resistance to creases, cellulose fibres' poor resistance

(Continued)

TABLE 3.3 (CONTINUED)

Fabric Properties and Their Applications

Category	Textile Parameter	Definition	Suitability
	Dimensional stability	The ability of the fabric to remain stable without change in its dimension after washing.	Woven and knitted fabrics
Comfort	Air permeability	The ability of the fabric to allow passage of air through its surface at a specified pressure difference over a certain period.	Dense woven fabric
	Moisture permeability	The ability of the fabric to allow moisture vapour to pass through its structure.	Fabric with finishes with certain coating for water penetration
	Thermal conductivity	The ability of the fabric to conduct heat (W/m.K); implies the quantity of heat that passes in unit time through a particular area and thickness.	Garments for outdoor wear
	Waterproof	The ability of the fabric to resist water penetration.	Fabrics with film/membrane coating
	Stretch and recovery	The ability of the fabric to stretch under deformation and recover to its original position after removal of deformation.	Knitted fabrics with elastane composition
Maintenance	Colourfastness to washing	It is a measure of how permanent a colour remains on the fabric during washing.	Mild to dark colour/shaded woven/knitted fabrics
	Colourfastness to rubbing	To determine the effect of colourfastness in wet and dry rubbing action.	Woven and knitted fabrics that are bright or dark coloured
	Colourfastness to light	To determine the effect of colourfastness during continuous exposure to light source.	Woven and knitted fabrics that are bright or dark coloured – outdoor/swimwear
	Colourfastness to perspiration	To determine the resistance of the textiles to human perspiration.	Woven and knitted fabrics that are bright or dark coloured worn next to the skin – socks, tights, etc.
Storage	Resistance to moths/insects	The ability of the fabric to resist from moth and insects during storage.	Fabrics with animal fibre composition
	Resistance to stains	The ability of the fabric to staining during long-term storage.	Woven/knitted fabrics

**CASE STUDY: FABRIC REQUIREMENTS
FOR A CYCLING ENTHUSIAST**

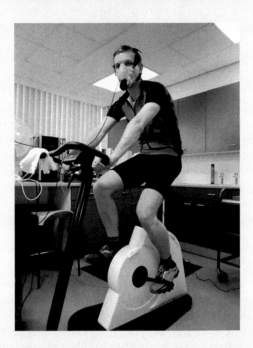

The cross-country cyclist illustrated in the picture (image reproduced with consent) is a regular cyclist who uses bike training twice a month. Each occasion can involve from 5 to 7 hours of intensive cycling. The garments that the cyclist uses include bib-shorts and tops. In addition, elbow support and shinbone protection are part of the kit. Some of the properties that the cyclist prefers in garments are

1. Lightweight
2. Economical
3. Durable
4. Washable
5. Easy to wear
6. Soft next to the skin
7. Breathable
8. Moisture absorbent
9. Practical (fit for purpose)
10. Economical

It could be noted that the cyclist, who is a professional, wants to be involved in the sport as a recreation as well as to keep fit. The cyclist pays significant attention to the clothing gear and desires to wear garments that allow free movement as well as protect from injuries.

3.3 Factors That Influence Fabric Behaviour

Functional apparel will be subjected to a wide range of end uses such that a garment will be affected by internal (fibres, yarn fineness, warp/weft or course/wale movement, fabric density, thickness, fabric count) and external (external environment – exposure to sunlight, wind, rain, cold weather conditions, fabric/human body interaction) factors. These factors affect the performance and behaviour of functional apparel, which are discussed in the following sections with examples.

3.3.1 Internal Factors Influencing Fabric Performance

Fabric is either interlaced with one or more sets of yarns or interconnected with loops of yarns, which are composed of fibres and filaments, and this is the fundamental element. In performance apparel a wide range of synthetic filaments such as nylon, polyester, elastane, acrylic and so forth are widely used. Fabric behaviour is affected by fibre blends and their fineness (McGregor and Naebe, 2013). Similarly, yarn quality (count) is pivotal in producing a uniform fabric texture. Some yarn quality parameters, such as yarn twist, number of folds and yarn count, affect fabric characteristics. For instance, a fine quality yarn made of fine denier filament will possess a supple and pliable fabric that has a low drape coefficient. On the other hand, a coarse yarn will produce a stiff fabric that has a high drape. Fibre type, yarn quality and fabric attributes have an effect on fabric performance. McGregor and Naebe (2013) reported the comfort properties of 81 single jersey knitted fabrics with varying fibre, yarn and fabric attributes. The research studied the comfort properties of wool knitted fabrics with 27 types of fibre blends, 16 types of yarn and 30 different fabrics. The authors reported that tighter fabrics were less comfortable and that progressive blending of cashmere with wool progressively increased comfort assessment. The tactile comfort properties were assessed using a Wool Comfort Metre (WCM). They further added that fabric thickness, yarn elongation and yarn quality (thick and thin places) affected fabric comfort properties (Figure 3.6).

It should also be noted that finer fibres (linear density) influence the behaviour of fabric and its performance, especially in sportswear. A fabric is also termed a micro-fabric when it is made of filaments whose fineness is less than 1 decitex. The term 'micro-denier fibre' is widely used in Asia and the United

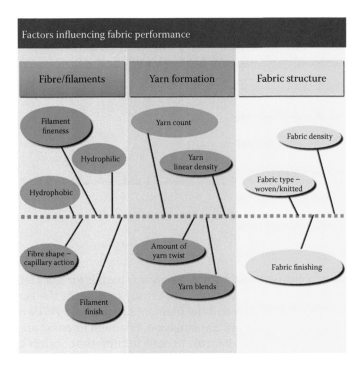

FIGURE 3.6
Internal factors affecting fabric performance.

States. Micro-denier fibres have excellent flexibility and yarns are of better regularity. This parameter enables yarns to be knitted and resulting fabric is soft and pliable (Chattopadhyay, 1997). In addition, knitted fabrics with micro-denier fibres have better dimensional stability and wick the moisture, resulting in better comfort. This makes the micro-fabric ideal for sportswear. Srinivasan et al. (2005) investigated the performance of polyester micro-denier knitted fabrics compared with normal polyester. The study reported that microfibre fabrics possessed excellent drape, moisture transmission property and wicking and were dimensionally stable. Such properties make it ideal for active sportswear fabric (Figure 3.7).

In addition, fibre fineness and yarn quality affect the fabric behaviour, particularly the comfort characteristics – wicking and moisture vapour transmission. Such a parameter is essential in maintaining comfort levels of an athlete wearing a base layer. Sampath, Mani and Nalankilli (2011) investigated the effect of filament fineness on comfort properties of knitted fabrics made of 150-denier polyester filament containing 34, 48, 108, 144 and 288 filaments. The fabrics were finished with a moisture management finish and were assessed for wetting, wicking and moisture vapour transmission. These researchers reported that when filament fineness increases, wicking rate increases to a certain level. The yarn made of 108 filaments had higher

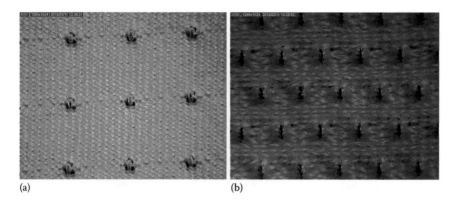

(a)   (b)

FIGURE 3.7
(a,b) Knitted fabric with vents.

wicking. The moisture vapour transmission was higher for finer fabrics than for fabrics made of coarser filaments. This study highlights the fact that the number of filaments in a yarn and the filament fineness should be at an optimum level to promote moisture transmission. Filament fineness and number of filaments in a yarn play a vital role in determining the comfort characteristics of micro-denier polyester knitted fabrics. Mori and Matsudaira (2000) highlighted that fabric density was an essential factor in determining the fabric handle of wool fabrics.

Fabric density greatly affects performance; for instance, a high fabric count has good abrasion resistance, fabric cover and dimensional stability. In addition, the fabric has excellent resistance to wind and reasonable strength. Such a property is widely preferred in work wear and trousers. On the other hand, low-count fabrics possess poor abrasion resistance, low fabric cover, low stability (leading to shrinkage) and low resistance to wind.

Matsudaira et al. (2009) investigated the effects of weave density, yarn twist and yarn count on fabric handle of polyester woven fabrics by using an objective evaluation method. Plain-woven fabrics made of polyester used for women's wear, such as taffeta, Dechine, Georgette 1, pongee, Yoryu and Georgette 2, were selected. Various fabric handle properties such as stiffness, antidrape stiffness, crispness, scroopy feel and flexibility with soft feel were studied. Stiffness increased with increase in weft yarn density (2000 to 5000 picks/metre) for all fabrics; anti-drape stiffness also increased with increase in weft yarn density (2000 to 5000 picks/metre). Fabric soft feel decreased as the weft density increased from 2000 to 5000 picks/metre. Hence, weft density is inversely proportional to fabric soft feel. Fullness and softness did not show any change with the variation in weft density. Larger weft density is needed to produce lower crispness. There was no difference in scroopy feel of the fabric with increasing weft density. The effect of yarn twist was noted in crepe de chine and Yoryu fabrics. In the case of crepe de chine the stiffness and anti-drape stiffness decreased with the yarn twist; however, in the case of Yoryu fabric, little change was noted in all the parameters. The effect of yarn count on taffeta and Georgette, where stiffness and antidrape stiffness decreased with increase in yarn count, was noted.

Prakash and Ramakrishnan (2013) explored the effect of fibre blend ratio, fabric loop length and yarn linear density on thermal comfort properties of single jersey fabrics. Three yarn count qualities (20s, 25s, 30s: Ne) were produced with blends of cotton and bamboo fibres. Investigators reported that thermal conductivity was reduced as the proportion of bamboo fibre increased; lowest thermal conductivity was observed with 100% bamboo yarns. For a given fabric of a certain composition, the air permeability increased as loop length increased. In addition, air permeability of 100% bamboo fabrics was 200% that of cotton fabric. Fabrics made of bamboo blended yarns had a lower thickness and fabric density than cotton fabrics. The water vapour and air permeability improved with the increase in the composition of bamboo fibre content. Water vapour permeability, the transmission of water vapour through fabric from the skin to the outer surface by diffusion and absorption–desorption processes (Das et al., 2009), determines breathability of the clothing material. As the yarn linear density increased, thermal conductivity decreased because fibres trapped more air. Finally, as the yarn linear density increased, the relative water vapour permeability increased particularly for bamboo blended fabrics. Researchers noted that increase in water vapour permeability can be attributed to lower fabric density and thickness. When the yarn count was coarser, the fabric density and thickness increased resulting in lower water vapour permeability. Karahan, Oktem and Ve Seventekin (2006) stated that natural bamboo fibre provided functional properties due to its excellent moisture absorption, quick evaporation and anti-bacterial properties. This shows that a number of internal factors, including fibres, yarn and fabric structure, affect the fabric characteristics, and designers/garment developers should pay particular attention in the selection of appropriate fabric. In the next section, evidence pertaining to external factors affecting the performance of the fabric is discussed.

3.3.2 External Factors Influencing Fabric Behaviour

Performance apparel is exposed to a wide range of external conditions, including sunlight, rain, wind and cold/warm weather conditions or during intense physical activity interaction with the human body. Generally, fabrics meet common requirements such as strength and durability; however, fabrics with special properties require them to be used in special applications intended for high-performance apparel. Typical performance apparel for activewear involves wearing clothing layers and various requirements of fabrics are outlined in Figure 3.8.

Wang et al. (2007) reported a wearer trial using clothing layers (four structures). They reported that the moisture management property of fabrics

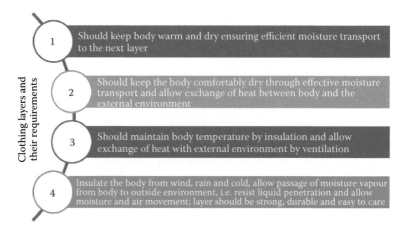

FIGURE 3.8
Clothing layers and their requirements.

significantly affected the moisture diffusion and temperature distributions in cold protective clothing systems and influenced the thermal and moisture sensations, when the wearer was exposed to a climate chamber maintained at −15°C.

Investigators Long and Wu (2011) reported that cotton fabric coated with nanoparticles made of a compound of titanium dioxide and nitrogen breaks down dirt and kills microbes when exposed to light. In the past, the invention was only applicable when fabrics were exposed to certain ultraviolet rays. But researchers reported that they intend to develop fabrics when exposed to ordinary sunlight. The fabric removes orange dye stain upon exposure to sunlight. Fabric coating is durable and remains intact after several washings and dryings.

Some of the other external factors that affect the wearer particularly when exposed to long and continuous amount of sunlight are discussed. Ultraviolet rays can cause damage to textiles as well as cause acute skin damage (Thiry, 2002). It should be noted that the Earth's atmosphere absorbs harmful wavelengths, and only 5% reach the Earth's surface (Rajendrakumar, Aggarwal and Bansal, 2006; Xin, Daoud and Kong, 2004). Researchers Sundaresan et al. (2011) recently investigated UV protection using nano-TiO_2 as well as the anti-microbial and self-cleaning properties of cotton fabrics. The fabrics treated with 12 nm particles exhibited higher ultraviolet protection factor (UPF) values than the fabric treated with 7 nm particles. The durability of the imparted function was in the range of 32 to 36 washes for antimicrobial activity and UV-protection property. In the case of self-cleaning activity, the smaller nanoparticle size TiO_2 derived using a sol-gel technique exhibited better self-cleaning activity as compared to large nanoparticles of nano-sol TiO_2.

When a garment is wet, the comfort is affected and this depends on thermal properties, moisture vapour resistance of clothing and the percentage of moisture accumulated inside the clothing. Hence, during wet conditions, the physiological comfort depends on the thermal resistance in the wet state and the active cooling resulting from the moisture evaporation from skin through the clothing and from direct evaporation of sweat from the fabric surface.

Onofrei, Rocha and Catarino (2011) reported the thermal comfort properties of elastic knitted fabrics produced with functional yarns (Coolmax® and Outlast®) that have thermoregulating effects. The measurements were made at dry and wet states, and the moisture transfer between the fabrics and a wet skin was assessed. Thermal parameters such as thermal resistance, thermal conductivity and absorptivity were reported. Two knitted fabrics were assessed: polyester (Coolmax)/elastane (Creora) and viscose (Outlast)/cotton/elastane (Creora).

3.3.2.1 Dry State Measurements

The air permeability of Coolmax fabric was lower than that of Outlast fabric due to the higher thickness of the Coolmax fabric. The higher surface area of the fibre increased the resistance to air flow, which resulted in lower air permeability. Thermal conductivity of a fabric represents the ability of the fabric to transport heat and results due to a combination of fibre conductivity and structural characteristics of the fabric. In a dry state, the Coolmax fabric exhibited higher thermal resistance, lower thermal conductivity and lower thermal absorptivity when compared with the Outlast fabric. Outlast knit fabric exhibited higher thermal absorptivity values, giving a cooler feeling at first contact with the skin.

3.3.2.2 Wet State Measurements

The rate of air flow through the wet fabric decreased when the fabric moisture increased. The authors reported a decrease in Outlast fabric porosity due to fibre swelling and replaced air spaces with moisture. Both fabrics in a wet state had higher values of thermal conductivity than in dry state. The thermal conductivity and thermal capacity of water is much higher than those of the fibres and air entrapped within the fabric, which leads to a higher heat conduction. For the Coolmax fabric, the effect of increasing moisture content above the 'sweating sensation' was not significant for thermal conductivity and resistance. In the case of Outlast, the increase in moisture content significantly raised thermal conductivity and decreased the thermal resistance. Researchers reported that Outlast fabrics were more prone to significant changes in thermal properties due to moisture uptake than the Coolmax fabrics were. In addition, 60% moisture content acts as a threshold to major changes on Outlast fabrics' thermal properties. Hence, it can be noted that the influence of moisture can affect the thermophysiological comfort of the wearer, and this is applicable to those functional yarns that have thermoregulating effects. Outlast is a thermally active material (phase change material) with viscose fibre structure and thermoregulating effect depending on heat absorption and heat emission of PCMs (Outlast: http://www.outlast.com). Thermoregulation of Coolmax depends on moisture management due to the shapes of fibres (multichannel), which depends on capillary theory – absorb sweat and moisture away from the skin and transport it to the fabric surface, leading to evaporation (http://www.coolmax.invista.com).

3.3.2.3 Commercial Examples – Fabrics for Outer Wear

Table 3.4 outlines some of the recent developments in fabrics with specific properties that prevent or protect the wearer from exposure to extreme weather conditions.

TABLE 3.4

Fabrics with Specific Properties

Specific Fabric Property	Description
Sunlight Protection	
Sun reflector – UV protection	Coldblack with a special finish reduces absorption of heat rays and provides a minimum of 30 UPF protection without
From Schoeller – Solar+ and coldblack	affecting appearance and sensation.
http://www.coldblack.ch	
Rain/Water Repellent	
Ecorepel for outdoors from Schoeller	Ecorepel is biodegradable and offers natural protection from stain and water. The fabric is breathable and soft and is made
http://www.ecorepel.ch	of fibres with long paraffin chains that wrap themselves like
Water repellent and breathable	spirals around filaments and reduce surface tension of water droplets and even mud runoff.
Water Resistance and Thermal Insulation	
Primaloft fibres – thermal insulation and water resistant	PrimaLoft fibres adsorb 100%–250% weight in water and insulate under wet conditions. They are resistant to wind and compressible. The fibres trap a large volume of air resulting in
http://www.primaloft.com	superior thermal performance.
Extreme Cold Weather	
Extremely cold conditions – Polartec Alpha	The fabric is made of three layers: a protective outer layer that is breathable, a middle lofty layer, which traps air, and a smooth inner lining fabric. It dries quickly to minimise heat loss and insulates even when wet. It is appropriate for a broad range of activities as a mid- or outer layer and is wind resistant and machine washable.

3.3.3 Fabric and Human Body Interaction

Clothing physiology is the interaction of clothing and the human body in various environment. Fabrics used for performance clothing should possess a reasonable fabric handle to facilitate the ease of wear. The interaction of fabric/clothing with the skin is regarded as a comfort factor by experts in this field. The human body remains comfortable at 37°C ± 0.5°C and under normal conditions produces heat continuously. Physical activity (exercise) may raise body core temperature by 3°C, whilst exposure to cold may lower core temperature by 1°C. Beyond this range, the human body is susceptible to *hypothermia* – where the skin or blood is cooled enough to lower the body temperature, slow down the metabolic and physiological process, reduce the heart rate and blood pressure, as a result consciousness is lost, which is life threatening.

Normally, when the core temperature increases higher than what it can disperse, the body reacts by producing perspiration from the skin. The evaporation of sweat results in a cooling effect. The body shell and core temperature in a warm environment remains at 35°C; however, in a cold environment the innermost core body temperature remains at 36°C. The boundary of core temperature shrinks to preserve the heat in the brain, thorax and abdomen; temperature of the limbs falls (Pocock and Richards, 2009), especially those in distal regions. Smith and Havenith (2011) investigated the sweat patterns of nine male Caucasian athletes during exercise in warm conditions. They reported that sweat rate increased significantly with exercise intensity in all regions except the feet and ankle. The lower back (posterior torso) consistently showed the highest sweat rates over the whole body. Highest sweat rates were observed on the central and lower back (posterior torso) and forehead whilst the lowest values were observed toward the extremities. These sweat pattern data would facilitate in designing thermoregulatory clothing (particularly for athletes), providing a comfortable microclimate through adequate ventilation, sweat absorption and moisture evaporation. When the surrounding temperature is below the core temperature, the body may lose heat and such heat losses be counteracted by wearing a garment with good insulation properties that can trap air and prevent heat loss.

Base layer clothing, which is often termed 'second skin', interacts with the body, and a number of parameters were investigated relating to physical, psychological and tactile sensation. These include fabric softness (Zhang et al., 2006), fabric–skin friction property (Wang, Liu and Wang, 2010) and tactile properties and subjective measures for next-to-skin knitted fabrics (Mahar, Wang and Postle, 2013). The fabric softness generally depends on fineness of filaments, finishes applied on fabric and the ability of fibres/filaments to transport or wick moisture away from the skin. Mahar et al. (2013) conducted a survey on descriptors of the fabric handle of approximately 50 next-to-skin knitted fabrics. The following bipolar descriptors were identified to describe the fabric handle:

- Smooth – rough (surface property)
- Soft – hard/harsh (flexural property)
- Heavy/thick – light/thin (mass/bulk property)
- Clean – hairy (surface property)
- Cool – warm (perceived temperature)

Yao et al. (2011) conducted an investigation of the interaction of skin and clothing using four cotton/polyester pyjamas (underwear) for three weeks under mild cold conditions. The wearer trial measured skin physiological parameters, subjective sensory response, stress level and physical properties of clothing fabric. Fabrics were treated with hydrophilic or hydrophobic treatment. Subjective sensations were obtained from a questionnaire by rating six sensations (dampness, coldness, itchiness, softness, breathability and overall comfort). A framework was established and the researchers reported that fabric properties influence skin physiology, sensation and psychological response. They further added that physiological effects from overall comfort sensations, sleep quality and stress also influenced skin physiology in terms of skin surface pH and sebum. They explained the relationships among fabric, skin physiology and psychology (Figure 3.9).

A garment plays a vital role in providing physical comfort to the wearer and offering a microclimate air cushion between the skin and garment. When a garment surface is not uniform, this may result in a tickling sensation or blistering in severe conditions where there is constant rubbing between the skin and garment. Pan et al. (2005) presented a simulated model for physical and psychological interactions between skin and fabric including contact and friction as well as during motion such as walking or running. They

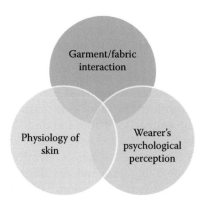

FIGURE 3.9
Fabric–human body interaction.

reported that the primary factors contributing to comfort and well-being of the wearer include fabric friction, fabric density, space between skin and fabric and elastic modulus of the fabric.

Investigators Qian and Fan (2006) reported the interaction of surface heat and moisture transfer from sweating/nonsweating mannequin under varying climatic conditions and walking speeds. Thermal mannequin simulates the human body temperature, the body movement and maintains a skin which is porous that allows the moisture vapour to permeate. The water is pumped to simulate body sweat. The researchers reported that surface thermal insulation was not affected by moisture transfer. The surface moisture vapour resistance at isothermal conditions was greater than those under nonisothermal conditions at wind velocity less than 2.0 m/sec.

The moisture accumulation due to condensation or absorption in sportswear clothing is a concern when the clothing is worn during cold conditions. Hence, it is necessary to know the underlying cause of moisture accumulation while constructing a multilayer assembly of garments, which can maximise moisture transmission and minimise condensation and moisture retained by clothing, as this will cause discomfort to the wearer. As this will cause discomfort due to chill effect. Wu and Fan (2008) established a theoretical model of heat and moisture transfer for multilayer assembly of fabrics. Two multilayer fabric assemblies were investigated: (1) GORE-TEX® inner fabric + multiple ply wool battings + multiple ply polyester battings + GORE-TEX outer fabric + (2) GORE-TEX inner fabric + multiple ply polyester battings + multiple ply wool battings + GORE-TEX outer fabric. It was reported that garment assembly which had wool batting in the inner region closer to the body and polyester batting in the outer region (away from the body) reduced moisture accumulation within and the total heat loss through the clothing assemblies. The information from this research is very useful to the design of multilayer padded jackets for cold weather conditions.

3.4 Fabric Structure and Characteristics

Structural variations of the fabric can affect the performance of fabrics, particularly in facilitating the thermoregulation of the human body. These include, yarn quality, fabric design, fabric layers, and coating on its surface. Özdil, Marmarali and Kretzschmar (2007) highlighted that knitted fabrics made from varying yarn count (linear density), yarn twist and processing (combed/carded) of fibres can affect the thermal properties of 1×1 rib fabrics. These rib knit fabrics are produced from finer yarns (yarn count) and have lower thermal conductivity and higher water vapour permeability values, as well as warmer feeling property. As the yarn twist increased, the thermal absorptivity and water vapour permeability also increased, resulting in a cooling effect. But the thermal resistance values decreased as the twist coefficient of yarn increased. Thermal resistance values of fabrics knitted with combed cotton yarns are lower than for fabrics knitted with carded cotton yarns. It was reported that fabrics knitted with combed yarns displayed higher thermal absorptivity values because the hairiness of this yarn was less and these samples gave the coolest feeling.

3.4.1 Structural Influence and Effect on Performance

Phase change materials have the ability to change their state within a temperature range intended for use in astronauts' space suits to provide thermal protection against extreme climate fluctuation (Mondal, 2008). PCMs absorb energy during the heating process as the phase change takes place, or this energy can be transferred to the environment in the phase change range during a reverse cooling process. Textiles containing PCMs react to the changes in the environment; when a rise in temperature occurs, the PCM microcapsules react by absorbing heat and storing this energy in the liquid state. When the temperature falls, the microcapsules release this stored heat energy and PCMs convert to solid state (Mondal, 2008). The PCMs are used in the thermoregulation of performance apparel where synthetic fibres containing PCM microcapsules are extruded (e.g. Schoeller®-PCM™). The melting point of PCMs is 28°C. PCMs are also incorporated in the form of lamination, coating, bicomponent fibres or foam. Researchers Wang et al. (2006) reported the effect of integrated application of PCMs and conductive fabrics in clothing using a bionic skin model. Clothing assembly of various thickness consisted of

1. Nonwoven fabric plus conductive fabric without heating
2. Nonwoven fabric coated with PCM plus conductive fabric without heating
3. Nonwoven fabric with conductive fabric with heating
4. Nonwoven fabric with PCM and conductive fabric with heating

Conductive fabric was controlled in such a way that it switched on when the temperature of the second layer (2) fell below 27°C; when temperature increased above 29°C it switched off. These researchers reported that the use of conductive fabric can enhance the thermal insulation of clothing and keep the wearer warm by preventing heat loss from the body while using clothing assembly coated with PCMs.

3.4.2 Special Multilayer Fabrics for Protection

Table 3.5 highlights that different types of fabric formation will have varying responses and outcomes under different climatic/weather conditions (Figure 3.10a–d).

TABLE 3.5

Waterproof Breathable Fabrics with Specific Fabric Properties

Densely woven fabrics	Fabrics constructed using hydrophobic and nonabsorptive fibre/yarns with high weave density and interyarn spaces should be small to protect against wind/rain.	Ventile, fabrics from microfibres
Microporous membranes and coating	Microporous membranes have pore size much smaller (2–3 μm) than the size of rain drops (100 μm) and water vapour (40 × 10^{-6} μm). Coated fabrics are composite materials whereby a polymer coating is applied to the fabric surface – polyurethane, polytetrafluroethylene (PTFE), acrylic, polyamino acids, whose pore sizes range from 0.1 to 50 μm. PU coating is widely preferred due to its flexibility, durability and ability of film to suit various end uses.	GORE-TEX, SympaTex
Hydrophillic membrane and coatings	This is based on chemical chain reactions between moisture molecules and nonporous film. The property is obtained by incorporating hydrophilic backbone in the coating material, which increases the affinity of polymer to the water molecule and with good breathability.	Durable, possess good strength, resistant to chemicals, can be designed for higher breathability than microporous material (e.g. SympaTex film)
Biomimetics	Mimicking the analogy of leaf stomata, which has the ability to open and close its structure, the moisture vapour transmission increases in open state and reduces in closed state. Stomatex is a closed foam made of neoprene with a series of convex domes vented by a tiny aperture at the apex. Domes mimic the transpiration process similarly to a leaf providing a controlled release of water vapour, enhancing comfort.	Stomatex is used in conjunction with SympaTex for waterproof insulating barrier.

Source: Mukhopadhyay, A. and Midha, V. K. (2008). *Journal of Industrial Textiles* 37 (3): 225–262.

The following list highlights some of the range of coated/laminated fabrics from the Frizza Group (2013), which are suitable for work wear and outer layer jackets. For instance, Venti Bi-stretch is a lightweight fabric with a dry film Teflon® lamination and has the ability to offer stretch for added comfort. Cordura 1100 is a woven, water-repellent fabric with Teflon lamination and is intended for work wear (Figure 3.11a–c).

1. Venti Bi-stretch – dry film S + Teflon 72 gsm; 65% nylon, 11% elastane and 24% polyurethane
2. Cordura 1100 – dry film bonded to woven fabric with Teflon coating – 410 gsm; 80% polyamide (nylon) and 20% polyurethane

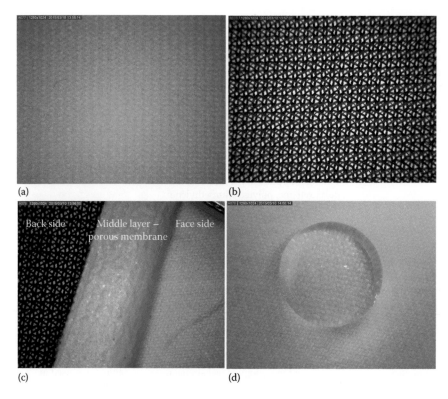

FIGURE 3.10
Microporous membrane – GORE-TEX. (a) Face side of the GORE-TEX fabric. (b) Backside of the fabric. (c) GORE-TEX de-laminated layers. (d) The face side of the fabric with a water droplet.

3. Polystretch – a micropile fabric with Teflon coating 346 gsm; 85% polyester, 14% polyurethane and 1% elastane (Figure 3.11d)

4. ASPEN – dry film + Teflon – 245 gsm, 82% nylon, 7% spandex and 11% polyurethane

3.5 Fabric Composition and Its Effect on Sportswear Performance

Fabric behaviour and characteristics are influenced by fabric composition. Some of the fabric properties that are affected are thermophysiological comfort in the form of water vapour permeability, moisture absorption, wicking and air permeability. The hydrophilic property of fibres also influences moisture absorption and transmission (wicking). The following section presents some of the recent findings along with commercial examples of natural and synthetic fibre blends.

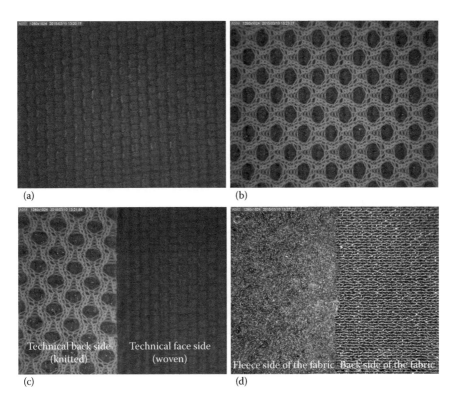

(a)

(b)

Technical back side
(knitted)   Technical face side
(woven)

Fleece side of the fabric   Back side of the fabric

(c)

(d)

FIGURE 3.11
(a–c) Multilayer fabrics. (d) Polystretch micropile fabric with Teflon coating.

3.5.1 Importance of Fabric Composition in Sportswear

Troynikov and Wardiningsih (2011) investigated the liquid moisture management properties of knitted fabrics made of different fibre blends – wool/polyester and wool/bamboo – of varying ratios for base layer sportswear. The researchers reported that blending wool fibres with polyester or wool with bamboo improved the moisture management properties of the fabrics in comparison to 100% wool and 100% bamboo fabrics.

Generally, blending fibres in clothing offers an optimum balance between durability and comfort properties. Investigators Das et al. (2009) explored the effect of fibre blends (polyester/viscose) on the moisture flow of eight different plain woven fabrics (varying the blend ratio of polyester/viscose). Air permeability, water vapour permeability, moisture regain, absorptive capacity and horizontal and vertical wicking were measured. They reported that, as the viscose proportion increased, the absorbency of the material increased linearly, similarly to the effect of moisture regain. They further added that as the fabric structure (cover factor, porosity) was similar among the test samples, the increase in the moisture absorption and regain can be attributed to the increase in the amount of hydrophilic fibres. Air permeability of all the test fabrics was similar, which confirms that the fabrics were similar in their construction. It was also reported that in the case of wicking, a small proportion of polyester fibre (hydrophobic – fibre does not bond with the absorbing group of water molecules) in the fabric increased the vertical wicking due to capillary flow. In the case of 100% viscose fabric, which is hydrophilic in nature (has good affinity to water; when the water molecule reaches it, it forms a bond with the absorbing group of fibre molecules), the wicking was less, due to the fact that fibres swell due to absorption resulting in the inhibition of capillary flow along its surface or between interfibre spaces. This resulted in less movement of moisture along the fabric surface.

This confirms the fact that when the hydrophilic proportion of fibres is increased, the fabric will offer quick absorption from the skin; however, the hydrophilic nature causes poor wicking, which accounts for moisture accumulation resulting in dampness and wet sensations next to the skin. Hence, in the case of activewear, where sweating levels are high, a higher proportion of polyester fibres will wick away the moisture and keep the wearer dry and a small proportion will facilitate quick absorption from the skin (typical fabric composition 80% polyester and 20% viscose).

3.5.2 Natural Fibres and Their Effects on Fabric Performance

Wool fibre has been used for outdoor apparel for its warmth, comfortable handle and superior performance. Some of the advantages of Merino wool include wrinkle recovery, water repellent, antistatic, durable and resilient. The Südwolle Group recently reported its high-performance fibre, Thermo°Cool®, offered by ADVANSA, a leading fibre producer in Europe. They added that for high-intensity sports activity the wearer can benefit from its Duoregulation™ fibre technology with Merino wool.

The combination of ADVANSA Thermo°Cool with Merino wool offers the possibility of light, breathable, exceptionally soft fabrics with a warm, natural, luxurious feel, and good comfort. ADVANSA Thermo°Cool is made with smart fibre cross sections. Fabrics from this blend do not absorb moisture as much as pure Merino wool and are well suited for aerobic sports. The blend of ADVANSA Thermo°Cool with Merino wool is suitable for warm and cold conditions and ideal for layering garments, socks and midlayer fleece. Wool fibres also provide natural odour control when sweat levels are high. The wearer feels fresh, as the fabric (ADVANSA Thermo°Cool) wicks the moisture away from the skin and dries quickly compared to 100% wool fabric. During intense activity, ADVANSA Thermo°Cool evaporates moisture faster than other materials and allows enhanced air circulation, focussing energy on the evaporation process. On the other hand, in the case of low activity, ADVANSA Thermo°Cool acts as a buffer which protects the wearer from changes in external temperature. This supports the wearer from postexercise cooling and allowing excess heat to dissipate into the air (Giebel and Lamberts-Steffes, 2013). Some of the brands by the Südwolle Group include

- Yarn in motion
- Biella yarn

Thermo°Cool has a blend of channelled fibre cross-section and hollow fibres, which creates interfibre spaces, which enables better circulation of air, and channelled filaments wick moisture away from the skin. The improved air permeability allows vapour to reach the exterior, preventing condensation, which affects evaporative cooling, and avoiding excess heat.

3.5.3 Synthetic and Smart Fibres Used in Sportswear

Dyneema®, from DSM, is a polyethylene fibre with ultrahigh molecular orientation which has high breaking strength and is reported to be 15 times stronger than steel (on a weight-for-weight basis) and resistant to cuts, as well as being lightweight and extremely tear resistant (Henssen, 2013).

Figure 3.12 reveals that Dyneema is approximately eight times as resistant to tearing as Cordura (high-tenacity nylon) and three times stronger compared to Kevlar (*para*-aramid).

Abrasion resistance results at various cycles of rub are presented for Kevlar and Dyneema in Figure 3.13. The fabric surface distortion is less for Dyneema at 8000 cycles compared to Kevlar, whereas Kevlar fabric reveals thread breaks at 180 cycles. The illustration in Figure 3.16 shows that Dyneema fabric is water repellent.

Finally, Figure 3.15 reveals that fabric made of Dyneema offers better thermal conductivity and comfort, helping to keep the body cool while it is exercising if a higher proportion of Dyneema is used in the fabric. Hence, fabric made of Dyneema is a high-performance material made of synthetic fibres which is resistant to tearing, lightweight, abrasion resistant, soft and supple

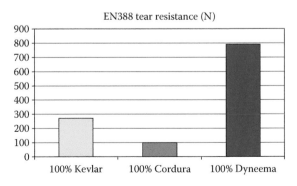

FIGURE 3.12
Tear resistance of fabric made of 100% Dyneema. EN388, European Standard for test requirements of safety gloves. (Source: Dyneema.)

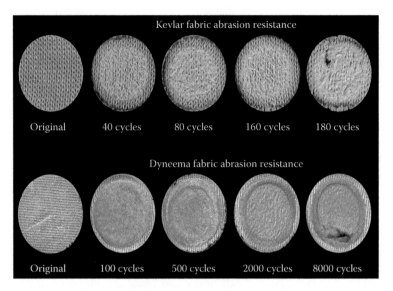

FIGURE 3.13
Abrasion-resistant fabric with Dyneema. (Source: Dyneema.)

FIGURE 3.14
Dyneema is hydrophobic. (Source: Dyneema.)

next to the skin, hydrophobic and fatigue or bending resistant; it offers better thermal conductivity, which makes it ideal to be used in high-performance sportswear and performance work wear.

3.6 Discussion

This chapter has highlighted the importance of fabric characteristics in the design and development of performance apparel, particularly for sportswear.

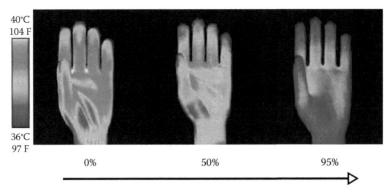

The more Dyneema, the cooler the hand.

FIGURE 3.15
Glove with Dyneema: high thermal conductivity. (Source: Dyneema.)

Recent developments in technology have resulted in the development of high-performance fibres/fabrics. Fabrics for performance apparel include waterproof or breathable fabrics, multilayer fabrics, knitted fabrics, fleece or brushed knits, smart fabrics and eco-friendly and hard-wearing fabrics for work wear. Fabric behaviour varies during use and manufacture. It should be noted that there are numerous requirements from a user perspective using high-performance apparel, and it is not possible to incorporate every possible need. In this context, this chapter has highlighted essential parameters that should be fulfilled in addition to desirable parameters. Widely played sports in the UK were chosen and various essential and desirable properties were presented. The case study also revealed that garments for cycling are multidimensional and functional. Market trends for sports garments are on the rise, as the number of adults involved in sports over the last few years has increased, as highlighted through the Mintel Group's report. There are a number of reasons behind this increased participation, but among many it is due to health benefits; for example, Sport England has stated that physically active adults have a reduced risk of chronic illnesses such as heart disease, diabetes or musculoskeletal concerns. Five major sports were identified as most preferred – swimming, running, football, cycling and golf. Most of these sports require specific clothing gear to support their performance.

A number of internal and external factors were highlighted with evidence and examples to facilitate a better understanding of fabric parameters. Some of the internal factors, including fibre type, blend, fineness, yarn quality, fabric density, thickness and structure, were presented. It was highlighted that in the case of wool-blend knitted fabrics, tighter knit fabrics were less comfortable. Fabrics made of synthetic microfibres were ideal for sportswear – particularly for athletics, as they had excellent moisture transmission properties, wicking and dimensional stability. In addition, the microfibre fabrics have excellent fabric handle, soft and pliable. The fineness of filaments influenced wicking and moisture transmission. In the case of woven fabrics, increase in the yarn density (weft) increased the stiffness of the fabric and the fabric was not smooth. The fabric stiffness decreased with the increase in the yarn fineness. It was also highlighted that fabrics with bamboo yarns had low thermal conductivity. Air permeability for knitted fabrics with bamboo fibres was much better than for fabrics with cotton fibres. Hence, it can be inferred that fabric structure affects its performance, particularly with regard to moisture- and/or thermally related comfort.

Various external factors were highlighted, including exposure to severe cold conditions, exposure to continuous sunlight and measurements of fabric properties in wet and dry states. The fabric properties differ when exposed to cold conditions, especially thermal and moisture sensation. Durable fabrics with TiO_2 nanocoating with self-cleaning ability and protection from UV rays and intended for high-performance outdoor apparel such as that used in mountain climbing or cross-country biking were also presented. It was also interesting to note the comfort measurements in the wet and dry states, and variations were presented. In the wet state, fabric porosity was reduced, resulting in low air permeability. A number of commercial examples were also presented with regard to UV protection, water repellency and thermal insulation.

The effect of fabric and human body interaction should be considered during the design and development of performance clothing, particularly for sportswear. It has been shown that a garment should be able to absorb sweat levels (particularly from the central and lower back), wick the moisture away from the skin, allow moisture evaporation, prevent excess heat loss from the body, avert condensation due to accumulation of moisture and body heat, provide an optimum microclimate between the skin and clothing and, most importantly, provide tactile comfort to the wearer. It was also highlighted that fabric properties influence the psychological well-being of the wearer and tactile sensation. These fabric properties are critical in the selection of appropriate fabrics – taking into consideration the fibre blend, quality, yarn structure, fabric density, fabric structure and lamination or coating techniques used to optimise thermoregulatory comfort of the wearer.

Various types of fabric structures used in sportswear were also presented with commercial examples and illustration. Materials such as PCMs, biomimetics, microporous coatings and membranes are getting popular among those interested in professional garments. Fabric composition in the form of fibre blends also influences the comfort properties, such as moisture regain, water vapour permeability, air permeability and wicking. The fabric moisture absorptivity is influenced by the proportion of hydrophilic fibre blended with hydrophobic fibres. Natural fibres such as Merino wool have also been reported to possess properties suitable for activewear. Pure Merino wool has been blended with varying fibre cross sections to regulate absorption, wicking, air circulation and insulation and to enhance the comfort of the wearer. Synthetic fibres with ultrahigh molecular orientation with superior performance (high breaking strength, lightweight, abrasion resistant, tear resistant) were also highlighted with illustration and their applications were presented. It is anticipated that the reader will benefit from this review of fabric properties, which has presented research relating to fabric properties for sportswear.

3.7 Summary and Conclusions

Fabric properties refer to a set of requirements fulfilled by a material intended for a specific application. Fabric knowledge is essential in the selection, design and development of a garment range for performance apparel and is highlighted in this chapter with examples and visual illustrations. Fabric properties and characteristics vary based on their texture, pattern, design and structure. Fabric knowledge is important for those involved in textile design, garment manufacturing, fashion design, product development, fashion buying, merchandising, quality control, fabric supply and sales. It also becomes a unique selling point for marketing and promoting various claims of a new technology. Garment manufacturing involves a combination of processes; for example, whilst sewing, various settings have to be carried out to prevent seam distortion or puckering. Fabric properties also affect cutting, laying-up of fabric layers and slip caused while handling fine-denier or microfibre fabrics. Understanding fabric behaviour and characteristics will enable technicians, technologists and designers alike to produce a garment fit for purpose and with the desired aesthetics.

Hence, if a company/designer prefers to design a range of base layer garments, the choice of a fabric will be based on a number of factors as highlighted in this chapter: internal and external factors, fabric structure, fibre blend ratio, fabric composition and, moreover, the interaction between the garment and skin. The typical fabric choice will be lightweight and made of microfibres, with a uniform yarn quality, and knitted fabric (medium to fine gauge), which has the ability to wick the moisture from the skin, allow air transmission and, at the same time, prevent excess body heat loss. However, for an outdoor garment, the fabric choice is heavyweight, coated or laminated or multilayered woven/knitted fabric (sometimes quilted) and possibly PCM, which has the ability to resist wind penetration, to repel water and, at the same time, to allow moisture vapour transmission. In addition, the designer should also assess the fabric–skin interaction under dynamic conditions (running, walking) and climatic conditions (wind/extremely cold/ warm) to verify its suitability. Therefore, it becomes mandatory for designers to have a wider understanding of fabric properties and characteristics as highlighted in this chapter so that they produce a performance garment ideal for the chosen application. It is anticipated that the reader, after reviewing this chapter, will be equipped with new ideas and be motivated to further his or her vision in designing functional and performance apparel for future generations.

References

Chattopadhyay, R. (1997). Spun yarns from micro denier fibre. *Man-made textiles in India* 40 (5):193.

Coolmax (http://www.coolmax.invista.com). Online resource accessed 4 June 2013.

Das, B., Das, A., Kothari, V. K., Fanguiero, R. and Araujo, M. D. (2009). Moisture flow through blended fabrics – Effect of hydrophilicity. *Journal of Engineered Fibers and Fabrics* 4 (4) (http://www.jeffjournal.org).

Frizza Group (http://www.frizzagroup.it). Online resource accessed 5 June 2013.

Giebel, G. and Lamberts-Steffes, E. (2013). Merino goes technical with ADVANSA Thermo°Cool. Performance Days Functional Fabric Fair, Munich, Germany, 16 May 2013.

Henssen, G. (2013). Comfortable protective apparel with Dyneema (http://www .Dyneema.com). Paper presented in the Performance Days Functional Fabric Fair, Munich, Germany, 16 May 2013.

Karahan, A., Oktem, N. and Ve Seventekin, T. (2006). Natural bamboo fibers. *Textile and Apparel* 4:236–240.

Long, M. and Wu, D. (2011). Cotton fabric cleans itself when exposed to ordinary sunlight. *Life Science Weekly*, 27 December:194.

Mahar, T. J., Wang, H. and Postle, R. (2013). A review of fabric tactile properties and their subjective assessment for next-to-skin knitted fabrics. *Journal of the Textile Institute* 104 (6):572–589.

Matsudaira, M., Nakano, K., Yamazaki, Y., Hayashi, Y. and Hayashi, O. (2009). Effect of weave density, yarn twist, yarn count on fabric handle of polyester woven fabrics by objective evaluation method. *Journal of the Textile Institute* 100 (3):265–274.

McGregor, B. A. and Naebe, M. (2013). Effect of fibre, yarn and knitted fabric attributes associated with wool comfort properties. *Journal of the Textile Institute* 104 (6):606–617.

Mintel Group. (2011). The sports clothing and footwear, Mintel Group Ltd.

Mondal, S. (2008). Phase change materials for smart textiles: An overview. *Applied Thermal Engineering* 28:1536–1550.

Mori, M. and Matsudaira, M. (2000). Engineered fabrics: Part 3. 29th Textile Research Symposium, Mt. Fuji, Japan, pp. 137–140.

Mukhopadhyay, A. and Midha, V. K. (2008). A review on designing the waterproof breathable fabrics. Part I: Fundamental principles and designing aspects of breathable fabrics. *Journal of Industrial Textiles* 37 (3):225–262.

Onofrei, E., Rocha, A. M. and Catarino, A. (2011). Investigating the effect of moisture on the thermal comfort properties of functional elastic fabrics. *Journal of Industrial Textiles* 42 (1):34–51.

Outlast (http://www.outlast.com), Outlast fabric. Online resource accessed 14 June 2013.

Özdil, N., Marmarali, A. and Kretzschmar, S. D. (2007). Effect of yarn properties on thermal comfort of knitted fabrics. *International Journal of Thermal Sciences* 46:1318–1322.

Pan, N., Zhong, W., Maibach, H. and Williams, K. (2005). Fabric and skin: Contact, friction and interaction. National Textile Centre, annual report, November (http://www.ntcresearch.org/current/FY2005?S05-CD04).

Pocock, G. and Richards, C. D. (2009). The regulation of body core temperature. In *The human body, an introduction for the biomedical and health sciences.* Oxford, UK: University Press.

Prakash, C. and Ramakrishnan, G. (2013). Effect of blend ratio, loop length, and yarn linear density on thermal comfort properties of single jersey knitted fabrics. *International Journal of Thermophysics* 34:113–121.

Qian, X. and Fan, J. (2006). Interactions of the surface heat and moisture transfer from the human body under varying climatic conditions and walking speeds. *Applied Ergonomics* 37:685–693.

Rajendrakumar, G., Aggarwal, A. K. and Bansal, C. P. (2006). UV protection evaluation, Ghaziabad, India. Northern India Textile Research Association, pp. 3–17.

Sampath, M. B., Mani, S. and Nalankilli, G. (2011). Effect of filament fineness on comfort characteristics of moisture management finished polyester knitted fabrics. *Journal of Industrial Textiles* 41:160–173.

Smith, C. J. and Havenith, G. (2011). Body mapping of sweating patterns in male athletes in mild exercise-induced hyperthermia. *European Journal of Applied Physiology* 111:1391–1404.

Srinivasan, J., Ramakrishnan, G., Mukhopadhyay, S. and Manoharan, S. (2005). A study of knitted fabrics from polyester microdenier fibres. *The Journal of The Textile Institute* 98 (1):31–35.

Sundaresan, K., Sivakumar, A., Vigneswaran, C. and Ramachandran, T. (2011). Influence of nano titanium dioxide finish, prepared by sol-gel technique, on the ultraviolet protection, antimicrobial, and self-cleaning characteristics of cotton fabrics. *Journal of Industrial Textiles* 41 (3):259–277.

Thiry, M. C. (2002). Here comes the sun. *AATCC Review*, June, pp. 13–16.

Troynikov, O. and Wardiningsih, W. (2011). Moisture management properties of wool/polyester and wool/bamboo knitted fabrics for the sportswear base layer. *Textile Research Journal* 81:621–631.

Wang, S. X., Li, Y., Hu, J. Y., Tokura, H. and Song, Q. W. (2006). Effect of phase-change material on energy consumption of intelligent thermal protective clothing. *Polymer Testing* 25:580–587.

Wang, S. X., Li, Y., Tokura, H., Hu, J. Y., Han, Y. X., Kwok, Y. L. and Au, R. W. (2007). Effect of moisture management on functional performance of cold protective clothing. *Textile Research Journal* 77 (12):968–980.

Wang, X., Liu, P. and Wang, F. (2010). Fabric–skin friction property measurement system. *International Journal of Clothing Science and Technology* 22 (4):285–296.

Wu, H. and Fan, J. (2008). Study of heat and moisture transfer within the multi-layer clothing assemblies consisting of different types of battings. *International Journal of Thermal Sciences* 47:641–647.

Xin, J. H., Daoud, W. A. and Kong Y. Y. (2004). A new approach to UV-blocking treatment for cotton fabrics. *Textile Research Journal* 74 (2):97–100.

Yao, L., Gohel, M. D. I, Li, Y. and Chung, W. J. (2011). Investigation of pajama properties on skin under mild cold conditions: The interaction between skin and clothing. *International Journal of Dermatology* 50:819–826.

Zhang, P., Liu, X., Wang, L. and Wang, X. (2006). An experimental study on fabric
softness evaluation. *International Journal of Clothing Science and Technology* 18
(2):83–95.

4

Fabrics for Performance Clothing

Tasneem Sabir and Jane Wood

CONTENTS

4.1 Introduction 87
4.2 Fabric and Its Application in Functional Clothing 88
4.3 Basic Structures and Their Influence on Sportswear Performance 89
    4.3.1 Conventional Woven Structures 89
4.4 Conventional Knitted Structures 90
    4.4.1 Classification of Knit Structures 90
        4.4.1.1 Weft Knitting 91
        4.4.1.2 Warp Knitting 92
    4.4.2 Properties of Warp- and Weft-Knitted Structures 92
4.5 Wovens versus Knits – A Case Study: The Speedo Story 93
4.6 Application of Fabric Structures in Sportswear 95
4.7 Future Developments in Woven and Knitted Fabrics 96
    4.7.1 Nanotechnology in Fabrics 97
    4.7.2 Fabric Developments 97
4.8 Conclusions 99
References 100

4.1 Introduction

The aim of this chapter is to provide an overview of the differences between woven and knitted structures and their application in sports garments. Furthermore, the chapter will provide the reader with basic terminology and structural details with their related properties. It further explains the application of woven and knitted structures using examples of some well-known sports brands. A case study on Speedo® explains the development of a swimsuit and the most innovative and controversial suit worn by the Olympic swimmer Michael Phelps. Finally, the chapter will go on to discuss the future of performance fabrics in the area of fibre and fabric developments.

4.2 Fabric and Its Application in Functional Clothing

There are many ways in which fabrics can be constructed; however, generally, fabrics can be categorised into three basic structures: woven, knitted and nonwoven (Eberle et al., 2008). Each of the structures has its own attributes and drawbacks and careful selection is required to ensure that optimum performance is attained for the end user.

In the field of sports and performance wear, woven fabrics dominated early developments as the yarns available were of natural origins and therefore relatively coarse in terms of yarn count. Silks, cottons and wools were all popular as they imparted both protection and performance properties in varying degrees alongside desirable tactile features. However, they were not without their drawbacks and many limitations were apparent (Kadolph, 2007). Swimwear was particularly difficult to produce as the natural fibres absorbed moisture readily, leading to garments that quickly sagged and moved out of shape (Gedeon, 2007).

Around the time of the Second World War, synthetic fibres were developed to such a degree that they became accessible to the mass market. The relatively high durability and low cost of the majority of synthetic fibres made them an appealing option to the consumer (O'Mahony and Braddock, 2005). Whilst most widely available synthetics lacked appeal in sportswear due to their poor breatheability and handle, there was one manufactured fibre that had a major impact on sportswear fabrics.

The creation of Lycra® revolutionised woven fabrics and saw the dawn of a new era in terms of woven close-fitting garments due to their newly found elastic properties (Invista, 2013). By the time of the 1972 Munich Olympics, swimwear had been developed using Lycra-containing knitted fabrics (Stefani, 2012). This sparked a trend in the widespread use of knitted fabrics with improved stretch throughout the sportswear industry.

As technologies developed and the synthetic fibre market became more advanced, the ability to produce extremely fine synthetic filaments encouraged the progression of knitting technology. In conjunction with the advances in synthetic fibres to enhance comfort properties, knitted fabrics became the popular choice for many sports apparel applications. Knitting is seen as the third largest fabric structure after woven and nonwoven structures. In recent times, advances in the technologies of fabric creation, alongside the use of innovative yarns and fibres, has seen interest refocus on woven fabrics, with many new applications being explored within the sports apparel market.

4.3 Basic Structures and Their Influence on Sportswear Performance

4.3.1 Conventional Woven Structures

Conventional woven fabrics are constructed from two sets of yarns: a warp (vertical) set and a weft (horizontal) set, which are interlaced at right angles to each other to form a sheet structure. The regularity with which the yarns interlace determines the nomenclature and influences some of the basic properties of the weave. The most basic woven structure is the plain weave, which involves the interlacing of alternate warp and weft yarns as shown in Figure 4.1.

This structure, although the most basic, is the most commonly found weave in apparel due to its relatively low cost and speed and ease of manufacture alongside its versatility (Elsasser, 2010). Twill and satin weaves, derivatives of plain weave, are also found in apparel and are used to give surface interest and to impart specific properties such as durability (twill) and drape (satin) to the structure.

It could be suggested that woven fabric construction itself does not have to be complicated in order to achieve fabrics displaying high performance. Fabrics created using a basic woven construction can be engineered for a multitude of end uses by clever selection of fibre, yarn and finishing techniques.

One of the earliest types of plain woven fabric known for its protection of the body against wind and moisture is known as Ventile. This fabric is a 100% long, staple-fibre cotton, densely woven basket-weave construction

FIGURE 4.1
Plain woven structure.

(a version of plain weave). When in contact with water, the cotton fibres swell, closing any gaps in the woven structure to such a degree that relatively high pressure is required for liquid to penetrate through to the skin. This imparts a degree of waterproofing to the fabric without any type of chemical finish. The gaps are still sufficient for water vapour to pass from the body to the outer environment, thus allowing the body to 'breathe' and be comfortable (Chaudhari et al., 2004).

Woven structures such as twill and satin have been developed for specific sportswear applications. Racing car driver suits were created by using a blend of carbon, polyester and cotton fibres (Abd El-Hady and Abd El-Baky, 2011). However, it was shown that the selection of the weave structure was of secondary importance to that of the fibre. In this case, the carbon fibre and excellent heat resistance, rather than the weave structure, imparted the specific performance properties required.

Similarly, in the case of performance swimwear garments, the weave structure was of secondary importance as the inclusion of a high proportion of elastane fibre was found to alter the shape of the body, rather than the weave structure itself.

4.4 Conventional Knitted Structures

Sports development encompasses improving the athlete's performance by seeking ways of studying the human body and engineered garments. By creating innovative fibres and fabrics, the textile industry has supported the development of functional clothing to enhance the level of an athlete's performance. The early adoption of knit structures was by luxury brands wanting to create figure-hugging garments. The early part of the 1980s saw the introduction of knits to the sportswear market (Abd El-Hady and Abd El-Baky, 2011). Many of the fabrics produced used high-quality fibres, yarns and complex structures, and these helped to improve the functional properties of garments for the sportswear market. In recent years, the development of both knitted and woven structures has appeared in the same garments, although many of today's highly engineered garments have opted for the knit structure.

4.4.1 Classification of Knit Structures

In contrast to woven fabrics, which can be defined as structures developed from the interlacing of yarns, knitted fabrics are those derived from the interlooping of yarns (Choi and Powell, 2005). Circular and flatbed knitting machines, using needles, distort yarns into loops, which are then interlooped with each other to form the knit structure.

Knit structures were once thought of as inferior to their woven counterparts due to their relative instability; however, innovations in both yarn and machine technologies have elevated knitted materials, in some cases, to have properties that far outweigh those offered by woven structures. Many brands in today's global market are utilising knit structures due to their advanced properties (Power, cited in Fairhurst, 2008). The rise and popularity of knitted fabrics in sportswear applications offer properties that lend themselves to casual leisure activities as well as to extreme performance sports. Knitted fabrics fall into two categories – warp and weft knitting, formed from a single yarn or from many yarns in either a weft-wise or warp-wise direction.

4.4.1.1 Weft Knitting

This form of knitting was commonly associated to hand knitting. Weft knitting is a method where a single yarn is used to generate a row of loops. The interlooping yarns are carried horizontally to form loops in rows as seen in Figure 4.2.

Weft knit structures can use multiple yarns to create complex pattern designs. Generally, the properties of weft-knitted fabrics are that they are soft and pliable and have good handle and drape. These materials have a tendency to unravel course by course and are extremely stretchy. General end use for weft-knitted garments includes socks, T-shirts and sweaters, cardigans and outwear garments (Gao, 2009). Most materials produced are in a tubular form.

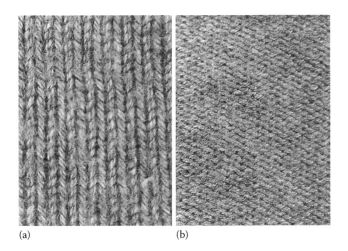

(a)					(b)

FIGURE 4.2
Weft knit structure: (a) face and (b) reverse.

4.4.1.2 Warp Knitting

In warp knitting, yarns are attached to the top of the machine, running vertically to create the knitted loops in a lengthwise direction, interlooping the yarns to form columns of loops. The formation of loops and properties differs significantly between each of the warp-knitted structures.

These materials have little stretch and are less likely to unravel. Warp-knitted structures are stronger and more stable than weft-knitted structures. End uses of these structures are technical applications, sportswear and underwear. Most warp-knitted materials are manufactured flat or in open width form.

4.4.2 Properties of Warp- and Weft-Knitted Structures

The properties of both knitted warp and weft structures are highly dependent on the fibres, yarn type and constructional details. Due to the mobility of the loops in weft-knitted structures, they excel in the thermal properties. The demand of many garments, in particular sportswear, requires the fabric to bend and stretch freely. This is achieved by the constructional details

TABLE 4.1

Weft Knit Structures and Applications

Weft knit structures offer plenty of movement and flexibility, although due to their unstable structures, they can have a limited appearance and can sag unless suitably laundered and stored. The basic weft knit structures are jersey knit (plain), rib knits, purl knits and interlock. These structures make up many of the fabrics in today's apparel market.	
Weft Knit Structure	**Properties and Applications**
Jersey knit (plain)	Jersey (plain) knit is the most economical structure to produce. These structures can be manufactured as lightweight to heavyweight fabrics. The structure allows the fabric to stretch both crosswise and lengthwise (more so in the crosswise direction) and has good drape. In sportswear, this structure is generally associated with polo-shirts, shorts, skirts and jackets.
Rib structure	The constructional details of these fabrics are complex, making its production slower than for plain single jersey. The properties of the fabrics include elasticity and stretch (considerably in the crosswise direction), which are important features in fashion garments. In sportswear, this structure is generally associated with collar, cuffs, neckline, shorts and socks.
Purl structure	Purl fabrics are considered the most expensive to manufacture of the basic knits as they require more production time. Purl fabrics have good stretch in all directions; however, due to the elasticity, the fabrics can be stretched out of shape easily. The fabrics can be quite decorative and are used heavily in children's wear.
Interlock structure	This is a stable fabric structure with limited crosswise stretch. Nowadays, this structure is rarely used in fashion garments (Power, cited in Fairhurst, 2008) but it has found a place in technical textiles. This structure has also been used in activewear, shorts and tops.

TABLE 4.2

Warp Knit Structures and Applications

Tricot and Raschel are two types of knitting machines which produce the vast amount of warp knitted structures in the textile industry. Many fabrics correspond to the two machines as seen in this table. Tricot is associated with plain tight structures, whereas Raschel lends itself to open, jacquard and fancy structures.

Warp Knit Structure	Properties and Applications
Tricot knits	Tricot knitted fabrics account for most of the warp knit structures. It is essential that high-quality, uniform filament yarns are used to create these structures. The characteristics of these structures are uniform in weight and appearance, displaying a tightly knitted structure. Tricot fabrics have little stretch.
Raschel knits	Raschel fabrics are similar to tricot, but are available in a variety of patterns and textural designs. The structure allows for heavy yarns to be used and creates open structures.

Note: In sportswear, these structures are generally associated with activewear, sportswear (seamless), sports shoes, swimwear, undergarments and compression garments.

provided by the knitted structures offering excellent elasticity, comfort and absorption ability (Chen, 2013). Warp-knitted fabrics are generally more stable than weft-knitted structures. However, weft-knitted structures often encounter problems with dimensional stability, as the mechanical stresses applied during washing can distort the fabric. Conversely, the application of finishes to the surface of the fabrics can minimise this behaviour. Tables 4.1 and 4.2 explain the knit structures.

4.5 Wovens versus Knits – A Case Study: The Speedo Story

Speedo is a manufacturer and distributor of swim apparel, with head offices in Nottingham, UK. The company has been a strong innovator of swimwear and sponsors several national swimming teams. Its innovations in swimwear began in 1932, focusing on garment design, with Arne Borg's racer-back swimsuit (Stefani, 2012).

Speedo became synonymous with swimsuits and the development of new fabrics which saw the reduction of drag and turbulence and the introduction of a swimwear bodysuit. Once the swimsuit had covered less; now the garments covered the entire body (Stefani, 2012).

Speedo created one of the most exciting swimsuits, known as the Speedo Fastskin™, which was designed to enhance performance of the swimmer by reducing the effects of friction drag in the water (Toussaint et al., 2007). The Fastskin 'sharkskin-based design' fabric was developed in conjunction with Fiona Fairhurst (a former competitive swimmer) and was a knitted base structure with features that were said to mimic sharkskin to improve the speed of the wearer in the water. It was marketed as 'the world's fastest swimsuit', as seen in Figure 4.3 (Stefani, 2012, p. 14).

The product was successful and saw improvements in racing times for the wearers. The Fastskin was considered a super lightweight suit and saw Michael Phelps beat his own 200-metre butterfly world record wearing the garment (Speedo, 2013). However, the Speedo development team realised that there was still room for further improvement.

A woven fabric 'LZR Pulse', which Speedo claimed to be 'the world's lightest woven swim fabric' was the next innovation, launched in 2008 (Speedo, 2013). The fabric was a densely woven microfibre nylon/elastane blend (Rodie, 2008). Containing two-way stretch, it was highly compressive and the areas of compression were concentrated in the suit at critical points to alter body shape and allow a more streamlined form in the water. Additionally the fabric was treated using nanotechnology to improve its water repellence/absorption, and the enhanced chlorine resistance meant that any degradation of properties such as compression or strength was kept to a minimum.

Clever garment engineering techniques, such as bonded seams, flowed with the contours of the body to further enhance the reduced drag resistance in the water. However, a major 'drawback' of the suit was that swimmers reported it took up to 20 minutes to dress due to the tightness of the fit. The Speedo LZR Pulse fabric was used in competition and swimmers wearing the suit broke a total of 46 world records, whilst at the Beijing Olympic Games, 94% of the swimming races were won by competitors wearing suits made from the

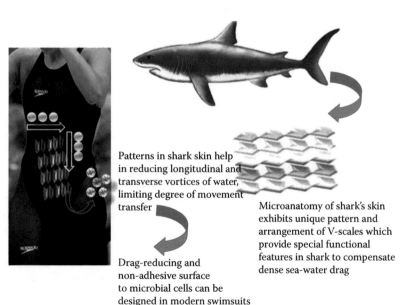

Patterns in shark skin help in reducing longitudinal and transverse vortices of water, limiting degree of movement transfer

Microanatomy of shark's skin exhibits unique pattern and arrangement of V-scales which provide special functional features in shark to compensate dense sea-water drag

Drag-reducing and non-adhesive surface to microbial cells can be designed in modern swimsuits

FIGURE 4.3
Sharkskin swimsuit design.

fabric (Stefani, 2012). This led to the international swimming federation (FINA) imposing a controversial international ban on the 'performance-enhancing' swimsuit (BBC, 2009; Marinho et al., 2012).

4.6 Application of Fabric Structures in Sportswear

The period of the 1980s saw a rise in modern sportswear (Abd El-Hady and Abd El-Baky, 2011). A single fibre or simply coating the fibre/fabric provided the functional characteristics for many garments in sporting activities. Over time, the creation of innovative fibres and complex structures has allowed for a wide variety of properties to be incorporated in one single garment. The application of knitted fabrics in sports garments has increased due to the demand for stretchable, wrinkle-resistant and snug-fitting garments. With advances in digital technology, this has opened the market to create innovative patterns, efficiently and practically. As the living standards of people have increased, their leisure activities have also seen growth where people are more health conscious. The demands for multifunctional clothing to incorporate comfort and health benefits are on the rise (Liu and Liu, 2012).

In any sporting activity, many factors need to be considered to perform at the optimum level. Three main attributes in the success or failure of a sport relate to (1) the athlete's ability, (2) equipment and facilities and (3) engineered clothing (Chowdhury et al., 2012; Yan et al., 2011). According to Feng and Liu (2012) and Onofrei et al. (2011), the most important characteristic of functional clothing is to create a stable microclimate close to the skin to support the body's thermoregulatory system in any physical environment. Tactility (hand) and aesthetics (appearance) are considered important qualities in garments (Emirhanova and Kavusturan, 2008). Understanding the fibre properties and the effects on the fabric is fundamental to the garment. Based on these characteristics, knitted fabrics are commonly preferred (Mikučioniené and Milašiene, 2013).

Some of the well-known engineered garments have amalgamated both science and technology to create functional sportswear. The intricate structures combined with material composition meet the demands of sportswear designs and performance. The structural details and mapping of the properties have led sports companies to concentrate on different aspects of the body. Nike (2013) launched a range of warp-knitted garments to provide breathable and cooling zones within their sports garments. The research and development team mapped the zone areas of females. The data generated allowed an engineered warp mesh structure to apply heated zones to certain parts of the torso.

X-BIONIC® was another recognized sportswear company specializing in incorporating aspects of nature in its most innovative and highly functional clothing range. In 12 years, X-BIONIC has revolutionised functional clothing (X-BIONIC, 2013). Most of the garments manufactured use intricate warp knit structures to create muscle control, support tissue and create partial compression, to name a few. The sophisticated mapping of the tightly knitted warp structures sends signals along the spine to the brain, enabling the body to cool through perspiration. Although this is considered a ground-breaking innovation in the apparel industry, very little research or published literature is available to verify the claims made concerning this clothing.

Woven structures have found a place in sports apparel as 'outer' or protective fabrics, rather than those lying in direct contact with the skin. Woven fabrics have proven particularly useful in protection against wind and rain whilst allowing breathability to enhance wearer comfort. However, it has been the development of fibres, yarns and finishes rather than the woven structure itself that has led to technological advances.

Microfibre woven fabrics are commonly used as 'soft shells' in sporting activities. The 'shell' is the garment forming the outermost layer of clothing, whilst the 'soft' is attributed to the tactility of the microfibre fabric. The development of microfibres has allowed the principle of the densely woven fabric structure to be further explored. Developments by companies such as Invista (2013) are typical examples:

Coolmax® is a microfibre polyester with a 'grooved' cross section which allows moisture to flow away from the body; THERMOLITE® is another polyester-based microfibre, this time with a hollow cross section which imparts insulation properties to the woven structure (ADVANSA, 2013). The microfibres themselves are pleasant to the touch whilst the polyester imparts strength and durability to the fabric. Additionally, the fineness of the filaments enables a tightly woven structure to be developed, which can impart properties such as windproofing without compromise to handle or breathability (O'Mahony and Braddock, 2005). Further fibre developments can be found in Chapter 2.

4.7 Future Developments in Woven and Knitted Fabrics

Integrating both woven and knit fabrics into a sports garment upgrades the wearability, function and design aesthetics of a garment (Chen, 2013). The textile industry is constantly seeking ways of improving garments to enhance performance levels. By combining material engineering and clothing, science has assisted in the physiology and physiological well-being of an athlete. The future developments in woven and knitted fabrics generally lie in its most basic element – the fibre and then further into the complex composites.

4.7.1 Nanotechnology in Fabrics

Developments in sportswear are to study and understand its core material, the 'fibre'. Variations in fibres are based on their dimensions (fineness/length), shape and constitution to increase functional properties required for the sports market including antibacterial, moisture regulation, comfort, breathable and soft and durable, leading to smart and functional designs. According to Nanostart (2013), nanotechnology brings about transformational change to the new era of sustainable energy. Engineered at the molecular level (1 to 100+ nm), the fabrics are manipulated to repel dirt, grease and oil (Wu and Li, 2004). Nanotechnology is being incorporated more into sportswear by reducing the stresses applied to the body or by improving comfort. Nanotechnology can improve textiles by creating a barrier against elements such as dirt, soiling and chemical attacks. Nanotechnology has also seen application in areas of medical and protective clothing. Nanoenhanced materials have incorporated silver to inhibit the growth of bacteria and reduce odour (Gorga, 2010). One of the most novel nanotech textiles was seen in the sharkskin swimsuit worn by the Olympic swimmer Michael Phelps. The suit included a nanoplasma layer, which significantly repelled water molecules and enhanced the swimmer's glide through the water (*Nanomagazine*, 2010). Nano-optimised particles have also been incorporated in the latest generation of clothing for skiing, water activities and golfing.

4.7.2 Fabric Developments

Variation in fabric constructions can be achieved by three methods. One method is to alter the weave or knit construction. This can take the form of complex structures of weaves and knits to more elaborate technical structures seen in 3D and spacer fabrics. 3D weaving is seen as the next big step in development of woven fabric structure. 3D weaving can be defined as 'a fabric, the constituent yarns of which are supposed to be disposed in a three-mutually-perpendicular-planes relationship' (Khokar, 2001, p. 196). This can be taken to mean any one of a variety of structures but it is commonly accepted that the structure is not of the 'flat' planar type usually associated with woven materials used for garments.

This type of woven structure has a noticeable 'depth' to its structure, in effect adding a third dimension to the fabric (Chen and Hearle, 2013). Another way of visualising a 3D woven structure is that of a preformed shape. This technology is predominantly used in highly engineered functions, such as air foils, fan blades and even car manufacture (Ceurstemont, 2011). It has also found use in ballistics protection and body armour (Kaufman, 2012) (Figure 4.4).

Many sportswear products are derived from sources that are associated with strong engineering backgrounds, such as the automotive or aeronautical industries. The development of the Speedo LZR Pulse fabric used techniques more commonly found in the development of cars to assess frictional

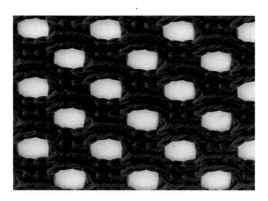

FIGURE 4.4
Warp knit military mesh.

drag of fluids. It is therefore reasonable to assume that 3D fabric could have potential in the sportswear apparel market. Ballistic protection has been explored in depth using 3D fabrics (Chen and Hearle, 2013); this could easily translate into protective wear such as shin pads for cricket, helmets for cycling or body armour type apparel used in American football (Marshall et al., 2002). Chen (2013) reports that even sports apparatus has been explored using 3D woven fabrics, with the development of lightweight golf clubs.

According to Sheikhzadeh et al. (2010), spacer fabrics can be described as two layers of knitted fabrics joined together by monofilament yarn. The fabric can also be referred to as a sandwich with the third layer tucked between the two other layers. The unique feature of the middle layer can take the form of tubes, pleats or engineered forms. The built-in pockets allow the zones to create layers of air, which act as insulation with thermoregulation effects. In the sportswear industry, warp- and weft-faced spacer fabrics can be found in applications such as functional clothing, sports shoes, shoulder pads and knee and elbow protectors (Chinta and Gujar, 2013). Figure 4.5 illustrates a

FIGURE 4.5
Warp knit spacer fabric.

spacer fabric constructed by Mayer & Cie. This is double jersey structure, which is very stable and has seen its application in shoe manufacturing.

'New science that has studied nature's creation has modelled or taken inspiration from these designs to solve human problems' (Benyus, 2002). Biomimetics is a science of using nature to solve human problems, creating innovative products. By observing nature we can mimic the living in sports clothing. A well-known example includes Velcro – a biomimetic example inspired by burs. Moisture management is extremely important in the sportswear market. The opening and closing of vents in clothing have mimicked pinecones. Nike introduced clothing incorporating the pinecone effect, the likes of which Maria Sharapova and Roger Federer wore at the 2006 US Open Tennis Championships.

Composite fabrics use a combination of different fibres or combine different fabrics' construction in one garment. Cloverbrook Fabrics is a leader in performance fabrics. A number of its garments include two-layer cellular construction where each layer is composed of a natural and synthetic fibre. Dry wool is one of the latest fabric developments from Cloverbrook; it uses a combination of Merino wool and synthetic fibres, leaving the body dry and comfortable.

4.8 Conclusions

Early sportswear was dominated by fabrics of natural origin like silk, cotton and wool woven into garments with a degree of performance. With the development of synthetic fibres came about the creation of sportswear using knitting technologies. Advances in comfort properties have led to the industry embracing knitting as a popular choice of constructional methods. However, interest has been refocused on woven fabrics and, using both constructions, many sportswear applications have improved performance.

Sport clothing has to consider not only the athlete but also the environment and activity. Therefore, the application of woven and knitted fabrics works well in this area. Woven materials are classified as the interlacing of yarns at right angles. Basic structures include plain, twill and satin. Each of the weaves has its own benefits and drawbacks. The latest generation of sophisticated clothing, using advanced knit structures have enabled controlled muscle tissue and compression, leading to increased oxygen and blood flow, and comfort. Similarly to woven fabrics, knitting has basic constructions in warp and weft directions. Many of the fabrics today are warp knitted and have been cleverly engineered to incorporate different degrees of tension to improve the optimum performance level of an athlete.

Some well known engineered garments in the marketplace have come from leading sportswear brands – for example, Nike, Speedo and X-BIONIC.

The evolution of technology and science has led to further advancements in fibres and fabrics where the incorporation of nature or a combination of complex structures has seen athletes perform at an optimum level. Woven and knitted structures both share benefits and limitations but the combination in one garment has led to the most successful and innovative garment in the sportswear industry: the Fastskin sharkskin swimsuit. The future in sports developments is to engineer garments that allow improved performance and functionality but remain aesthetically pleasing.

References

Abd El-Hady, R. A. M., and Abd El-Baky, R. A. A. (2011). Enhancing the functional properties of sportswear fabric based carbon fibre. *Asian Journal of Textiles*, 1 (1): 14–26.

ADVANSA. (2013) Coolmax fabrics. [Online] 27 April 2013 (http://www.advanced fibres.eu/).

BBC. (2009). Hi tech suits banned from January. [Online] October 2012 (http://news .bbc.co.uk/sport1/hi/other_sports/swimming/8161867.stm).

Benyus, J. M. (2002). *Biomimicry: Innovation inspired by nature.* New York: Pernnial.

Ceurstemont, S. (2011). Giant 3D loom weaves parts for supercar, *New Scientist* TV. [Online] June 2013 (http://www.newscientist.com/blogs/nstv/2011/02 /giant-3d-loom-weaves-parts-for-supercar.html).

Chaudhari, S. S., Chitnis, R. S. and Ramkrishnan, R. (2004). Waterproof breathable active sports wear fabrics. *Man-made Textiles in India* 5: 166–171.

Chen, S. (2013). An application research into the different-material insertion in knitting clothing design. *Advanced Material Research* 753–755: 1591–1594.

Chen, X. and Hearle, J. (2013). 3D woven fabrics for functional textiles, *Proceedings of the International Conference*: *Advances in Functional Textiles*, Textile Institute, Manchester, UK, July 2013.

Chinta, S. K. and Gujar P. D. (2013). Significance of moisture management for high performance textile fibres. *International Journal of Innovative Research in Science, Engineering and Technology* 2 (3): 814–819.

Choi, W. and Powell, N. (2005). Three dimensional seamless garment knitting on v-bed flat knitting machines. *Journal of Textile and Apparel, Technology and Management* 4 (3).

Chowdhury, H., Alam, F., Mainwaring, D., Beneyto-Ferre, J. and Tate, M. (2012). Rapid prototyping of high performance sportswear. *Procedia Engineering* 34: 38–43.

Eberle, H., Hornberger, M., Menzer, D., Hermeling, H., Kilgus, R. and Ring, W. (2008). *Clothing technology: From fibre to fashion*, 5th ed. Haan-Gruiten, Germany: Europa-Lehrmittel.

Elsasser, V. (2010). *Textiles.* New York: Fairchild Books.

Emirhanova, N. and Kavusturan, Y. (2008). Effects of knit structure on the dimensional and physical properties of winter outerwear knitted fabrics. *Fibres and Textiles in Eastern Europe* 16 (2): 69–74.

Fairhurst, C. (2008). *Advances in apparel production.* Cambridge, UK: Woodhead Publishing Limited.

Feng, L. and Liu, Y. (2012). Development of Coolsmart functional knitwear fabrics. *Advanced Material Research* 503–504: 498–502.

Gao, J. (2009). Design on knitwear fashion. *Asian Social Science* 5 (1): 128–130.

Gedeon, J. (2007). Succeeding in swimwear. *The Beaver: Exploring Canada's History* Aug.–Sept. 2007: 13.

Gorga, R. E. (2010). Nanotechnology in textiles. *Textile World* 160 (6): 33.

Invista. (2013). http://www.invista.com/en/brands/lycra.html.

Kadolph, S. J. (2007). *Quality Assurance of Textiles and Apparel,* 2nd ed., New York: Fairchild Books.

Kaufman, J. (2012). An introduction to 3D-weaving. *Textile World* July/August 2012.

Khokar, N. (2001). 3D-weaving: Theory and practice. *Journal of the Textile Institute* 1(2).

Liu, P. and Liu, Y. (2012). Development of hydroscopic and fast-dry sportswear fabrics. *Advanced Material Research* 503–504: 178–181.

Marinho, D. A., Mantha, V. R., Vilas-Boas, J. P., Ramos, R. J., Machado, L., Rouboa, A. L. and Silva, A. J. (2012). Effect of wearing a swimsuit on hydrodynamic drag of swimmer. *Brazilian Archives of Biology and Technology* 55 (6): 851–856.

Marshall, S., Waller, A., Dick, R., Pugh, C., Loomis, D. and Chalmers, D. (2002). An ecologic study of protective equipment and injury in two contact sports. *International Journal of Epidemiology* 31: 587–592.

Mikučionienė, D. and Milašiene, D. (2013). The influence of knitting structure on heating and cooling dynamic. *Materials Science* 19 (2): 174–177.

Nanomagazine. (2010). Nanotechology and textiles. [Online] 15 November 2013 (http://www.nanomagazine.co.uk/index.php?option=com_content&view=article&id=149:editorial--nanotechnology-and-textiles&catid=44:issue-9&Itemid=151).

Nanostart. (2013). Solving Problems with Nanotechnology. [Online] 15 November 2013 (http://www.nanostart.de/index.php/en/nanotechnology/solving-problems-with-nanotechnology).

O'Mahony, M. and Braddock, S. (2005). *Techno textiles 2.* London: Thames and Hudson.

Nike. (2013). Nike Pro Elite Knit: Seamless, breathable and lightweight. [Online] August 2013 http://www.innovationintextiles.com/nike-pro-elite-knit-seamless-breathable-and-lightweight/#sthash.OLLkHBEy.dpuf).

Onofrei, E., Rocha, A. M. and Catarino, A. (2011). The influence of knitted fabrics' structure on the thermal and moisture management properties. *Journal of Engineered Fibres and Fabrics* 6 (4): 10–22.

Rodie, J. (2008). Ultratech, ultraspeed. *Textile World* [Online] May/June 2008 (http://www.textileworld.com/Articles/2008/May_2008/Departments/QFOM.html).

Sheikhzadeh, M., Ghane, M., Eslamian, Z. and Pirzadeh, E. (2010). A modeling study on the lateral compressive behaviour of spacer fabrics. *The Journal of the Textile Institute* 101 (9): 795–800.

Speedo. (2013). Speedo: Behind the technology: Fastskin LZR suit. [Online] 24 November 2013 (http://www.speedousa.com/technology/popup.jsp?technologyId=Fastskin%20LZR%20Racer).

Stefani, R. (2012). Olympic swimming gold: The suit or the swimmer in the suit? *Significance* 9 (2): 13–17.

Toussaint, H., Truijens, M., Eelzinga, M., Van de Ven, A., De Best, H., Snabel, B. and De Groot, G. (2007). Effect of a Fast-skin™ 'Body' suit on drag during front crawl swimming. *Swimming; Sports Biomechanics* 1 (1): 1–10.

Wu, C. E. and Li, Y. (2004). The influence of nanotechnology toward sports. Pre-Olympic Congress. Sport Science through the Ages: Challenges in the New Millennium, Athens, 2004.

X-BIONIC. (2013). https://www.x-bionic.com/labs/materials.

Yan, Y., Yi, X., Tao, J., Lin, M. and Li, R. (2011). Research on the influence of the fabric organization to wind-resistant property of sportswear. *Advanced Materials Research* 331: 646–649.

5

Composite Fabrics for Functional Clothing

Jane Ledbury and Emma Jenkins

CONTENTS

5.1 Introduction 104
    5.1.1 Laminates and Coatings 105
    5.1.2 Microencapsulation 105
5.2 Application of Composite Fabrics 106
    5.2.1 Composite Fabrics for Protection and Survival 106
    5.2.2 Chemical and Biological Protection 106
    5.2.3 Physical and Mechanical Protection 107
    5.2.4 Flame and Heat Protection 107
    5.2.5 High Visibility 108
    5.2.6 Outdoor Clothing and Sportswear 108
5.3 Human Physiological Response and Functional Requirements
    of Composite Fabrics 108
5.4 Composite Fabrics for Functional Outerwear 109
    5.4.1 Technologies 110
    5.4.2 Functionality Considerations 111
5.5 Measurement Techniques and Comparison of Performance 113
    5.5.1 Water Resistance 113
    5.5.2 Water Vapour Permeability 114
    5.5.3 Performance Comparisons 114
5.6 Innovation in Composite Fabrics for Functional Outerwear 114
5.7 Case Study: Jackets for Hill Walkers 116
    5.7.1 Performance of Composite Fabrics for Hill Walking Jackets.... 116
    5.7.2 Consumer Requirements for a Hill Walking Jacket 117
        5.7.2.1 Consumer 1 117
        5.7.2.2 Consumer 2 118
        5.7.2.3 Consumer 3 118
        5.7.2.4 Consumer 4 120
        5.7.2.5 Consumer 5 120
    5.7.3 A Ventilation Design Feature for Jackets for Hill Walking 121
    5.7.4 Evaluation of a Prototype Hill Walking Jacket 121

5.8 Composite Fabrics: Considerations for Garment Design
and Development 130
    5.8.1 Design Features 131
    5.8.2 Assembly Techniques 132
5.9 Case Study: Survival Clothing for Military Flight Crew 133
    5.9.1 Current Clothing Analysis 133
    5.9.2 The Design Brief 134
5.10 Challenges of Designing Cold Weather Clothing Systems
for Aircrew 135
    5.10.1 Environmental Considerations 135
    5.10.2 Human Factors Affected by Cold Weather 137
    5.10.3 Thermal Balance and Layered Clothing Systems 138
        5.10.3.1 Base Layer 139
        5.10.3.2 Midlayer 139
        5.10.3.3 Outer Protective Shell 139
5.11 Design Process 139
5.12 Composite Materials for Cold Weather Survival Clothing 141
    5.12.1 Selection of Fabrics and Materials 141
        5.12.1.1 Protection Layer 143
        5.12.1.2 Comfort Layer 144
        5.12.1.3 Insulation and Ventilation 144
        5.12.1.4 Mobility 146
        5.12.1.5 Fit 146
        5.12.1.6 Construction 146
5.13 Conclusions 147
Acknowledgements 148
References 148

5.1 Introduction

Composite fabrics comprise two or more materials which are bonded together to produce a structure with entirely new characteristics. Composite fabrics are composed of a textile base, strengthening resin, a stabilising filler and chemical additives, which provide additional or improved functional properties (Lawler and Wilson, 2002). Materials are combined that are dissimilar and which have distinct mechanical properties, in order to produce the best possible blend of high-performance properties for a range of applications/functions. Composites are constructed utilising methods used in the production of coated and laminated fabrics, such as surface coating, impregnation, and lamination (Fung, 2002). Composite, laminated and coated fabrics are used for a broad range of applications in functional and protective clothing, and are often referred to as 'technical textiles'. Protective apparel, which includes military clothing, personal protective equipment (PPE), personal protective systems (PPS) and sportswear, represents a substantial portion of the technical textiles market (Fung, 2002).

5.1.1 Laminates and Coatings

An important development in waterproof breathable clothing was when GORE-TEX®, a microporous polymeric film made from polytetrafluoroethylene (PTFE), was introduced to the market in 1976 (Shishoo, 2002). The microporous film provides a barrier to water droplets (rain) whilst allowing the diffusion of moisture vapour through the film. GORE-TEX film is used in both two-layer (fabric bonded to one side of a fabric) and three-layer (fabric bonded to both sides of the film) construction in order to produce high-performance materials (Shishoo, 2002).

Microporous composite fabrics, constructed from PTFE laminates, have the greater share of the outdoor clothing market. PTFE laminates generally prevent rain penetration, whilst allowing moisture to be transferred from the inner to the outer environment. In addition to microfibrous membranes such as GORE-TEX, microporous coatings (coated fabrics) and hydrophilic membranes (laminated fabrics) offer similar functions, preventing the passage of rain through the textile, whilst allowing the transmission of water vapour to the outer environment.

According to Shishoo (2002), developments in multifunctional coated and laminated fabrics have resulted in new and innovative (smart) textiles for protective clothing which offer waterproof, breathable and heat insulation, as well as heat protection or antistatic properties for functions such as firefighting and military applications. Schoeller®Textiles developed NanoSphere®, which allows water and oil to run off the surface of the fabric and prevents staining. NanoSphere is a finishing technology which mimics the self-cleaning effect and water repellence of plants; dirt cannot adhere to the finely structured surface of the fabric (Schoeller, 2014).

5.1.2 Microencapsulation

Thermal comfort plays an important part in performance for both flight crew and hill walkers, therefore micro-encapsulation will be considered briefly within this chapter. Phase change material (PCM) used in outdoor clothing and sportswear was originally developed by Outlast® Technologies for NASA. Phase change materials, which can change from a solid to liquid state, are microencapsulated and enclosed permanently in a polymer shell (Outlast, 2014), which can be coated onto fabrics or incorporated into a polyurethane foam. The microencapsulated PCM absorbs and stores surplus body heat and releases excess heat back to the body in order to regulate the body's microclimate (Outlast, 2014). PCMs are useful for sportswear and work wear applications, where phases of physical exertion, which generate heat, are followed by periods of inactivity, resulting in rapid cooling.

Other microencapsulated materials used in sportswear applications include antimicrobial, moisturising and deodorising chemicals, fragrances and aromatherapy oils. Additionally, nanocomposites provide antisoiling properties for a range of work wear and sportswear applications (Fung, 2002). Micro-encapsulated materials, such as PCMs, originating from NASA in the development of apparel for space travel, remain crucial considerations in the design of clothing for military applications.

5.2 Application of Composite Fabrics

5.2.1 Composite Fabrics for Protection and Survival

'Protective clothing refers to garments and other fabric-related items designed to protect the wearer from harsh environmental effects that may result in injuries or death' (Adanur, 1995, cited in Zhou, Reddy and Yang, 2005). The application of technical textiles for survival is concerned with protection and ranges from complex military apparel to work wear such as personal protective equipment (PPE), which may include flame, chemical and biological protection. Protective clothing can be uncomfortable, bulky, heavy and restrictive (Fung, 2002); however, significant improvements have been made with the development of new composites enabling designers and product developers to satisfy increasing demand for protective clothing that meets user needs and is comfortable. Wearers of protective clothing may demand garments which offer protection against extreme and harsh environments (both hot and cold), contaminants, physical hazards and impact. The designer must source materials which satisfy the need for protection as well as comfort.

Hazards are generally categorised as chemical, biological, physical/mechanical, radiological, flame and thermal. These categories are used to establish national and international standards and performance specifications. The selection of textiles which are fit for purpose is critical to the function, use and care of protective and survival clothing. An appreciation of the working environment and functional requirements, together with knowledge of textiles, is crucial to the successful development of PPE and apparel for first responders against chemical, biological, radiological, nuclear, and explosive (CBRNE) hazards (Shaw, 2005).

5.2.2 Chemical and Biological Protection

According to Shaw (2005), textiles for chemical protection can be woven or nonwoven, laminated to membranes, coated, bonded or laminated with plastic films or rubber. In order to minimise risk to the wearer, materials are selected that provide resistance to penetration or permeation of chemical contaminants, thereby preventing dermal exposure. Demand for barrier materials for clothing in the health industries has increased over recent years due to risk of exposure to blood-borne pathogens (HIV and hepatitis B). Barrier materials for protection include laminated nonwovens and microporous polytetrafluoroethylene (PTFE) films laminated to nonwovens (Fung, 2002).

5.2.3 Physical and Mechanical Protection

Physical and mechanical hazards encompass stab, ballistic and abrasion protection. Antistab materials require a dense construction, which provides resistance to penetration from sharp objects. Alternatively, a protective coating of hard particles (silicon carbide) may be added to the fabric. The purpose of the particles is to blunt sharp objects, adding to the penetration resistance of the material. Recent developments to protect against abrasion include small domes of strong resin, which are bonded to the outer surface of the fabric or garment (Buckley, 2005). X-BIONIC® has produced apparel for road racing which uses a similar technique with a base of boron carbide for its 'Black Diamond' cycle road-racing suit.

Ballistic protective vests comprise many layers of fabric; however, some garments include a composite plate to reduce bulk. The plate is designed to spread the area of impact and reduce soft-tissue damage. Despite effective protection, composite plates are rigid and impermeable, which may have an impact on thermal burden and comfort for the wearer (Buckley, 2005).

5.2.4 Flame and Heat Protection

Flame and heat thermal protective clothing requires materials that have intrinsic flame-resistant properties; additionally, a multilayer construction is used, with each layer fulfilling a particular purpose (Shaw, 2005). Current firefighters' clothing in the United States is constructed of three layers laminated together with an aramid outer layer, a barrier midlayer of neoprene-coated aramid or a GORE-TEX film laminated to nylon and a thermal insulating layer. Whilst the layered garment has proved efficient, the possible deterioration of the barrier film is a cause for concern, as this would leave the wearer vulnerable to risk (Fung, 2002). Where a finish is applied in order to protect the wearer, the durability of the finish is of prime importance as it may ultimately affect the performance of the product (Shaw, 2005). Military and fire services currently use FR aramid or aramid/FR viscose blend yarns, which have essential flame-resistant properties in materials for protective outerwear, as the fabrics are both strong and durable (Buckley, 2005). According to Shishoo (2002), it is possible to combine the functions of different fibre qualities such as *p*-aramid, which provides high strength, and *m*-aramid, which offers flame and heat resistance, to form the basis of protective clothing with exceptional strength and temperature resistance which can be used by the fire service and for military applications.

5.2.5 High Visibility

Other applications of textile coatings for PPE include high-visibility material. Two types of high-visibility material can be produced by textile coating, either by incorporating pigment into resin, which is applied as a coating, or by a preprepared film containing pigments that may be laminated onto fabrics. A third type of high visibility is produced through reflective microprisms, which are integrated into a film to reflect light. The emergency services and first responders, as well as sportsmen and -women, commonly use reflective material such as Scotchlite® in order to enhance road safety (Fung, 2002).

5.2.6 Outdoor Clothing and Sportswear

The use of composite fabrics for outdoor clothing is concerned with protection against the elements, primarily wind and precipitation. In tandem with these primary considerations, it is necessary to take into account the thermophysiological comfort of the wearer. Composite fabrics used in sportswear may include protection against impact or other hazards and may incorporate aspects of performance. A commonly used composite textile for sportswear is neoprene, which is used predominantly for wetsuits in a range of water sports.

5.3 Human Physiological Response and Functional Requirements of Composite Fabrics

Heat generated by the body during exercise is proportionally similar to oxygen consumed (Johnson, Benjamin and Silverman, 2002). This is because the conversion of chemical energy into mechanical energy during muscle contractions is an inefficient process. Even in optimum conditions 75% of energy is released as heat, causing the participant to start warming up (Cheuvront and Haymes, 2001). In response, the hypothalamus region of the brain redirects blood flow to the skin, where heat can be transferred to the surrounding environment by conduction, convection and radiation (Robinson, 2000). However, this is dependent on a favourable thermal gradient and the presence of convection currents, which can be inhibited by clothing. Therefore, the principal means of heat exchange in humans is by sweating.

Evaporation is the process by which molecules change from a liquid to a gaseous state. This requires heat to raise the kinetic energy of molecules and break the bonds connecting them (Cengel, 1997). The latent heat for vaporisation of sweat is taken from the body surface, providing a substantial cooling effect. Low relative humidity and convection currents speed up the process by increasing vapour diffusion away from the skin (Cheuvront et al., 2004). Outerwear garments must therefore have high air and water vapour permeability. If the fabric or design prevents this, sweat and water vapour may accumulate within the underlying clothing layers and limit evaporative power by increasing local relative humidity (Spencer-Smith, 1978; Havenith, 1999; Chen, Fan and Zhang, 2003). This leads to tactile sensations of discomfort related to the buildup of heat and moisture, such as clamminess and prickliness (Yoo and Barker, 2005).

Water collected in the underlying clothing layers also displaces trapped still air, which would ordinarily provide insulation. This significantly reduces thermal resistance by conducting heat away from the skin along the path of water flow (Brownless et al., 1995). The net reduction in insulation is related to volume of water accumulated (Hall and Polte, 1956; Chen et al., 2003). Therefore, the outer garment must also prevent rain, snow and drizzle entering the clothing system.

Insulation is also diminished by wind. Heat is conducted to the cooler air and transferred away from the body by the moving stream. This forced convective cooling is intensified by lower air temperatures and higher wind speeds. Another function of the outer layer must therefore be to provide a shield against the effects of wind chill.

5.4 Composite Fabrics for Functional Outerwear

Composite fabrics are engineered to protect the clothing system from penetration by rain and wind, but to allow water vapour from the evaporation of sweat to diffuse to the external environment. Historically, this has been achieved through application of close-woven fabrics constructed from highly processed yarns in densely packed plain weave structures (Lomax, 1991; Holmes, 2000). However, modern composites have now all but replaced these in the marketplace due to their improved water resistance, breathability, weight, durability and versatility.

Composite fabrics have become a staple for outerwear performance where a number of specific functional attributes are required from one garment. This is particularly true for outdoor sports where protection from the weather and maintenance of comfort are paramount for both safety and comfort. Recent market research indicates that customers would invest in clothing designed specifically for their chosen sporting activity and are prepared to spend more money on innovative fabrics to improve their comfort and performance (Mintel Group, 2011, 2013). This is supported by the European Outdoor Group (2013), which suggests that the annual turnover of outdoor goods across Europe in 2012 exceeded €10 billion, with 53% of this value being attributed to clothing (as opposed to footwear and other accessories). Furthermore, the market specifically for waterproof breathable fabrics is expected to grow by approximately 5% per annum until 2016 (Anon, 2013a).

Products in the UK outdoor sportswear market can be broadly categorised into garments for mountaineering, hill walking, fast and light activities (such as mountain biking, fell running and adventure racing) and snow sports. The primary function of all these garments is to maintain a comfortable microclimate for the wearer. However, their performance specifications change according to variables in the use environment and the heat and moisture generated by the intensity of the activity performed. Moreover, the shape and design of the garment are also partially dictated by the type of activity being undertaken. For example, jackets for mountaineering will generally be extremely water and abrasion resistant and the fit may be slightly longer and wider to allow for layering. In contrast, the jackets used for fell running will tend to be highly breathable with a slimmer fit, and the incorporation of design features is kept to an absolute minimum to save on weight.

The conventional approach to achieving and maintaining comfort in all these sports is by active adjustment of a layered clothing system. This consists of a base layer worn next to the body and engineered to rapidly wick moisture away from the skin, insulation midlayers designed to trap air and retain warmth and an overlying outer layer for protection from the elements. This outer layer is predominantly composed of composite materials. In this chapter, application of composite fabrics is presented in the context of functional apparel. Recent innovations in the area of composite fabrics are described in relation to human physiology responses to outdoor sports. Focus group responses from five outdoor wearers are presented to highlight specific requirements for their chosen activity. An in-depth analysis of hill walking jackets is discussed with specific user requirements, and appropriate assembly techniques are discussed with regard to whole garment performance and functionality. The case study on survival clothing for military crews examines the design and development of a low-burden, high-performance survival jacket which forms part of the cold-weather clothing system for military flight crews. It further highlights the requirements of air crews that are unique, which demands a structured approach with user-centred clothing design to ensure satisfying user needs as well as facilitating the survival of the user in cold weather conditions, and offers to carry out their primary task.

5.4.1 Technologies

Composite fabrics are constructed from a base fabric with a polymer bonded to the reverse side, next to the skin. The polymer confers the performance characteristics to the base fabric and can be coated, or laminated in place. Methods of coating allow the liquid polymer to be directly applied to the base fabric and include knife systems (direct coating), transfer from release paper and rotary screens. In a laminated fabric, the polymer is incorporated into a film or membrane, which must then be joined to the base fabric using an adhesive, with heat and pressure (Fung, 2005).

The structure of the base fabric must be selected for its durability. During normal use, it can be expected that the garment will be constantly abraded by the repetitive flexing action of the exercise and against rucksacks, harnesses and rock faces. It is also at risk from snagging by thorns and other hazards in the use environment. The fibres must therefore be strong, dimensionally stable and resistant to rotting, and the fabric structure must be densely woven.

Contemporary waterproof and breathable fabrics are broadly classified as microporous or hydrophilic. Microporous fabrics incorporate polymers manufactured from expanded polytetrafluoroethylene (ePTFE) or cheaper polyurethane (PU) (Mukhopadhyay and Midha, 2008). They add a system of micropores and channels, which are permeable to water vapour and impermeable to liquid water (Lomax, 1991; Holmes, 2000).

Hydrophilic fabrics are manufactured using polyurethane, or polyester polymers. These do not have pores and water vapour is diffused through them from the inside to the outside face by a molecular process. Very thin hydrophilic layers can be added to the face of microporous fabrics to improve water and contaminant resistance, without compromising breathability (Lomax, 1991). This is often referred to as a bicomponent, olephobic coating or membrane.

5.4.2 Functionality Considerations

Several studies have shown that waterproof breathable fabrics can be ranked according to their capacity to transfer water vapour to the ambient environment. These agree that ePTFE laminated membanes are most permeable, followed by hydrophilic and PU microporous laminates and then PU coatings (Holmes, Grundy and Rowe, 1995; Ruckman, 1997a,b). Laminated fabrics also provide increased flexibility with regard to the structure of the base fabric (e.g. they can be incorporated into stretch fabrics) and present superior handle and drape (Fung, 2005). However, coated fabrics are cheaper to manufacture and estimated to account for approximately 80% of the total waterproof breathable fabric market (Anon, 2013a).

Membranes and, to some degree, coatings are both relatively fragile and must be protected from repeated abrasion and contamination from oil and dirt (Weder, 1997). This can take the form of a loose lining in the garment, or additional materials incorporated into the laminated structure. Carefully selected, these can increase the functionality of the composite (see Table 5.1).

> **Two-layer laminates:** Two layer laminates are prevalent in the middle market for skiing, hiking and fashion applications. They are currently being redefined as new 'basics' that have the potential to

TABLE 5.1

Composition, Function and Application of Laminated Composite Fabrics

Classification	Composition	Function	Application
2-layer	Base fabric + membrane	Waterproof, windproof and breathable	Middle-market hiking, mountaineering, skiing and fashion
2.5-layer	Base fabric + membrane + raised printed backing	Lightweight, compressible, waterproof, windproof and breathable	Fast and light, high-intensity activities
3-layer	Base fabric + membrane + soft tricot backing	High durability, flexibility, waterproof, windproof and breathable	Premium hiking, mountaineering and skiing
Soft shell	Base fabric (+ membrane or coating) + soft backing	Insulation, flexibility, windproof, breathable and water resistant	Multifunctionality and high-intensity activities
4-layer	Base fabric + membrane + insulation + soft tricot backing	Insulation, waterproof, windproof and breathable	Hiking, skiing, mountaineering and fashion
Drop lining	Membrane + lining	Waterproof, windproof and breathable	Middle-market hiking, skiing and fashion

combine high performance with a directional look. For example, canvas, herringbone and denim effects can be incorporated into the base fabric. Loose linings must be added to these garments and tend to be warp-knitted mesh constructions to promote airflow. Brushed finishes also add insulation and improve hand feel. Taffeta is commonly used in the sleeves and trouser legs, to increase the ease of donning and doffing the garment, and adding a PU coating at the hem and cuff will prevent rain wicking into the interior.

2.5-layer laminates: Another method of protecting the laminate membrane is to print it with a raised pattern, to provide a barrier between the skin or other clothing layers (see Figure 5.9 later in this chapter). This is called a 2.5-layer construction and removes the requirement for a loose lining. Therefore, subsequent garments are much lighter and more compressible and applicable to fast and light activities, such as fell running and mountain biking.

Three-layer laminates: Three-layer systems sandwich the membrane between the base fabric and a soft tricot scrim backing. It produces a heavier fabric, with reduced drape, but increased durability (Mukhopadhyay and Midha, 2008). It also allows the designer to incorporate a level of elasticity, which would otherwise be impaired by use of a fixed lining. Performance garments incorporating these fabrics are currently very important in the outdoor sportswear market and promoted as exclusive high-performance.

Soft shells: The three-layer construction also encompasses another class of composite, referred to as 'soft shell'. These integrate the warmth, flexibility and high breathability of fleece, with wind and water resistance. Densely constructed stretch materials are selected for the base fabric and thermal, or high wicking, materials for the reverse. Membranes and coatings can be sandwiched in between to give waterproof performance. However, durable water-repellent (DWR) finishes are more frequently used so that the highest level of breathability can be maintained. Recent innovations include the use of Merino wool for its hygroscopic properties and soft handle. Garments often have an active fit that is cut close to the body, to maximise the function of the inside face. They are designed for multi-functionality and high-intensity activity.

Four-layer laminates: Four-layer fabrics extend the soft shell concept by adding a layer of insulation. They comprise a base fabric, waterproof breathable membrane, thermal padding and soft backing. Manufacturers claim that the construction also stabilises the thermal batting without conventional quilting so that it does not migrate or bunch (Fratitex, 2015).

Drop liners: Drop liners are a lining fabric laminated with a waterproof, breathable membrane. This allows the shell fabric to be entirely separate and selected for its appearance and drape, which is ideal for more fashion-focused applications. The lining is positioned so that the membrane is inherently protected by its orientation within the garment construction (Holmes, 2000).

5.5 Measurement Techniques and Comparison of Performance

5.5.1 Water Resistance

The British Standard method for measuring the water resistance of composite fabrics uses the hydrostatic head test (BS EN 20811:1992). The fabric is clamped in a rubber gasket and water is applied under pressure to the face of the specimen. The water pressure is increased until penetration occurs in three places. Other tests (Bundesmann and WIRA shower test) may replicate rainy conditions more accurately, but the hydrostatic head test has the benefit of being simple and fast. It can also be used to gauge the effect of loaded rucksack straps, sitting or the pressure of elbows on the fabric (Holmes, 2000).

5.5.2 Water Vapour Permeability

The main techniques for measuring the vapour permeability of fabrics are the upright cup, inverted cup and guarded sweating hot plate methods. However, the values obtained from each method are not comparable due to differing test conditions, which drive vapour transfer at dissimilar rates and favour particular fabric types. Therefore, one method must be selected and adhered to (McCullough, Kwon and Shim, 2003).

The British Standard (BS 7209:1990) employs the upright cup method. The fabric to be tested is sealed over a cup of water and the mass lost over a given time period is equated to vapour diffused through the fabric. The cup is rotated on a turntable to maintain a favourable vapour pressure difference and sustain diffusion through the fabric. The premise for the inverted method is similar. However, the cup is upturned, causing the water to touch the fabric specimen. This simulates a wet clothing layer and is therefore favourable to hydrophilic fabrics. The principle of the guarded sweating hot plate is different. A water-saturated porous plate is heated to skin temperature to imitate a sweating human body. The power required to maintain the plate at a constant temperature whilst water evaporates from the surface and permeates a fabric specimen is measured. This value is used to calculate the resistance to vapour transfer (Saville, 1999).

5.5.3 Performance Comparisons

Anecdotally, it is accepted that manufacturers select vapour transfer test methods that favour the performance characteristics of their product. Therefore, as intertest results are not comparable, it is notoriously difficult to assess the effectiveness of different branded fabrics. However, it is important to note that increasing the water resistance of a fabric reduces the corresponding permeability to water vapour (Lomax, 1991).

5.6 Innovation in Composite Fabrics for Functional Outerwear

ePTFE: GORE-TEX produced the first composite fabric manufactured from an ePTFE membrane. It revolutionised waterproof, breathable fabric performance when released in the 1970s and continues to innovate its product range, offering 2-, 2.5- and 3-layer laminate options. Its most recent product, called GORE-TEX® Pro, is claimed to yield a further 10%–28% increase in breathability (Anon, 2013b; GORE-TEX, 2014b).

GORE-TEX continues to dominate the market for performance clothing. However, alternative fabrics are now much more visible.

For example, eVent®, manufactured by General Electric, is also constructed with an ePTFE membrane. However, individual ePTFE filaments are finished with an oleophobic coating rather than a film, which the manufacturers claim allows for Direct Venting™ of moisture to the outside fabric face and superior performance (Anon, 2012; eVent, 2014).

Biomimetic fabrics: Biomimetic fabrics are designed to mimic the structure and function of naturally occurring materials and processes. Plant mechanisms have been of particular interest to researchers developing fabrics to improve clothing microclimate comfort. Schoeller Technologies have developed a waterproof breathable membrane called c-change™ modelled on the natural response of pinecones to increased heat and humidity. It is claimed that hot ambient conditions or periods of high activity induce the polymer structure to open and facilitate higher rates of water vapour transfer. In contrast, the structure will close in cooler conditions. The company is also responsible for a water-repellent finish called Ecorepel®, which is said to imitate the oily secretions produced by water fowl to keep their feathers dry (Schoeller Technologies, 2014).

UV absorbing and reflecting fabrics: Finishing treatments that utilise, or block the warming effect of ultraviolet (UV) radiation, without compromising the breathability of composite fabrics, are available from a number of manufacturers (Anon, 2013b,c). However, PolychromeLab has incorporated the capacity for both into one material. It is a reversible three-layer, waterproof, breathable fabric comprising a silver face, membrane and dark matt face. The manufacturers claim that the silver side can reflect up to 28% of radiating light, whilst the matt side can absorb approximately 98% of radiating light, providing a buffering effect dependent on the orientation of the fabric (PolychromeLab, 2014).

Thermally adaptive fabrics: Thermoregulatory fabrics adapt to temperature changes and produce a buffering effect. However, this is temporary (approximately 12.5–15 minutes) and they are therefore most suitable for sports that are characterised by bursts of high activity and rest (e.g. skiing; Shim, McCullough and Jones, 2001). The market leader is a PCM called Outlast, which can be incorporated into waterproof breathable fabrics using a range of methods and is described in detail in Chapters 2 and 6.

Sustainability: Many of the coatings, membranes and finishing treatments conventionally used to create waterproof breathable performance in composite fabrics contain fluorochemicals that are potentially carcinogenic to humans. In response to programmes designed to eliminate their use, manufacturers have altered their composition, using fluorocarbons that are considered less damaging. However, these have reduced performance implications. Therefore, it is thought that demand for fluorine-free performance will drive innovation in this sector (Anon, 2013a). This is supported by the growing trend for sustainable products, with high-profile prizes being awarded to brands that source fabric responsibly (Anon, 2013d).

5.7 Case Study: Jackets for Hill Walkers

Hill walking is defined as a recreational activity to traverse summits, requiring little or no specialist mountaineering skills and equipment. Areas popular with hill walkers are likely to be in national parks with concentrations of high peaks such as the Cairngorms, Loch Lomond; the Trossachs, Snowdonia; the Lake District and the Peak District. These regions offer a diversity of terrain typified by rocky outcrops and ridges, upland fells, peat bogs, heather moors, woodlands, river gorges, lochs, lakes and fertile valleys.

Weather conditions in these parks may vary seasonally between average temperatures of below freezing during the winter to approximately 17°C in the summer. Increased rainfall and wind speeds are associated with active weather patterns due to proximity to the Atlantic. Topography can also further affect daily weather conditions. Ambient temperatures decrease by 0.5°C for every 100 metres ascended and, whilst wind accelerates over peaks and exposed high ground, the moist air carried with it can cool and condense to form cloud and rain.

Even in optimal conditions the conversion of chemical energy into mechanical energy during muscle contractions is inefficient and generates heat (Ainslie et al., 2002; Minetti et al., 2002). Climbing hills requires muscles to lift the body against gravity and the work done is increased by traversing yielding terrain such as bog, peat moor-land or high vegetation, which necessitates changes to gait and stride patterns (Givoni and Goldman, 1971; Creagh, Reilly and Nevill, 1998; Lejeune, Willems and Heglund, 1998).

Therefore, during a hill walk a range of exertion is required over a considerable period of time, requiring occasional stops for navigation, resting and refuelling. This presents unique challenges for thermoregulation of the clothing system, which is further complicated by rapidly changing climatic conditions.

5.7.1 Performance of Composite Fabrics for Hill Walking Jackets

Studies agree that PTFE laminates are most permeable – capable of transferring 4400–5550 g/m^2 of water vapour in 24 hours. This can be improved if a favourable vapour pressure difference is maintained between each fabric

TABLE 5.2

Approximate Work Rates, Perspiration Rates and Water Vapour Transfer Rates Associated with Hill Walking

Activity	Work Rate (watts)	Perspiration Rate (g/24 h)	Rate of Water Vapour Transfer Required (g/m²/24 h)[a]
Active walking carrying a light pack on the level	400	15,200	6080
Active walking carrying a heavy pack on the level	500	19,000	7600
Active walking carrying a heavy pack in the mountains	600–800	22,800–38,400	9120–15,360

Source: Keighley, J. H. (1985). *Journal of Coated Fabrics* 15:89–104.
[a] Based on assumption that a man's medium jacket will use approximately 2.5 m² of fabric and will cover the torso, head and arms, from which most water vapour will be lost.

face (Holmes et al., 1995; Ruckman, 1997a,b). However, when compared with Table 5.2, which shows approximate work and sweat rates associated with hill walking, it is clear that water vapour will not be completely diffused, even at the lowest activity rate. Furthermore, rain reduces performance because vapour pressure above the fabric surface is increased (Keighley, 1985; Holmes et al., 1995). Fabrics may stop breathing altogether in prolonged precipitation as pores on the outer fabric surface can become blocked (Ruckman, 1997c). In addition, condensation may form at low temperatures, which also blocks the micropores (Ruckman, 1997a; Bartels and Umbach, 2002). Therefore, capacity for moisture vapour transfer through PTFE composite fabrics is limited in the circumstances when a raincoat would be required during a typical hill walk in the UK.

5.7.2 Consumer Requirements for a Hill Walking Jacket

The performance limitations of waterproof breathable fabrics are further illustrated by qualitative data collected in focus groups with hill walkers, mountaineers, fell runners and members of the mountain rescue organisation. Content analysis revealed key customer needs, which should be addressed to optimise the function of hill walking jackets:

5.7.2.1 Consumer 1

The jacket will allow the clothing system to remain free from moisture. With the exception of fell runners, all participants were concerned with protecting their clothing systems from rain. A water-resistant hard shell was always taken on the hill, regardless of the weather. However, most were ultimately disappointed with the water vapour permeability of these garments and claimed that 'you just get soggy from the inside out, so there's no point in it'. The fell runners demonstrated the extreme of this argument. They chose to wear a windproof shell because they believed the heat and moisture they generated were beyond the moisture management capability of a fully water-resistant garment. One fell runner described the experience of wearing a water-resistant garment as follows: 'You can't tell if the water's coming from the outside or inside'.

5.7.2.2 Consumer 2

The jacket will allow the wearer to make effortless, sensitive adjustment to the comfort of the clothing system. Jackets were primarily modified by operation of the centre front zipper on the outer shell. This was described as a rapid and simple way of ventilating the clothing system, with a degree of fine-tuning. Some participants owned smock-style jackets. However, most felt that these retained more heat than their full-zip counterparts, which allowed the entire front section of the subsequent garment layer to be exposed to ambient air. Half zips were preferred on the mid- and base layers to permit cooler air directly on the skin of the neck and upper torso if required.

Other types of zipped ventilation were also considered advantageous in outer shells. However, again it was very important that these should be easy to use and swiftly effective. Mesh vents on the torso and zipped side seams were regarded as useful. One hill walker described how, when he was hot, he would unzip the side seams of his jacket and then use the hip belt of his rucksack to hold open the hemline, allowing ambient air into his clothing system. Some participants also owned jackets with zips to open the armpit region of the garment (pit-zips). However, these were judged difficult to reach and operate. Consequently, most tended not to use them and relied upon the ventilation provided by the centre front zipper instead.

Other means of adapting clothing to improve thermal comfort included loosening and tightening cuff tabs and waist and hem cords, putting the hood up or down, pulling the sleeves down over the hands or pushing them up to expose other clothing layers or the skin to cooler air and tucking or untucking the clothing layers covering the upper body into trouser layers. Participants understood the benefits of ventilating their clothing systems, but it was clear this action did not keep them sufficiently comfortable whilst working hard. Ultimately, most resorted to removing and carrying a clothing layer. One participant described how 'if it's not raining I'd take my jacket off if I was getting too hot. If it is then I keep it on and take a different layer off'.

5.7.2.3 Consumer 3

The jacket will allow the clothing system to remain comfortable in relation to the variable metabolic demands of hill walking in the UK. Participants identified diversity of terrain as the fundamental reason for changing energy expenditure and understood that this placed considerable comfort demands upon their clothing. They reported that climbing hills produced increased amounts of heat and sweat, but described subsequent ridge and descent sections as less physically challenging. This caused a drop in heat generation and made them feel cooler, which is illustrated by the following conversation:

You are destined to be a sweaty mess by the time you get to the top.
And that's when you start cooling down again.

In fact, several participants recounted similar experiences when stops had not been made for fear of getting cold. These actions initiated cognitive mistakes due to tiredness and consequently forced changes in pace and stops, which were associated with more intense feelings of cold. The mountaineers proposed that less fit people required longer and more frequent stops and suggested that this caused problems in mixed-ability groups, when fitter people were more likely to get cold.

All participants layered their clothes to keep themselves comfortable. They accepted that their sport involved varying levels of heat generation and therefore selected several lightweight wicking and insulating layers and a water-resistant, windproof shell. Additional midlayers were habitually carried in case participants began to feel uncomfortably cold. These were high-insulation, lightweight, compressible garments to cover the torso (e.g. a wool jumper, fleece gilet, or down- or synthetic-filled gilet or jacket). They claimed that they used this layered system to easily regulate their temperature. One participant stated, 'I don't really mind getting hot; I just take off a layer'.

However, in practice it appears that the hill walkers, fell runners and mountaineers were reluctant to stop and change their clothing layers and the mountain rescue team did not have the opportunity. Compromises were often made to maintain a comfortable clothing microclimate. Male participants reported how they would deliberately underdress for conditions at the beginning of a session on the hill. This was to prevent their becoming uncomfortably hot later. They commented that 'if you're right for that first 5 to 10 minutes, you're going to be wrong for the rest of it, guaranteed'.

Most frustrations related to water-resistant shells. Some participants stated that they preferred to start an outing with these garments on, because they did not want to 'faff' about later. Others would wait until it began to rain and then put the jacket directly over the clothes they were already wearing, even if they were previously comfortable. This was because they did not want to get wet and cold while they stopped and removed and stored layers. However, the subsequent increase in heat and moisture generation would then exceed the moisture vapour permeability of the water-resistant shell, as described by the following quote:

Well, to be honest, those team jackets are better than any other shell waterproof I've come across for breathability. But most of them, you're

> working hard and you've got a lot of kit with you and then you stop and
> it just all runs down the inside and you just get soggy from the inside
> out, so there's no point in it.

Some participants did not don their waterproofs until it was too late or not
at all.

Because of this perceived problem with water-resistant shells, some participants had invested in soft shells, which they wore as single, warm, breathable,
windproof and shower-proof layers. However, all those who owned these also
took a traditional water-resistant jacket with them in case of heavier rain, further
increasing the amount of kit they carried.

5.7.2.4 Consumer 4

The jacket will protect the wearer from the effects of wind chill. All participants had
experienced the effects of wind chill. They understood that this was a consequence of the terrain they were exercising in. One hill walker explained, 'If
you're going reasonably high, there's always likelihood that the wind will be
quite high and the temperature a lot lower'. Therefore, a windproof top was
regarded as essential kit and all members of the mountain rescue team were
automatically equipped with one. This item of clothing was typically taken on
the hill despite the season or weather conditions at the beginning of an outing.

5.7.2.5 Consumer 5

*The jacket will allow the clothing system to remain comfortable within the range
of average daily temperatures expected in mountainous regions of the UK.* The
changeable nature of the British weather was a frustration to all participants.
They found it difficult to select garments which would allow them to remain
thermally comfortable in all the weather conditions encountered. This was
considered most problematic during the summer, when the risk of getting
wet and cold was balanced against the weight of the kit carried.

The fell runners and mountain rescuers tended to make compromises with
their clothing. They often judged these incorrectly and three of the four fell
runners admitted they had been dangerously cold due to poor choice of kit.
The hill walkers and mountaineers were much more cautious. They tended
to check a reputable weather forecast and then took clothing to account for
all the weather conditions they were likely to confront. The following quote
illustrates the dissatisfaction these groups felt:

> I think it's more annoying packing for the summer because, you know,
> especially in Scotland, you never know if it's going to rain. Well, I mean
> you can assume that it's going to rain at some point. So you know, I end
> up packing my bag and thinking, 'Why am I taking all this stuff'? This
> is just as much as I take in winter, but it's all stuff you might need at
> some point.

In addition, participants suspected that summer clothing was inferior. They tended to use the same water-resistant jacket year round, but indicated this was uncomfortable during the warmer months: 'I think that the water-proof and windproof clothing in the summer is so … is much less comfortable'.

5.7.3 A Ventilation Design Feature for Jackets for Hill Walking

It is clear that waterproof, breathable fabrics do not fully meet the customer needs described. An alternative solution is to increase the ventilation inside hill walking jackets and improve evaporation of sweat by increasing convective heat loss and lowering relative humidity (Cheuvront et al., 2004). Diaphragm pumps have been identified as a viable method of achieving this by drawing air from the external environment and forcing it through the clothing microclimate. They also have the benefit of being cheap, simple and reliable with no moving parts to add unnecessary weight and complexity (Karassik et al., 2001).

The example represented in Figure 5.1 has a series of small diaphragm pumps arranged in parallel inside the back of the jacket. It utilises the combination of rucksack movement against the counter-rotation of the hips and shoulders to compress the pumps and drive fluid flow (Rose and Gamble, 2006). This allows for autonomous operation proportional to the speed of walking. Valves prevent backflow and direct external air into the centre back of the jacket. A vent positioned above the pumps and opening directly through the jacket shell allows the warmed, moisture-laden air to be forced out as the pumps are compressed by the rucksack (Fourt and Hollies, 1970).

5.7.4 Evaluation of a Prototype Hill Walking Jacket

A prototype hill walking jacket incorporating the diaphragm pump arrangement has been tested in a controlled laboratory wearer trial based on framework established by other researchers (Reischl and Stransky, 1980; Ruckman, Murray and Choi, 1999). The following physiological and microclimate parameters were measured during an experimental protocol that included a 30-minute acclimatisation period, 35 minutes of brisk level walking on a motorised treadmill and a 10-minute recovery period:

- Temperature and relative humidity of the clothing microclimate were measured using a Signatrol SL54TH data logging button suspended in a mesh bag pinned to the jacket lining.
- Mean skin temperature was measured using thermistors positioned on the skin at the chest, abdomen and upper and lower back.
- Amount of sweat retained within the clothing system was calculated from the difference in clothing weight immediately before and after the protocol.

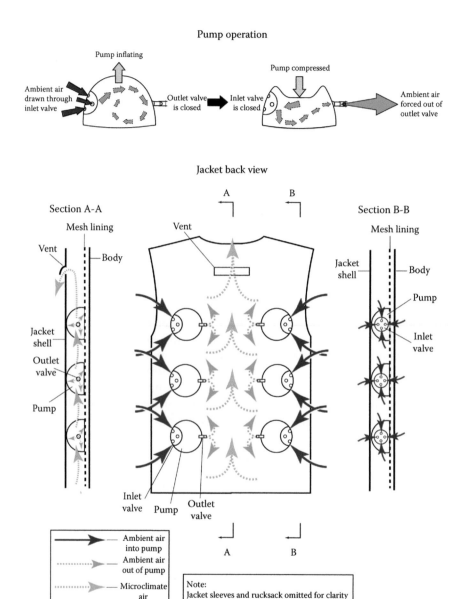

FIGURE 5.1

Schematic diagram showing pump operation and integration with prototype hill walking jacket.

A 20-litre rucksack weighing approximately 2 kg was used during the exercise period to operate the pumps. Two waterproof, breathable composite shell fabrics were analysed in the experiment: PTFE laminate and PU coated nylon. These were both tested with and without the pump arrangement. Six males participated in the wearer trial and a within-subjects design was used to address variations in fitness. The experimental results were analysed using two-way analysis of variance for a within-subjects repeated measures design.

The graph shown in Figure 5.2 and analysis of variance shown in Table 5.3 demonstrate that the pumps significantly reduced the rate of increase in air temperature during the warm-up period. They continued to do so during exercise, indicating that cooler air was drawn from the external environment and into the garment microclimate. The air temperature then remained stable during recovery. This was a significant effect, which correlates with the results for relative humidity during this period (Figure 5.3 and Table 5.4). It suggests that the air trapped in and around the pumps increased thermal insulation and also presented resistance to moisture vapour transfer through the jacket shell (Spencer-Smith, 1977). This argument is maintained by the results for mean skin temperature at the upper back, shown in Figure 5.4 and Table 5.5.

At the chest (Figure 5.5 and Table 5.6), where the jacket is compressed against the body by the rucksack straps, the rate of temperature increase is significantly lower in the PTFE fabrications. This demonstrates that superior moisture vapour transfer through the jacket shell is most important for temperature regulation in this area. The pattern continues at the abdomen, although Figure 5.6 and Table 5.7 indicate that pumps did have a small impact on the rate of temperature increase here. It is proposed that the pumps were able to stimulate airflow around the side seams in the lower portion of the jacket (Vokac, Kopke and Keul, 1973). This would promote dry heat exchange and subsequent evaporation of sweat and moisture transfer through the hemline.

At the lower back (Figure 5.7 and Table 5.8), where airflow is not restricted, the pumps significantly reduced the rate of temperature increase during

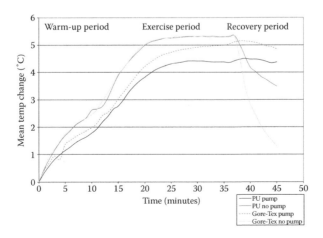

FIGURE 5.2
Mean change in air temperature measured inside all jackets.

TABLE 5.3

Analysis of Variance for Change in Air Temperature inside Jackets
during Warm-Up, Exercise and Recovery Periods

Source	Type III Sum of Squares	df	Mean Square	F	Sig.
Warm-Up Period					
Fabric	0.936	1	0.936	0.490	0.515
Error (fabric)	9.561	5	1.912		
Pump	26.634	1	26.634	11.772	0.019
Error (pump)	11.312	5	2.262		
Fabric * pump	0.336	1	0.336	0.349	0.580
Error (fabric * pump)	4.810	5	0.962		
Exercise Period					
Fabric	6.059	1	6.059	0.885	0.390
Error (fabric)	34.219	5	6.844		
Pump	73.516	1	73.516	5.571	0.065
Error (pump)	65.977	5	13.195		
Fabric * pump	5.701	1	5.701	0.953	0.374
Error (fabric * pump)	29.905	5	5.981		
Recovery Period					
Fabric	6.064	1	6.064	1.558	0.267
Error (fabric)	19.457	5	3.891		
Pump	65.867	1	65.867	12.273	0.017
Error (pump)	26.833	5	5.367		
Fabric * pump	51.959	1	51.959	11.521	0.019
Error (fabric * pump)	22.549	5	4.510		

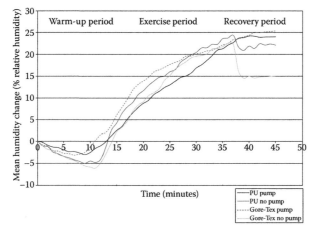

FIGURE 5.3
Mean change in relative humidity measured inside all jackets.

TABLE 5.4

Analysis of Variance for Change in Relative Humidity inside Jackets during Warm-Up, Exercise and Recovery Periods

Source	Type III Sum of Squares	df	Mean Square	F	Sig.
Warm-Up Period					
Fabric	0.625	1	0.625	0.003	0.955
Error (fabric)	906.920	5	181.384		
Pump	343.826	1	343.826	5.249	0.071
Error (pump)	327.531	5	65.506		
Fabric * pump	45.753	1	45.753	0.417	0.547
Error (fabric * pump)	548.358	5	109.672		
Exercise Period					
Fabric	145.002	1	145.002	0.134	0.730
Error (fabric)	5426.058	5	1085.212		
Pump	8.835	1	8.835	0.010	0.926
Error (pump)	4571.221	5	914.244		
Fabric * pump	903.215	1	903.215	1.783	0.239
Error (fabric * pump)	2533.019	5	506.604		
Recovery Period					
Fabric	398.713	1	398.713	1.377	0.293
Error (fabric)	1447.466	5	289.493		
Pump	1287.881	1	1287.881	5.966	0.058
Error (pump)	1079.376	5	215.875		
Fabric * pump	640.071	1	640.071	1.414	0.288
Error (fabric * pump)	2263.090	5	452.618		

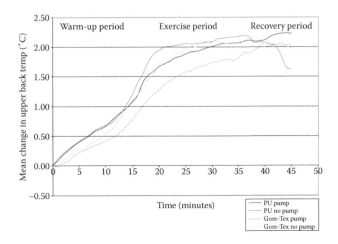

FIGURE 5.4
Mean change in skin temperature measured at the upper back.

TABLE 5.5

Analysis of Variance for Change in Skin Temperature at the Upper Back during Warm-Up, Exercise and Recovery Periods

Source	Type III Sum of Squares	df	Mean Square	F	Sig.
Warm-Up Period					
Fabric	2.862	1	2.862	3.843	0.107
Error (fabric)	3.724	5	0.745		
Pump	0.462	1	0.462	0.461	0.527
Error (pump)	5.016	5	1.003		
Fabric * pump	0.245	1	0.245	0.486	0.517
Error (fabric * pump)	2.526	5	0.505		
Exercise Period					
Fabric	6.268	1	6.268	1.787	0.239
Error (fabric)	17.535	5	3.507		
Pump	7.714	1	7.714	2.613	0.167
Error (pump)	14.758	5	2.952		
Fabric * pump	0.932	1	0.932	0.318	0.597
Error (fabric * pump)	14.649	5	2.930		
Recovery Period					
Fabric	0.163	1	0.163	0.061	0.815
Error (fabric)	13.375	5	2.675		
Pump	0.257	1	0.257	0.083	0.784
Error (pump)	15.391	5	3.078		
Fabric * pump	0.776	1	0.776	0.260	0.632
Error (fabric * pump)	14.948	5	2.990		

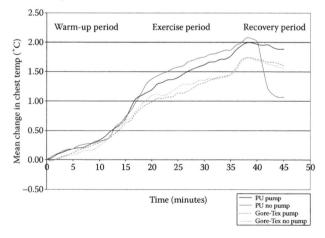

FIGURE 5.5

Mean change in skin temperature measured at the chest.

TABLE 5.6

Analysis of Variance for Change in Skin Temperature at the Chest during Warm-Up, Exercise and Recovery Periods

Source	Type III Sum of Squares	df	Mean Square	F	Sig.
Warm-Up Period					
Fabric	0.441	1	0.441	7.052	0.045
Error (fabric)	0.313	5	0.063		
Pump	0.004	1	0.004	0.016	0.905
Error (pump)	1.275	5	0.255		
Fabric * pump	0.011	1	0.011	0.016	0.904
Error (fabric * pump)	3.429	5	0.686		
Exercise Period					
Fabric	11.532	1	11.532	22.854	0.005
Error (fabric)	2.523	5	0.505		
Pump	0.945	1	0.945	0.539	0.496
Error (pump)	8.769	5	1.754		
Fabric * pump	0.133	1	0.133	0.164	0.702
Error (fabric * pump)	4.072	5	0.814		
Recovery Period					
Fabric	0.858	1	0.858	0.298	0.609
Error (fabric)	14.398	5	2.880		
Pump	1.726	1	1.726	0.486	0.517
Error (pump)	17.766	5	3.553		
Fabric * pump	1.388	1	1.388	0.643	0.459
Error (fabric * pump)	10.799	5	2.160		

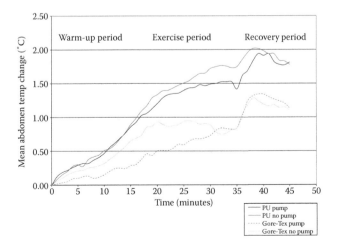

FIGURE 5.6

Mean change in skin temperature measured at the abdomen.

TABLE 5.7

Analysis of Variance for Change in Skin Temperature at the Abdomen during Warm-Up, Exercise and Recovery Periods

Source	Type III Sum of Squares	df	Mean Square	F	Sig.
Warm-Up Period					
Fabric	3.896	1	3.896	2.381	0.183
Error (fabric)	8.182	5	1.636		
Pump	0.621	1	0.621	0.684	0.446
Error (pump)	4.540	5	0.908		
Fabric * pump	0.556	1	0.556	0.713	0.437
Error (fabric * pump)	3.901	5	0.780		
Exercise Period					
Fabric	56.170	1	56.170	3.829	0.108
Error (fabric)	73.354	5	14.671		
Pump	4.563	1	4.563	0.904	0.385
Error (pump)	25.241	5	5.048		
Fabric * pump	0.176	1	0.176	0.059	0.818
Error (fabric * pump)	14.927	5	2.985		
Recovery Period					
Fabric	26.129	1	26.129	3.921	0.105
Error (fabric)	33.320	5	6.664		
Pump	0.015	1	0.015	0.013	0.912
Error (pump)	5.564	5	1.113		
Fabric * pump	0.459	1	0.459	0.194	0.678
Error (fabric * pump)	11.796	5	2.359		

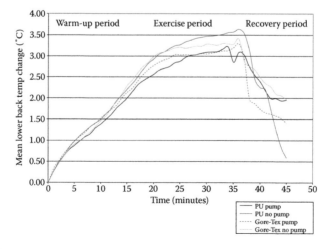

FIGURE 5.7

Mean change in skin temperature measured at the lower back.

TABLE 5.8

Analysis of Variance for Change in Skin Temperature at the Lower Back
during Warm-Up, Exercise and Recovery Periods

Source	Type III Sum of Squares	df	Mean square	F	Sig.
Warm-Up Period					
Fabric	0.095	1	0.095	0.586	0.479
Error (fabric)	0.812	5	0.162		
Pump	0.163	1	0.163	0.226	0.655
Error (pump)	3.601	5	0.720		
Fabric * pump	0.452	1	0.452	0.935	0.378
Error (fabric * pump)	2.416	5	0.483		
Exercise Period					
Fabric	2.083E-5	1	2.083E-5	0.000	0.997
Error (fabric)	5.648	5	1.130		
Pump	9.075	1	9.075	9.452	0.028
Error (pump)	4.801	5	0.960		
Fabric * pump	1.055	1	1.055	0.248	0.640
Error (fabric * pump)	21.305	5	4.261		
Recovery Period					
Fabric	0.117	1	0.117	0.050	0.832
Error (fabric)	11.675	5	2.335		
Pump	1.442	1	1.442	0.738	0.430
Error (pump)	9.765	5	1.953		
Fabric * pump	8.588	1	8.588	1.071	0.348
Error (fabric * pump)	40.094	5	8.019		

exercise. This area of the jacket is loose and unaffected by the rucksack and
a bellows effect created by the slack fabric may enhance the airflow generated from the pumps. In addition, the proximity to the hemline facilitates a
ready exchange of microclimate and environmental air. Therefore, dry heat
exchange, sweat evaporation and moisture vapour transfer out of the clothing system are all enhanced. All jackets experienced a large drop in skin
temperature at the lower back during the recovery period. This is associated
with the rapid evaporation of moisture. It is indicative of inadequate provision for moisture vapour transfer from the back of all garments, leading to
collection of sweat around the sensor (Ruckman et al., 1999).

The graph shown in Figure 5.8 and the analysis of variance shown in
Table 5.9 demonstrate significantly less sweat collected in the jackets fabri-
cated from the PTFE composite. However, the pumps reduced accumulation
by an average of 7.5% in the PU jacket and by 10.7% in the PTFE jacket and
confirmed that airflow generated by the pumps also facilitated moisture
vapour transfer from the clothing.

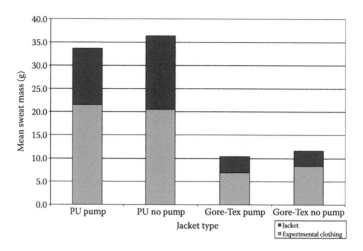

FIGURE 5.8

Mean mass of sweat accumulated within the clothing system after the experimental protocol.

TABLE 5.9

Analysis of Variance for Mass of Sweat Accumulated in the Clothing System after the Experimental Protocol

Source	Type III Sum of Squares	df	Mean Square	F	Sig.
Fabric	3468.251	1	3468.251	20.933	0.006
Error (fabric)	828.417	5	165.683		
Pump	23.860	1	23.860	0.890	0.389
Error (pump)	134.066	5	26.813		
Fabric * pump	3.383	1	3.383	0.118	0.745
Error (fabric * pump)	143.185	5	28.637		

5.8 Composite Fabrics: Considerations for Garment Design and Development

Functional outerwear must be waterproof and facilitate the evaporation of sweat and dispersal of water vapour for maintenance of user comfort. Performance composite fabrics prevail as the main commercial solution. However, these have operational limitations, leading to some customer dissatisfaction.

Pumps incorporated into functional outerwear can improve microclimate and user comfort where air may freely flow. However, it is clear that better integration between garment or component design and appropriate fabric selection is key to optimising complete product performance.

5.8.1 Design Features

Other design features important to functional outerwear for outdoor sports are shown in Figure 5.9 and Table 5.10. These include secure pockets that are large enough to store essentials such as a compass, map, whistle, hat, gloves and energy snacks. Pockets must be easy to reach and operate whilst on the move, but not look bulky.

Fixed or removable hood designs are both acceptable. However, they be must carefully shaped and adjustable to maximise head coverage, whilst maintaining a maximum field of vision at all times. Participants of some sports, such as climbing and skiing, will also require the hood to fit over a helmet and goggles. In fact, the position of fasteners, vents and pockets must always be considered in respect of other equipment which may be used by the customer (e.g. rucksacks and climbing harnesses).

Other features commonly found on functional outerwear include snow skirts, high-cut collars, storm flaps and adjustable cuffs. It is important that the design of these components is inherently water resistant, to prevent unintentional passage of moisture into the clothing system. Some composite fabric brands will only seal their products for production after the entire garment assembly has been lab tested to standard performance criteria.

Special accessories augment performance and add design interest. Waterproof zippers, cord stoppers, pullers, eyelets and reflective trim are available in a huge variety of colours, patterns and textures. However, these must be applied with regard to safety, ensuring they do not hang in the field of vision, or snag during garment use.

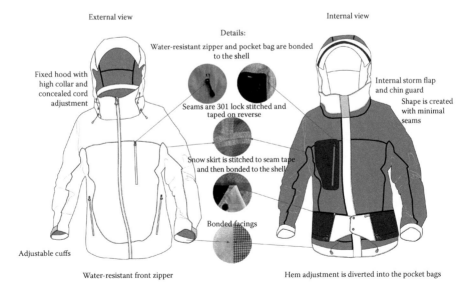

FIGURE 5.9
The anatomy of an outdoor hill walking jacket.

TABLE 5.10

Design Features Commonly Found on Functional Outerwear

Design Feature	Position	Function
Pockets	External and internal chest, side hips	Storage of map, whistle, snacks, compass, etc.
Hood	Fits over the head and around the face	To protect the head from wind, rain and snow and to prevent water entering the jacket through the neck opening
Ventilation features	Under armpits, chest pockets	To control airflow into and out of the jacket
Snow skirt	Internal – seals around the waist and hips	To prevent wind and snow entering the jacket through the hemline
High collar	Around the neck and chin	To control airflow into and out of the jacket and prevent rain entering the garment through the neckline
Cuff adjustment	At the sleeve hem (typically a self fabric or moulded rubber tab with hook and loop fastening)	To control airflow into and out of the jacket and prevent rain entering the garment through the cuff
Hem adjustment	At the jacket hem (typically an elastic, eyelet and cord-lock system)	To control airflow into and out of the jacket and prevent rain entering the garment through the hemline
Storm flap	Internal, external or combination – covering the centre front zipper	To prevent rain entering the jacket through the zipper tape, or teeth

5.8.2 Assembly Techniques

The majority of composite fabrics are reasonably easy to sew using a 301 lock stitch and superimposed seam construction. Therefore, complex shapes are routinely incorporated into functional garments at an acceptable cost. Lap felled seams can also be used where additional strength is required in a straight run (e.g. shoulder seams).

Polyester or core spun polyester thread with a water-resistant finish is commonly used for sewability, strength and durability that matches the expected life span of the composite fabric. The thread must also be resistant to mould and mildew due to potential issues relating to the use and storage conditions. Coats Epic thread is a good example; however, demand for higher specifications is also driving innovation in this sector and GORE-TEX has recently marketed an innovative thread manufactured from the ePTFE fibres, called Tenara®. This is claimed to be highly resistant to UV light and extreme weather conditions, extending the expected life of the seam and contributing to the continued performance of the garment (GORE-TEX, 2014a).

The needle should be the smallest diameter possible, with an acute, round point to ensure the resulting stitch hole is as small as possible, thus reducing the potential for damage to the fabric. The coating or laminate on the composite fabric may stick to the needle and create friction. This will dull the point relatively quickly and create burrs, which may ultimately damage the fabric, creating a weaker seam. An antiadhesive coating such as Teflon® and frequent needle changes will address this problem. Use of a Teflon presser foot will also reduce friction and maintain correct movement of fabric plies through the sewing machine.

Waterproof tape applied to the reverse side of the seam maintains the water resistance of the fabric by sealing the stitch holes. The tape typically comprises two or three layers, incorporating an adhesive film, a waterproof breathable membrane and a backing layer. A patterned tape printed to match a 2.5-layer laminated fabric is detailed in Figure 5.9. The tape is applied using a hot air sealing machine and should be positioned centrally and overrun the end of the seam to prevent leaks. Additionally, any thread ends must be cut flush and folds in the fabric or the tape should be avoided to prevent the formation of channels, which may allow water to enter the garment. Seam tape can also be used to attach design features without stitching through the shell of the garment. For example, snow skirts and gaiters can be stitched to the tape and then subsequently bonded in place (see Figure 5.9).

Recently, technologies have become available which facilitate the joining of composite fabrics without compromising their water resistance. Hot air, hot wedge or ultrasonic energy may be utilised to melt thermoplastic elements within the fabric, or an added thermoplastic film. Subassemblies are compiled using narrow lapped or butted constructions and then pressure is applied over a given time period to weld or bond the seams together. The effect is a low-profile seam, which is also flexible and lightweight to provide a performance edge (Tyler, Mitchell and Gill, 2012). Examples of bonded zippers, pocket bags and facings are shown in Figure 5.9.

5.9 Case Study: Survival Clothing for Military Flight Crew

This section examines the design and development of a low-burden, high-performance survival jacket, which forms part of a cold weather clothing system for military flight crew. The needs of aircrew are unique and a structured approach to user-centred clothing design is required in order to ensure that the garment will not only satisfy the user's need, but will also facilitate the survival of the wearer and make possible the continuation of the primary task.

5.9.1 Current Clothing Analysis

Thermal burden is a considerable problem when survival garments are worn in warm conditions prior to flying to colder regions, where thermal insulation is required and when the aircrew are required to perform more physical activities. Current cold weather clothing systems can be bulky and uncomfortable to wear for long periods of time, particularly when seated and strapped into the confines of an aircraft cockpit (Oliver, 2012). As indicated in the preceding information from Oliver, garments supplied to aircrew are heavy, bulky and can be restrictive when in motion. Additionally, the current system comprises up to seven layers of clothing, which significantly increases thermal burden for the wearer.

5.9.2 The Design Brief

New lightweight, high-performing materials and technologies are currently emerging and the design brief was to exploit these new technologies in the development of an improved lightweight, cold weather clothing system, which would introduce and integrate 'new to market' materials and manufacturing techniques into conceptual garments.

The cold weather clothing system can consist of a single garment, or, more traditionally, a two-piece garment (upper and lower). The system design may consist of a layered approach, with underlying thermal garments designed to optimise their interface with other garments within the system to introduce flexibility in order to accommodate all environmental conditions.

Flight clothing for military aircrew is complex and must be considered alongside flight equipment and the confines of the cockpit, and as seen in Figure 5.10, additional survival garments such as 'anti-G' trousers and body armour carrier systems, may be worn within the system, which can have a significant impact upon thermal burden and the comfort of the wearer.

Flight clothing must be comfortable whilst the wearer is seated in the cockpit or rear of the aircraft, be durable enough to survive possible ejection from the aircraft and provide protection in hostile environments, post evacuation. In addition, outer layer materials must be flame retardant, underlying garments must be nonmelt and all materials must be antistatic in order to avoid the generation of explosive sparks. The set of primary design criteria in the

FIGURE 5.10
Example of pilot's flight clothing. (Courtesy of Ariel Bravy/Shutterstock.com.)

following list was established in consideration of the needs of the body, the impact of the environment and the requirements of activities undertaken both in flight and outside the aircraft.

Extreme cold weather survival system for military aircrew key ranking criteria based on user needs (see list below):

1. Minimise/eliminate water ingress
2. Minimise/eliminate wind penetration
3. Improved thermal insulation with lower weight penalties
4. Reduce bulk, improve mobility
5. Introduction of lighter weight, breathable, durable materials
6. Flame retardant outer garments
7. Nonmelt materials for underneath garments
8. Antistatic materials to avoid explosive spark
9. Layering systems to enable flexibility in the levels of protection provided
10. Lowered heat stress and thermal burden during physical exertion by the introduction of air channelling and ventilation within the clothing system
11. Ergonomic considerations – improved mobility and comfort
12. Ergonomics considerations – predominant posture (seated in one position for extended periods of time)
13. Ease of donning and doffing
14. Survival clothing to integrate with other garments and equipment which may be worn by aircrew
15. Equipment stowage pockets
16. Urination/toilet provision – particularly during remote operations and escape and evasion scenarios
17. Use by both male and female aircrew
18. Consider legislation and standards

5.10 Challenges of Designing Cold Weather Clothing Systems for Aircrew

5.10.1 Environmental Considerations

Military operations require flexibility, dependability and a wide range of performance from their clothing and equipment (Scott, 2009). Flight crew must wear, use and transport a wide range of apparel to accommodate survival in a variety of harsh environments, varying from arctic conditions to mountainous regions and high altitudes to hot, dry deserts (Scott, 2009).

Thermal burden presents unique challenges for the design of cold weather clothing systems for aircrew; in addition, these may vary between types of aircraft, such as fast jet (FJ) and nonfast jet (NFJ). Fast jet aircraft may take off in hot environmental conditions, with the crew enduring cold conditions during flight operations and with possible extremely cold weather post ejection from the aircraft, during survival on land. Nonfast jet flight conditions vary according to the crewmember; during flight, the pilot is predominantly sedentary, whilst rear crew require more mobility to carry out operational activities (Oliver, 2012).

Additionally, there is a constant change between physical work and sedentary periods of inactivity (e.g. seated during transit); this is further complicated by doors being open during flight operations with the consequent wind chill factor, as well as a rotor downdraft. The designer must therefore accommodate a wide variation in conditions, ranging from extreme heat to extreme cold through the careful selection of materials and innovative design.

Essentially, operational range of conditions can include:

- Cold, wet conditions and effects of wind chill
- Cold, dry conditions and effects of wind chill
- The cold, wet environment – rain, snow and sleet
- Extremes of environment – hot to cold (Oliver, 2012)

The cold weather survival suit must ultimately be able to protect against extreme environments in harsh conditions. Environmental considerations must include protection against wind chill, water penetration and extreme cold and provide thermal protection, comfort and mobility, camouflage and concealment, as well as having low noise signature. Further considerations for escape and evasion, after aircraft abandonment, include an ability to vent garments during periods of intense activity and the facilitation of urination/toilet provision with the minimum of doffing. Garments must be quick drying, whilst offering head, hand and foot protection as well as adequate storage for survival equipment (Oliver, 2012).

Further challenges for the designer are presented through conflicting requirements of utility and performance, an example of which, might be reduced weight and bulk and greater insulation. Although insulation materials have become thinner and less bulky over recent years, the designer must look to innovative methods of providing insulation in order to minimise both weight and bulk. Additionally, the wearer must carry essential survival equipment and rations in equipment stowage pockets, which increases both weight and bulk.

FIGURE 5.11
Flight gear in the confines of a cockpit. (From Stocktrek Images, Inc./Alamy.)

According to Scott (2005), the requirement for lighter weight materials in turn conflicts with the need for fire retardance as lighter weight materials such as polyester and nylon cannot easily be made fire retardant. Generally, cold weather clothing requires that the outer protective layer be loosely fitted, in order to avoid compression of the insulation (mid) layer. This conflicts with the necessity to design closer fitting garments for flight crew, in order to prevent snag hazards due to the equipment present within the confines of the cockpit (Figure 5.11).

The need for air permeability is compromised through wind and waterproof materials; therefore, newly developed materials and design innovation needed to be exploited in order to minimise conflicts in order to deliver the best possible product to the user.

5.10.2 Human Factors Affected by Cold Weather

Cold presents a danger to health and can affect body function and performance and potentially life (Holmér, 2005). Cold weather apparel must protect the wearer from loss of body heat resulting from exposure to dry cold, wind, rain, snow or sleet. A layered clothing system provides flexibility and insulation in order to counteract the damaging effects of cold temperatures. It was necessary for the designer to find a clothing solution in order to balance heat loss and heat produced during exercise, as heat production may increase by a factor of 10 during intense activity (Thwaites, 2008), resulting in possible heat stress, despite prevailing cold conditions.

Cold weather can primarily be divided into three types:

1. Wet cold: +10°C down to −10°C
2. Very cold: 10°C down to −30°C
3. Extremely cold: 30°C and below, down to −60°C

The body core temperature is generally 37°C. With the onset of cold, blood flow to the extremities is reduced in order that the body core temperature can be maintained (Thwaites, 2008). Loss of blood flow and subsequent cooling can result in discomfort, reduced manual dexterity, vagueness and eventually frostbite and hypothermia (Bougourd and McCann, 2009). As the body can lose heat through convection, radiation, conduction and evaporation, it was incumbent on the designer to design apparel which could prevent the body losing heat faster than it could be produced. This might be achieved by reducing heat loss through the provision of optimum insulation and the facilitation of controlled moisture transmission produced by perspiration (Hu and Murugesh Babu, 2009).

5.10.3 Thermal Balance and Layered Clothing Systems

Thermal balance can be maintained through layered clothing systems, which offer flexibility due to the donning and doffing of garments as required. Layered clothing enables air to be trapped between garments; the air trapped between layers often provides greater insulation than the layers themselves (Hu and Murugesh Babu, 2009). Although this case study is concerned only with the development of a protective outer-layer jacket, the garment was considered as part of an integral clothing system.

The cold weather clothing system is a complex layering system that aims to protect the body from a range of conditions and has evolved from military combat clothing to provide thermal protection from varying activities, environmental conditions and climatic pressures (McCann, 2005). The assembly becomes an interface for the body to maintain protection through the physical and physiological interactions in the microclimate between the body, clothing and environment. The main purpose of the clothing system is to provide protection from the environment and to compensate for any adverse changes that occur in the microclimate during the interactions (Pan, 2008).

Thermal comfort is dependent on the thermophysiological properties of textile materials as the body regulates its temperature by sweating during periods of exercise and activity; which facilitates evaporative cooling. Clothing presents a barrier to moisture dissipation, which in turn can lead to condensation resulting in wet garments and loss of thermal insulation (Scott, 2009). The cold and inactivity necessitate additional insulation, whilst exercise and activity require that clothing allow heat to escape through ventilation or doffing in order to maintain thermal comfort.

Layering within the clothing system can optimise the comfort of aircrew by enabling clothing combinations to be selected in accordance with operational conditions and mission requirements to be encountered.

A typical clothing ensemble will require:

- Base layer/underwear
- Mid (insulation) layer

- Outer protective shell
- Accessories

5.10.3.1 Base Layer

The primary requirements of the base layer are to provide breathability, insulation and wicking in order to provide moisture management and to be comfortable next to the skin. The specific needs of aircrew include nonmelt materials, to avoid drip hazard, and biostatic to prevent explosive sparks. The base layer is a sanitary layer and therefore antimicrobial properties and ease of care (for frequent laundering) were desirable.

5.10.3.2 Midlayer

The midlayer of the clothing system is primarily concerned with providing insulation and should be lightweight and breathable. The particular needs of aircrew comprise flame-retardant materials, low bulk and a low noise signature. A modular or scalable approach to vary insulating characteristics based on conditions and operational requirements was necessary.

5.10.3.3 Outer Protective Shell

Clothing engineered for protection against the cold is dependent upon the selection of textile materials and carefully considered design characteristics (Kapsali, 2009). The primary functions of the outer shell are to offer protection against precipitation (rain, snow and sleet) by preventing water ingress, to retain heat provided by the insulation layer and to manage moisture generated by the wearer. Whilst design features are important in order to retain heat and also to allow rapid periodic ventilation of heat through openings and vents, composite textile materials have a key role in managing the user's personal microclimate.

5.11 Design Process

A structured user-centred approach to design (De Jong, 1984) was repeated by Niessing (2012) in the development of a cold weather survival jacket for military flight crews. The first stages of research and analysis examined current clothing provision, the needs of the body, the requirements of the activity and the impact of the environment, which enabled the designer to gain a broad view of the design situation. Design criteria were developed based upon exploration of the design problem through analysis of research in the initial stages of the design process (Table 5.11).

TABLE 5.11

Design Process

General Request
Extreme Cold Weather Jacket for Aircrew

Exploration of the Design Situation

General Objective	**Problem Definition**
• Design and evaluate extreme cold weather jacket for aircrew	• Input from industry partner as military clothing provider • Primary research

Review of Literature

Literature Search	**Market Analysis**
• Thermal regulation • Ventilation	• Evaluation of current garments on the market

Review of Literature
• The design process
• Aircrew task requirements
• Aircrew safety and comfort

Problem Structure

User Input	**Materials Analysis**
• The design process • Aircrew task requirements • Aircrew safety and control	• Compliance with Tencate as fabric provider – based on company visit

Design Specifications

Safety/Protection	**Comfort**	**Durability**	**Production**
• Weather protection • Heat and flame resistant • Antistatic for explosion risk • Water, oil and petrol repellent	• Low weight • Air permeable • Water vapour permeable • Moisture management • Thermally insulating • Fit and mobility • Feels soft and supple • Good handle and drape	• Long life cycle • Physical properties (strength) • Abrasion resistance • After wash appearance • Resistance to pilling • Colourfastness	• Acceptable cost • Low maintenance and easy care • Components readily available • Repairable • Ease of production

Interaction of design criteria established

Prototype development

Design development

Source: Adapted from De Jong (1984) by Niessing, L. (2012). Cold weather jacket. Unpublished coursework, MSc dissertation, Manchester Metropolitan University.

5.12 Composite Materials for Cold Weather Survival Clothing

As described earlier, the needs of aircrew are unique and further complicated by the range of environment, which might be encountered on a single mission. It was therefore important to select appropriate materials in order to meet the needs of aircrew. Current product analysis revealed that survival clothing provided to military aircrew met essential criteria such as protection against the elements, heat and flame. However, in order to provide essential protection the garments were often both heavy and bulky and therefore compromised comfort and mobility.

5.12.1 Selection of Fabrics and Materials

A review of textiles indicated that composite materials, which offered the combined properties of protection against the elements and high levels of breathability, as well as defence against heat and flame, would provide the most suitable area for selection. It was established that criteria 1–8 from the 'Criteria for Extreme Cold Weather Survival System for Military Aircrew' chart would be satisfied through appropriate fabric selection.

Material requirements for the cold weather survival jacket were identified and ranked as follows (see list below).

1. Windproof – moisture vapour permeable (MVP) membrane – 100% windproof
2. Breathable – MVP material, water vapour, air permeable
3. Flame retardant – inherently flame retardant or improved flame-retardant surface finishes
4. Antistatic – carbon fibres in weave construction to reduce electrostatic discharge (ESD)
5. Lightweight – approximately $180g/m^2$
6. Camouflage/colour – multiple dying processes: producer coloured to improve colourfastness and reduce infrared reflective (IRR)
7. Low noise signature – flexible materials with good drape and handle
8. Maintainable – easy to mend or repair in the field Figure 5.12

Following research, it was established that there were limited materials fit for purpose that met all of the established design criteria. A range of MVP membranes were available; however, lightweight fabrics with improved FR properties were limited. Additionally, some of the MVP membranes did not fulfil the requirement for low noise signature which, although a lower ranking criterion, is crucial for operational manoeuvres. TenCate: Defender M9180 was identified as being the composite fabric which offered the best

FIGURE 5.12
Example of current clothing for FJ aircraft. (Courtesy of Matt Cardy/Stringer/Getty Images.)

properties. It was considered that the fabric selected met design criteria 1, 2, 5, 6 and 8 and acknowledged that user trials would confirm whether the textiles performed to the required standards and therefore satisfied the needs of the wearer; however, this was outside the scope of the small-scale research project, as the design brief only required an initial prototype. FR materials are selected with regard to the limiting oxygen index (LOI) test method, which provides an indication of the flammability of materials. Importantly, the selected fabric, Defender M9180, met the ISO 13506-2008 standard for burn injury prevention tested on a thermal mannequin, with aramid solution predicted to be 28% compared to Defender M at 8% (TenCate, 2012); which indicates that the chosen fabric provides extra seconds of protection and a lower risk of burn injuries.

Following selection of the outer shell materials, consideration was given to providing insulation for protection against extreme cold. Variable insulation was investigated as preflight comfort in hotter climates needed to be considered. Initial exploration included an 'air vest', based on a concept developed by W. L. Gore and marketed as 'Air Vantage'. A gilet was subsequently developed by BMW for motorcycle clothing. Insulation was provided by air pockets, captured between two laminated fabrics, in a series of small cells. The wearer could fill the pockets, by blowing air through a tube in the collar of the garment, to inflate the vest when required and it could be released or emptied of air through a valve when cooling was needed. Further research indicated that although this method could provide variable insulation, the nature of the laminates (in order to trap and retain air) could compromise breathability of the jacket. It was established that as this compromised textile criteria 2, this would be an area for further research, and alternative methods of variable insulation such as body mapping were to be investigated.

TABLE 5.12

Fabric Selection Based upon Selection Criteria and Testing

	Fabric Selection: Two-Layer System			
Area	**Protection Layer**	**Comfort Layer**	**Comfort Layer**	**Comfort Layer**
Product name	Defender M–DM9180 fabric	Defender M 5.5 156 g/m² (±5%) stretch mesh knit	Polartec Thermal FR	Polartec Power Stretch FR
Company	TenCate	TenCate	Polartec	Polartec
Composition	64% Lenzing FR 24% *Para*-aramid 10% Polyamide 2% Antistatic	64% Lenzing FR 24% *Para*-aramid 19% Polyamide 2% Antistatic	93% Meta-aramid 5% *Para*-aramid 2% Carbon	68% Modacrylic 29% Rayon 3% Spandex
Weight	180 g/m² (±5%)	187 g/m²	237 g/m²	396 g/m²
Width	163 cm (+2/–1 cm)		142 cm	147 cm
Construction	2/1 Twill	Mesh	Double velour	Jersey/grid
Flame resistance	ISO 15025:2000 No flaming debris; no flaming edge; no hole formation After flame: >2 sec After glow: >2 sec	ISO 15025:2000 No flaming debris; no flaming edge; no hole formation After flame: >2 sec After glow: >2 sec	ISO 14116 (EN 533) Index 3/ After flame: <2 sec No flame spread to top or side/ no hole formation	ISO 14116 (EN 533) Index 3/ After flame: <2 sec No flame spread to top or side/ no hole formation

Source: Niessing, L. (2012). Cold weather jacket. Unpublished coursework, MSc dissertation, Manchester Metropolitan University.

The cold weather survival jacket was designed on a two-layer integrated principle consisting of a protective layer and a comfort layer, and fabric was selected in consideration of fabric properties, test results and specification as can be seen in Table 5.12.

- Protection layer: outer shell to provide protection against the weather and other hazards
- Comfort layer: inner shell to control moisture and airflow and to provide insulation (Niessing, 2012)

5.12.1.1 Protection Layer

Military clothing demands high standards of safety and durability; therefore, textile tests were carried out on five selected fabrics (Figure 5.13). TenCate Defender M9180 was deemed the optimum fabric for the protection layer of the jacket. Defender M9180 is intrinsically flame resistant, self-extinguishes, does not shrink, melt or drip and offers protection equivalent to that of Kevlar but with a lesser risk of burn injury (Niessing, 2012). This

Co/PES   Aramid fabrics   TenCate Defender M^{TM}

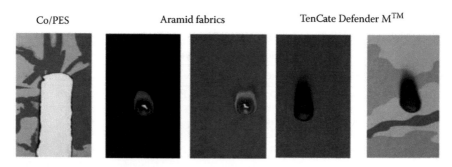

FIGURE 5.13
Flame resistance – EN 15025. (Courtesy of Niessing, 2012. With permission.)

was preferred to flame-retardant treatments, which, according to Sudhakar, Krishnaramesh and Brightlinigstone (2008), are not permanent and therefore not fit for purpose. Additionally, Defender M9180 is available in a laminate version, which offers foul-weather protection, has a low noise signature and allows camouflage print with high standards of colourfastness (TenCate, 2012). Subsequent textile testing for flame resistance established that protection offered by Defender M9180 was equivalent to that offered by fabrics such as Kevlar and aramid, with a lower risk of burn injuries (Niessing, 2012).

5.12.1.2 Comfort Layer

The comfort layer was developed based on body mapping, which identified areas in need of insulation and areas requiring venting and airflow. The body needs thermal protection in areas surrounding vital organs. The fabric selected was a Polartec Thermal FR R2206 flame-resistant fleece fabric that provided insulation through a velour construction, which traps air and retains body heat. Polartec Powerstretch FR 2400 was chosen for the side panels for flexibility. The Powerstretch was a midweight, meta-aramid Nomex fibre, which offers excellent insulation and is inherently nonflammable. TenCate Defender M9180 (180 g/m^2) mesh for ventilation was selected in order to provide behaviour consistent with the protection layer. Pattern construction for the comfort layer was mapped according to Figure 5.14 and in addition was smaller than the outer layer in order to reduce bulk and enhance airflow within the jacket. A mesh construction formed a bridge between the protection layer and the lining in the shoulder, sleeve and upper torso area (Figures 5.14 and 5.15).

5.12.1.3 Insulation and Ventilation

As previously described, the cold weather survival jacket for aircrew must accommodate both sedentary and intense activities; insulation provided within the mapped construction traps air and thereby conserves heat. It was

Cold weather system - construction inside

Standing collar for weather
and hazard protection

Anti slip lining in shoulder and sleeve area
for improving donning and doffing

Round shape in the
front for comfort

FR mesh in back part and
armwhole for breathability and
venting

Inside pocket for personal stuff

FR fleece in front and back
for protection of organs

FR stretch on sides for
improved mobility

Rib at the back - no rib in front for
comfort in sitting position

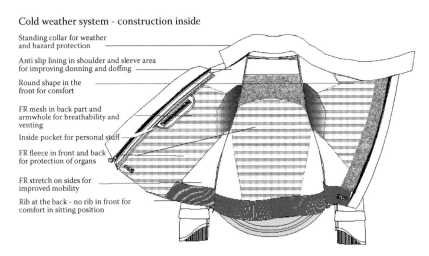

FIGURE 5.14
Description of inner jacket construction. (Courtesy of Niessing, L., 2012.)

Cold weather system - thermal layer

TenCate:
Defender
M9180

Fleece
FR thermal
(comfort)

Fleece
FR stretch
(mobility)

Fleece + lining
FR and thermal
(anti-stick)

Mesh
FR + antistatic
(ventilation)

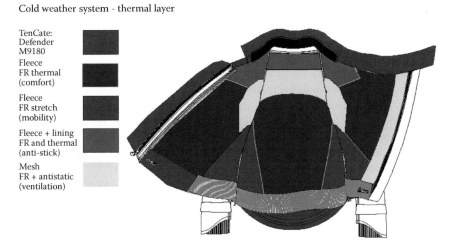

FIGURE 5.15
Fabric mapping of the jacket inner. (Courtesy of Niessing, L., 2012.)

necessary to facilitate ventilation for the release of heat during periods of activity in order to maintain a safe body temperature and comfort. As the jacket is primarily worn in the seated position and with an inertia harness, opening the jacket at the neck for ventilation could be difficult. Consequently, adjustable openings are a more efficient way to control heat and moisture than by passive diffusion through garment layers. According to Niessing (2012), during exercise, airflow between the shells aids ventilation and heat released through the mesh construction within the comfort layer and vented to the atmosphere via adjustable openings in the front upper sleeves. Following evaluation of the initial prototype, an additional vent was added to the chest area as, according to Ruckman et al. (1999), openings near the chest may be most effective in creating ventilation.

5.12.1.4 Mobility

Taking into account the complexities previously described, a range-of-motion study was conducted through indirect observation (video footage and film) and indirect interview, through a leading manufacturer of military apparel. It was established that the survival clothing for aircrew must facilitate a wide range of movement, including full abduction of the limbs, and allow free manipulation of the joints (knees and elbows) without restriction. Concepts were developed accordingly, which provided for predominant posture whilst seated in the aircraft as well as articulation at the elbow joints to accommodate the range of activities undertaken whilst in the aircraft as well as in a survival situation post ejection. A stretch FR fleece was used to enhance mobility and accommodate body movement in the back and side panels. Clothing insulation changes with motion (Lotens and Havenith, 1992, cited by Niessing, 2012), so the mesh 'bridge' between the outer and lining was designed to accommodate movement, without affecting the protection layer and consequently the insulation.

In addition to range of motion, consideration was given to speed of donning and doffing the garment in order to optimise safety and comfort, as time can be critical in emergencies. According to Niessing (2012) and Ashdown (2007), the main considerations for donning and doffing are pattern construction and fabric selection. The initial design proposed a diagonal zipper to allow a larger opening; this was later evaluated and modified. Easy-glide lining was placed inside the shoulder and sleeve areas in order to reduce frictional drag and ease speed of donning and doffing.

5.12.1.5 Fit

As previously described, in providing insulation within a layered system, it is important that the outer layer does not compress the inner garments, resulting in the effective insulation of the clothing assembly. The inner layer must be able to provide insulation and wick away moisture. As the comfort layer was incorporated into the outer shell, a closer, more ergonomic fit was achieved; this provided insulation and moisture transfer capability, whilst effectively reducing the risk of snag hazards presented by loose-fitting garments.

5.12.1.6 Construction

Design criteria had established that the jacket offer protection against wind and rain and therefore sewn seams required sealing with impermeable or microporous tape. Despite continuous development in waterproof taping, sealed seam tape inevitably adds weight to a garment and therefore alternative methods of joining were initially explored.

The laminated version of Defender M9180 selected for the protective layer was heat and flame retardant and offered potential for ultrasonically welded seams. The first prototype was constructed using traditional sewn joining techniques; it was proposed that a subsequent prototype would employ ultrasonic welding, which would be tested for strength and durability. Construction of the comfort layer utilised traditional sewn seams using a 401 chain stitch and a twin-needle 406 cover stitch for elasticity, durability and strength, whilst creating a flat seam.

5.13 Conclusions

In this chapter, consumer requirements for functionality have been discussed in relation to composite fabric construction and performance, with particular reference to outer layer garments for outdoor sports and military survival situations. This illustrates well the 'technology transfer' that occurs between military and commercial uses of composite fabrics in apparel design and development.

High-performance outerwear jackets are essential to protect the end user from prevalent weather conditions whilst they enjoy safe participation in their sport. However, to remain comfortable, these garments must facilitate the evaporation of sweat and dispersal of water vapour in relation to the increased metabolic demand generated by the associated terrain and activity.

Composite fabrics provide a solution to these requirements, and developments in recent years have become increasingly innovative. However, their performance can be limited in the unique atmospheric conditions created through the interaction of outdoor sporting activities with the use environment, which can lead to dissatisfaction amongst end users.

Designers and product developers should select and position high-performance composite fabrics in relation to physiological demands specific to the sporting activity and consider the use of ventilation design features to optimise comfort. In addition, the application of specialist assembly techniques is essential for supporting the integrity of composite fabrics and maintenance of complete garment performance.

The case study on a military flight crew demonstrates the application of a user-centred approach, which better meets user needs in terms of thermal burden, weight and clothing comfort. Protective garments insulate the wearer from adverse weather, heat and flame, but can retain heat and moisture buildup, compromising the body's ability to maintain thermal balance and resulting in heat stress and the 'plastic wrap effect' (Sudhakar et al.,

2008:323). In response to user needs, a cold weather jacket was designed to maximise airflow and ventilation between the layers and maintain thermal comfort, thereby minimising risk of heat-related illness.

Lightweight, waterproof, breathable and (additionally to the application for outdoor clothing) durable FR materials and the TenCate composite laminated version of Defender M9180 provided optimum properties for meeting military needs. Polartec thermal FR R2206 flame-resistant fleece provided lightweight insulation, whilst Polartec PowerStretch FR offered insulation and enhanced mobility. Challenges within the design process include finding the most appropriate innovative materials, with a fire-retardant ability, which meet stringent military needs and which reduce both bulk and weight.

As can be seen in the case studies presented in Chapter 5, the advancement of composite materials has led to the provision of clothing with additional properties, which can enhance both comfort and performance.

Acknowledgements

With gratitude, the authors would like to acknowledge Paul Oliver, chief designer, Survitec Group, for his generosity in sharing his expertise and Linda Niessing, Product Design and Development, X-Technology Swiss Research & Development AG, for her kind contribution.

References

Adanur, S. (1995). Cited in Zhou, W., Reddy, N. and Yang, Y. (2005). Overview of protective clothing. In *Textiles for protection*, Scott, R. A. (ed.). Cambridge, UK: Woodhead Publishing Ltd., pp. 3–30.

Ainslie, N., Campbell, I. T., Frayn, K. N., Humphreys, S. M., Maclaren, D. P. N. and Reilly, T. (2002). Physiological and metabolic responses to a hill walk. *Journal of Applied Physiology* 92: 179–187.

Anon. (2012). Product developments in innovations. *Textile Outlook International* 158: 40–63.

———. (2013a). Waterproof breathable fabrics: Demand for comfort is driving innovation. *Performance Apparel Markets* 3rd quarter, pp. 23–60.

———. (2013b). Product development and innovations in textiles and apparel. *Textiles Outlook International* 164: 54–95.

———. (2013c). Temperature control fabrics: Optimising wearer comfort. *Performance Apparel Markets* 1st quarter, pp. 27–63.

———. (2013d). Fast track: Trends in performance apparel at ISPO Munich 2014. *Performance Apparel Markets* 4th quarter, pp. 4–11.

Ashdown, S. P. (2007). Sizing in clothing – Developing effective sizing systems for ready-to-wear clothing. In *Cold weather jacket*, Neissing, L. (ed.). Cambridge, UK: Woodhead Publishing. Unpublished coursework, MSc dissertation, Manchester Metropolitan University.

Bartels, V. T. and Umbach, K. H. (2002). Physiological water vapour transport through protective textiles at low temperatures. *Textile Research Journal* 72 (10): 899–905.

Bougourd, J. and McCann, J. (2009). Factors affecting the design of cold weather performance clothing. In *Textiles for cold weather apparel*, Williams, J. T. (ed.). Cambridge, UK: Woodhead, pp. 152–195.

British Standards Institution. (1990). Specification for water vapour permeable apparel fabrics. BS 7209:1990, London: BSI.

———. (1992). Textiles – Determination of resistance to water penetration – Hydrostatic pressure test. BS EN 20811:1992, London: BSI.

Brownless, N. J., Anand, S. C., Holmes, D. A., Rowe, T. and Silva, A. D. (1995). The quest for thermophysiological comfort. *Journal of Clothing Technology and Management* 12 (2): 13–23.

Buckley, R. (2005). Surface treatments for protective textiles. In *Textiles for protection*, Scott, R. A. (ed.). Cambridge: Woodhead Publishing Ltd., pp. 196–216.

Cengel, Y. A. (1997). *Introduction to thermodynamics and heat transfer,* int. ed. New York: McGraw–Hill.

Chen, Y. S., Fan, J. and Zhang, W. (2003). Clothing thermal insulation during sweating. *Textile Research Journal* 73 (2): 152–157.

Cheuvront, S. N. and Haymes, E. M. (2001). Thermoregulation and marathon running. *Sports Medicine* 31 (10): 743–762.

Cheuvront, S. N., Carter, R., Montain, S. J., Stephenson, L. A. and Sawka, M. N. (2004). Influence of hydration and airflow on thermoregulatory control in the heat. *Journal of Thermal Biology* 29: 471–477.

Creagh, U., Reilly, T. and Nevill, A. M. (1998). Heart rate response to 'off-road' running events in female athletes. *British Journal of Sports Medicine* 32: 34–38.

De Jonge, J. O. (1984). Forward the design process. In *Clothing the portable environment*, Watkins, S. M. (ed.). Iowa, Ames, IA: Iowa State University Press, pp. vii–xi.

European Outdoor Group. (2013). The state of trade survey 2012. October 2013. [Online] accessed 23 April 2014 (http://www.europeanoutdoorgroup.com /market-research).

eVent. (2014). Technology. [Online] accessed 23 March 2014 (http://eventfabrics .com/technology/).

Fourt, L. and Hollies, N. R. S. (1970). *Clothing comfort and function.* New York: Marcel Dekker, Inc.

Fratitex. (2015). Technical fabrics and softshell. [Online] accessed 23 March 2014 (http://www.fratitex.it).

Fung, W. (2002). *Coated and laminated textiles.* Cambridge, UK: Woodhead Publishing Ltd.

———. (2005). Coated and laminated textiles in sportswear. In *Textiles in sport*, Shishoo, R. (ed.). Cambridge, UK: Woodhead Publishing, pp. 134–176.

Givoni, B. and Goldman, R. F. (1971). Predicting metabolic energy cost. *Journal of Applied Physiology* 30 (3): 429–433.

GORE-TEX. (2014a). GORE-TEX Tenara sewing thread. [Online] accessed 23 March 2014.

GORE-TEX. (2014b). Outerwear technologies. [Online] accessed 23 March 2014 (http:// www.gore-tex.co.uk/remote/Satellite/content/outerwear-technologies#).

GORE-TEX. (2014c). https://www.gore.com/en_xx/products/fibers/sewingthread outdoor/sewingthreadoutdoor.html.

Hall, J. F. and Polte, J. W. (1956). Effect of water content and compression on clothing insulation. *Journal of Applied Physiology* 8 (5): 539–545.

Havenith, G. (1999). Heat balance when wearing protective clothing. *Annals of Occupational Hygiene* 43 (5): 289–296.

Holmér, I. (2005). Protection against the cold. In *Textiles in sport,* Shishoo, R. (ed.). Cambridge, UK: Woodhead, pp. 262–286.

Holmes, D. A. (2000). Waterproof breathable fabrics. In Horrocks, A. R. and Anand, S. C. (eds.). *Handbook of technical textiles.* Cambridge: Woodhead Publishing.

Holmes, D. A., Grundy, C. and Rowe, H. D. (1995). The characteristics of a waterproof breathable fabric. *Journal of Clothing Technology and Management* 12 (3): 142–158.

Hu, J. and Murugesh Babu, K. (2009). The use of smart materials in cold weather apparel. In *Textiles for cold weather apparel,* Williams, J. T. (ed.). Cambridge, UK: Woodhead, pp. 84–112.

Johnson, A. T., Benjamin, M. B. and Silverman, N. (2002). Oxygen consumption, and muscular efficiency during uphill and downhill walking. *Applied Ergonomics* 33: 485–491.

Kapsali, V. (2009). Biomimetics and the design of outdoor clothing. In *Textiles for cold weather apparel,* Williams, J. T. (ed.). Cambridge, UK: Woodhead, pp. 113–130.

Karassik, I., Messina, J. P., Cooper, P. and Heald, C. C. (2001). *Pump handbook,* 3rd ed. New York: McGraw–Hill.

Keighley, J. H. (1985). Breathable fabrics and comfort in clothing. *Journal of Coated Fabrics* 15: 89–104.

Lawler, B. and Wilson, H. (2002). *Textiles technology.* Oxford, UK: Heinemann Educational Publishers.

Lejeune, T. M., Willems, P. A. and Heglund, N. C. (1998). Mechanics and energetics of human locomotion on sand. *Journal of Experimental Biology* 201: 2071–2080.

Lomax, G. R. (1991). Breathable, waterproof fabrics explained. *Textiles* 20 (6): 12–16.

Lotens, W. A. and Havenith, G. (1992). A comprehensive clothing ensemble heat and vapour transfer model. In *Cold weather jacket,* Neissing, L. (ed.). Soesterberg: TNO-Institute for Perception. Unpublished coursework, MSc dissertation, Manchester Metropolitan University.

McCann, J. (2005). Material requirements for the design of performance sportswear. In *Textiles in sport,* Shishoo, R. (ed.). Cambridge, UK: Woodhead, pp. 44–69.

McCullough, E. A., Kwon, M. and Shim, H. (2003). A comparison of standard methods for measuring the water vapour permeability of fabrics. *Measurement Science and Technology* 14: 1402–1408.

Minetti, A. E., Moia, C., Roi, G. S., Susta, D. and Ferretti, G. (2002). Energy cost of walking and running at extreme uphill and downhill slopes. *Journal of Applied Physiology* 93: 1039–1046.

Mintel Group. (2011). Sports clothing and footwear: UK. August 2011. [Online] accessed 23 March 2014 (http://academic.mintel.com).

———. (2013). Sports goods retailing: UK. July 2013. [Online] accessed 23 March 2014 (http://academic.mintel.com).

Mukhopadhyay, A. and Midha, V. K. (2008). A review on designing the waterproof breathable fabrics. Part 2: Construction and suitability of breathable fabrics for different uses. *Journal of Industrial Textiles* 38 (1): 17–41.

Niessing, L. (2012). Cold weather jacket. Unpublished coursework, MSc dissertation, Manchester Metropolitan University.

Oliver, P. (2012). Design approach to cold weather clothing systems. *Proceedings of the SAFE Symposium 2012*, Merseyside Maritime Museum, Albert Docks, Liverpool.

Outlast. (2014). http://www.outlast.com/en/technology (last accessed 9 September 2014).

Pan, N. (2008). Sweat management for military applications. In *Military textiles*, Wilusz, E. (ed.). Cambridge, UK: Woodhead, pp. 137–157.

PolychromeLab. (2014). Alta quota. [Online] accessed 23 March 2014 (http://poly chromelab.com).

Reischl, U. and Stransky, A. (1980). Assessment of ventilation characteristics of standard and prototype firefighter protective clothing. *Textile Research Journal* 50 (3): 193–201.

Robinson, D. (ed.) (2000). *Temperature and exercise*. Milton Keynes: The Open University.

Rose, J. and Gamble, J. G. (2006). *Human walking*, 3rd ed. Philadelphia: Lippincott Williams and Wilkins.

Ruckman, J. E. (1997a). Water vapour transfer in waterproof breathable fabrics. Part 1: Under steady-state conditions. *International Journal of Clothing Science and Technology* 9 (1): 10–22.

———. (1997b). Water vapour transfer in waterproof breathable fabrics. Part 2: Under windy conditions. *International Journal of Clothing Science and Technology* 9 (1): 23–33.

———. (1997c). Water vapour transfer in waterproof breathable fabrics. Part 3: Under rainy and windy conditions. *International Journal of Clothing Science and Technology* 9 (2): 141–153.

Ruckman, J. E., Murray, R. and Choi, H. S. (1999). Engineering of clothing systems for improved thermophysiological comfort – The effect of openings. *International Journal of Clothing Science and Technology* 11 (1): 37–52.

Saville, B. P. (1999). *Physical testing of textiles*. Cambridge, UK: Woodhead Publishing.

Schoeller. (2014). http://www.schoeller-textiles.com/en/technologies/nanosphere .html (last accessed 9 September 2014).

Schoeller Technologies. (2014). Passion for technology. [Online] accessed 23 March 2014 (http://www.schoeller-tech.com/en).

Scott, R. A. (2005). Textiles for military protection. In *Textiles for protection*, Scott, R. A. (ed.). Cambridge, UK: Woodhead, pp. 597–621.

———. (2009). Cold weather clothing for military applications. In *Textiles for cold weather apparel*, Williams, J. T. (ed.). Cambridge, UK: Woodhead, pp. 305–328.

Shaw, A. (2005). Steps in the selection of protective clothing materials. In *Textiles for protection*, Scott, R. A. (ed.). Cambridge, UK: Woodhead Publishing Ltd, pp. 90–116.

Shim, H., McCullough, E. A. and Jones, B. W. (2001). Using phase change materials in clothing. *Textile Research Journal* 71 (6): 495–502.

Shishoo, R. (2002). Recent developments in materials for use in protective clothing. *International Journal of Clothing Science and Technology* 14 (3/4): 201–215.

Spencer-Smith, J. L. (1977). The physical basis of clothing comfort. Part 3: Water vapour transfer through dry clothing assemblies. *Clothing Research Journal* 5 (3): 82–100.

————. (1978). The physical basis of clothing comfort. Part 6: Application of the principles of the design of clothing for special conditions. *Clothing Research Journal* 6 (2): 61–67.

Sudhakar, P., Krishnaramesh, S. and Brightlinigstone, K. S. (2008). New developments in coatings and fibers for military applications. In *Military textiles,* Wilusz, E. (ed.). Cambridge, UK: Woodhead Publishing.

Tencate (2012). In *Cold weather jacket,* Neissing, L. (ed.). Unpublished coursework, MSc dissertation, Manchester Metropolitan University.

Thwaites, C., W. L. Gore & Associates, UK Ltd. (2008). Cold weather clothing. In *Military textiles,* Wilusz, E. (ed.). Cambridge, UK: Woodhead, pp. 158–182.

Tyler, D., Mitchell, A. and Gill, S. (2012). Recent advances in garment manufacturing technology: Joining techniques, 3D body scanning and garment design. In *The global textile and clothing industry: Technological advances and future challenges,* Shishoo, R. (ed.). Cambridge, UK: Woodhead Publishing, pp. 131–170.

Vokac, Z., Kopke, V. and Keul, P. (1973). Assessment and analysis of the bellows ventilation of clothing. *Textile Research Journal* 43: 474–482.

Weder, M. (1997). Performance of breathable rainwear materials with respect to protection, physiology, durability and ecology. *Journal of Coated Fabrics* 27 (2): 146–168.

Yoo, S. and Barker, R. L. (2005). Comfort properties of heat resistant work wear in varying conditions of physical activity and environment. Part II: Perceived comfort response to garments and its relationship to fabric properties. *Textile Research Journal* 75 (7): 531–541.

6

Smart Materials for Sportswear

Jane Wood

CONTENTS

6.1 Introduction 153
6.2 The Definition of Smart Materials 155
    6.2.1 Power Supplies 155
    6.2.2 Conductive Yarns 156
    6.2.3 Integrated Sensors 157
6.3 The Influence of Smart Materials on Sportswear Performance 158
    6.3.1 Health and Performance Monitoring 158
    6.3.2 Automatic Adjustments 159
        6.3.2.1 Shape Memory 159
        6.3.2.2 Phase Change 160
        6.3.2.3 Chromic Effects 163
6.4 Specific Applications in Sportswear 163
    6.4.1 Athletics 163
    6.4.2 Ski Wear 164
6.5 Future Developments in Smart Materials 166
6.6 Discussion 166
6.7 Summary 168
References 168

6.1 Introduction

'Smart' and 'intelligent' textiles are terms that are frequently used interchangeably. The first intelligent clothing systems comprised a power source, wiring and electronic devices being concealed within the garment's construction. This was merely a meeting of two technologies, rather than a true amalgamation. Although this gave an element of convenience to the wearer through clever garment construction and ease of operation of the electronic device, it could not be seen as a true development of technology. Garments such as the Philips ICD+ jacket are examples of technologies being used in this way for the mass market, and although at their launch much media interest was created, the consumer quickly lost interest in the novelty value of such garments (Tuck, 2000). Since this starting point, much research has been undertaken to seamlessly merge fabrics, clothing and information technology concepts, resulting in truly smart materials. The 'wearable motherboard' garment, developed primarily for use by the US military, is considered one of the first smart garments as it incorporated the majority of electronic components within the fabric structure and was described as 'The Wearable Motherboard™: The first generation of adaptive and responsive textiles' (Gopalsamy et al., 1999:152).

From these beginnings, the scope for the potential use of smart and intelligent fabrics has widened. The development of smart textiles in the area of healthcare to aid the monitoring of soldiers upon injury quickly translated into a domestic healthcare setting, with products being developed to enable patients to be monitored remotely. These developments enabled products to be targeted toward the mass market, enabling consumers to monitor their own health through their clothing.

Sport has become more popular as a leisure activity and the promotion of health and well-being are a high priority on government agendas (Mintel Group, 2013). However, research has also shown that whilst participation in sports such as cycling and athletics showed strong growth in the years leading up to the London Olympic Games in 2012, this pattern is now in decline. Various reasons can be outlined for this decline in the market, one being that motivation can be a limiting factor. It is suggested that visual representations of the benefits of physical activity can be great motivators and this is an area in which smart textiles are finding a rapidly growing market. Consumers can use technologies incorporated into their clothing not only to see their progress during the activity, but also to set personal goals which can act as a strong driver to improve commitment to the sport (Mintel Group, 2013). It is therefore not surprising that the major sports brands, such as, for example, Nike and Adidas, have developed footwear and clothing housing inbuilt sensors that can be linked to wristbands and mobile phone apps to enable wearers to track their progress during physical activities or intensive training.

The aim of this chapter is to provide an overview of current smart and intelligent textile technologies. It commences with a broad definition of smart textiles and then explores how such garments can be powered and how information can be collected and transmitted. The chapter will then further present how smart textiles can respond to the changes within the body and adapt to these to enhance comfort and performance, with specific reference to sportswear applications. A case study will be used to illustrate the applications of smart textiles in the sportswear market. Finally the chapter will go on to discuss the future for smart textiles and the limitations of the technologies.

6.2 The Definition of Smart Materials

It is generally considered that the term 'smart material' refers to a material that is able to sense changes in its surroundings and react to these changes accordingly. However, Tao (2001) suggests that smart materials can be sub-categorised into three divisions as follows:

- Passive smart materials can be thought of as sensory devices; that is, materials of this type can sense changes in the environment, but do not have the capabilities to react to these changes.
- Active smart materials are able to both sense external stimuli and react accordingly.
- Very smart materials are able to sense, react and adapt themselves according to the changes in their surroundings.

There has been much work on the development of smart materials in the medical industry, where major innovations have been noted in the monitoring of patients. Similarly, the military have found smart materials extremely useful in the monitoring of soldiers in combat, enabling medical aid to be targeted to those whose need is greatest (Park and Jayaraman, 2003). The major part of these developments can be used and adapted to enhance the use of smart materials in the sportswear industry.

As the potential applications of smart materials are so diverse, there are many solutions offered to the issues of power supplies, conductive yarns and integrated sensors – each of which will now be discussed in turn.

6.2.1 Power Supplies

A limiting factor in the development of smart textiles has been the amount of power required to allow the textile 'system' to operate. Early smart garments were simply garments engineered to allow wires to be hidden within the garment structure so that, for example, the wires connecting the headphones to an MP3 player could be concealed within the garment structure. Later developments, particularly in ski wear, such as jackets produced by Spyder™, used Eleksen™ 'softswitch' technology (Peratech, 2013). These jackets concealed the electronics required to connect the fabric keyboard to the MP3 player, with the output via either headphones or an inbuilt speaker systems within the structure of the garment collar. Another garment produced collaboratively by Phillips and Levis was the ICD+ jacket (Van Langenhove and Hertleer, 2004). This jacket allowed the wearer access to his or her mobile phone and MP3 player via an inbuilt microphone and hidden wires within the garment structure (Tuck, 2000). However, in all cases the issue of power was met by the batteries of the MP3 player or mobile phones themselves.

As technology has developed beyond entertainment systems, the issue of power source has been a difficult one. Smart garments require a source to power integrated systems and, in some cases, to transmit the data generated by these systems.

Traditional mains power sources from an AC/DC supply have been explored. These can easily meet the demands of the smart garment system, but require hard wiring to the power source. In a medical environment, where the patient is confined to bed, this is not considered an issue. However, even in a medical context, restriction of movement causes discomfort to the wearer, which renders this method totally unsuitable for the sportswear garment.

Lithium ion batteries, such as those used in MP3 and mobile phone technologies, have been successfully used in smart garments. The problem with this type of power source is the relatively limited life span and need for frequent recharging due to the power demands of the systems being supplied. This issue caused researchers to try to find alternative power supplies (Lam Po Tang and Stylios, 2005).

Solar energy harvested using photovoltaic (PV) cells has been an area of much interest. Traditional materials for PV cells, such as crystalline silicon, gave hard, brittle structures that were not suitable for a garment end use. Developments in nanotechnologies have meant that materials based on silicon can be used to produce thin films or even be spun into fibres, thus incorporating the PV component into the textile substrate itself. The advantage of this sustainable power source is that it has limited impact on the environment, with the minimum of harmful by-products being produced. An additional advantage is that wearers in remote locations can still have access to this abundant source of power (Taieb, Msahli and Sakli, 2009). This also leads to a drawback: Those locations which do not have a large amount of strong sunlight may have difficulty in charging the cells to the degree required to power the smart textile systems.

6.2.2 Conductive Yarns

In order for smart garments to progress from those which cleverly hide wiring within the garment structure, the textile itself needed to become the 'wiring system'. Conductive yarns, such as those illustrated in Figure 6.1, appeared to offer the solution to this problem, but were not without initial problems of their own. Early work focused on metallic fibres and yarns, such as copper, steel, nickel and silver, which were known for their conductive properties. However, issues were quickly found with the flexibility of these fibres, thus impacting on the drape and handle and ultimately the comfort of the textile. Additionally, cost was a prohibitive factor in the development of these fabrics due not only to the price of the raw materials themselves, but

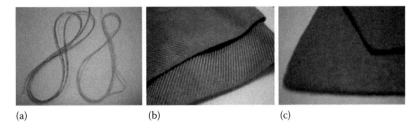

FIGURE 6.1
Metal conductive fibres: (a) spun metal and metal polyester blend, (b) woven spun metal yarns and (c) nonwoven metal fibre fabric. (From Lam Po Tang, S. and Stylios, G., 2005. *International Journal of Clothing Science and Technology* 18 (2): 108–128, p. 120.)

also to the additional costs of the excessive damage sustained by machinery during fabric production (Lam, 2005).

Further developments involved polymer-based filament yarns. These yarns can be created by the traditional method of melt spinning, with the active conductive polymer component being added at the doping stage. Coating yarns using conductive polymers was also explored, with some success being found by coating polyethylene Dyneema® yarns with polyaniline (PANI) salts (Devaux et al., 2009). Polypyrrole (Ppyr) has also been used as a successful conductive coating for polyester in fabric form, thus expanding the possibilities for the development of truly integrated electronic systems (Mokhatari and Nouri, 2012).

Another approach to integrating the electronic circuit into the textile is the use of embroidery techniques. Various researchers have explored this field and have used the yarns for embellishment as well as functionality. Researchers at Nottingham Trent University have used silver-coated yarns to embroider antenna onto garments to facilitate wireless megahertz radio frequency data transmission, with some success (Cork et al., 2013).

Another possible solution to the issue of integrated conductivity is that of conductive ink printing. This allows specific circuits to be printed as required, but there is still much development required in this field to ensure such circuits are completely effective and reliable (Moonen, Yakimets and Huskens, 2012).

6.2.3 Integrated Sensors

Sensors as part of a garment need to be innocuous, particularly in sportswear, so as to not distract wearers from their activity, which could be detrimental to performance. Developments in textile technology mean that the textile itself can now be considered as the sensor, rather than trying to discreetly house an external sensor within the body of the garment or textile structure.

Researchers such as Coyle et al. (2009) used existing moisture wicking fabrics as a means of collecting sweat from the athlete during activity. Using a colorimetric pH indicator, an LED detector and a wireless transmitter, the researchers were able to monitor the rate of fluid loss from the body. Such a system could be useful for athletes to monitor hydration levels (which have a profound effect on performance), particularly in environments where the temperature and humidity are markedly different from that in which the athlete normally trains and competes.

Respiratory rate is considered a key indicator of performance in sports activity. Fabrics that can measure pressure strain can be used in monitoring the movement of the ribcage and therefore give an indication of respiratory rate. Piezoelectric fibres (those which generate an electrical potential when under strain) are considered the best types of fibres for this end use. Natural fibres such as silk can display these properties to some degree, but synthetic polymers such as polyvinylidene fluoride (PVDF), polyproplylene (PP) and polyethylene terephthalate (PET) can be engineered to display piezoelectric properties (Vatansever et al., 2011). Typically, these textiles need to be incorporated as a strap within the garment structure, positioned firmly around the ribcage, to give the best respiratory measurements. Additionally, piezoelectric fibres can be used to monitor the movement of the athlete during an activity. Traditionally, the movement of the athlete and his or her posture during training and competition is monitored by the coaching team and discussed watching postperformance video footage. Smart textile monitors can be used to give real-time feedback to both the athlete and the coaches, enabling adjustments to posture and technique to be made whilst the athlete is undertaking activity. For example, using piezoelectric sensors in shoes has enabled feedback to be gained on walking and running gait; similarly, piezoelectric wristbands have enabled tennis players to adjust their racket grip and wrist posture (Coyle et al., 2009).

Garments with integrated sensors are becoming more commonplace. Companies such as Smartlife (Smartlife, 2013) have developed a commercially available garment, the Smartlife healthvest®, which contains integrated sensors to monitor heart rate, respiratory rate and temperature with the potential to also provide electrocardiogram (ECG) readings if required. The technology allows users to download the information to their personal smartphone or computing device, allowing personal monitoring of vital signs.

6.3 The Influence of Smart Materials on Sportswear Performance

6.3.1 Health and Performance Monitoring

Health monitoring has been a key area for the development of smart textiles. Placing textiles directly against the skin and hard wiring them to a monitor has been a technique used for several years in the monitoring of vital signs such as heart and respiratory rates. However, more recent developments have seen advancements in wireless technologies, eliminating the need for hard wiring to an external piece of equipment and thus increasing the potential for mobility of the patient, in turn improving patient comfort and therefore, as some research suggests, decreasing recovery times (Zheng et al., 2007).

These technologies are easily transferrable to the field of sportswear. Incorporating the sensors within the body of the garment and wirelessly transmitting information can provide critical information on vital signs and thus the performance of the athlete. This can provide valuable information for the athlete during the preparation phase of the competitive season and training sessions can be tailored to enhance performance, without impeding activity. Similarly, the performance of the athlete can be monitored during competition with analysis of data providing a platform with which to build a strategy for future events. However, it is critical that the sensors cannot be detected by the athlete as this could cause distraction or discomfort and ultimately cause a reduction in performance.

The limitation in this type of technology is the mode by which the wireless data are transported. Traditional wireless protocols consume large amounts of energy and thus the size of the battery to support this proved prohibitive. Recent developments in wireless protocols ANT™ (Dynastream, 2013; Stylios, 2013) require a much reduced amount of energy for operation and therefore enable power sources to be small enough to be incorporated into the garment and frequency of recharging reduced to an acceptable level.

6.3.2 Automatic Adjustments

The definition of a true smart material is that it is one able to detect and respond to external stimulus, adapting itself accordingly. There are various types of materials that could fall into this category.

6.3.2.1 Shape Memory

Shape memory textiles exhibit the ability to be deformed by external stimuli (usually temperature) into a temporary form and then return to their original shape (Kim and Lewis, 2003). Such materials are manufactured from polymers that are heated to a specific temperature (the temperature of deformation) at which they are set into shape. The material is then allowed to cool. During use, if the material is heated to its temperature of deformation, it will then lose its set shape and thus change the properties it imparts. This process is entirely repeatable as the changes are within the morphology of the structure (Figure 6.2) and not due to the degradation of the polymer (Hu, 2007).

In sportswear applications, such textiles can be useful in the thermoregulation of the body. Schoeller® c-change materials are biomimetic structures based on the movement of the opening and closing of pinecones due to environmental conditions. The fabric structure opens as the body temperature

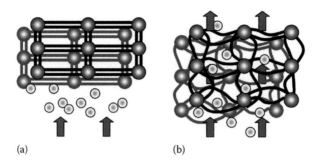

(a)   (b)

FIGURE 6.2
Schematic diagram of shape memory effects: (a) closed structure at low temperature and (b) open structure at high temperature. (From Lam Po Tang, S. and Stylios, G. 2005. *International Journal of Clothing Science and Technology* 18 (2): 108–128, p. 112.)

rises, allowing heat and moisture to travel away from the body and facilitating cooling. As the environment and body cool, the structure closes, thus trapping air between the garment and the body and allowing thermal insulation to occur (Schoeller, 2013).

6.3.2.2 Phase Change

Phase change materials are those which have the ability to change state when absorbing or releasing thermal energy, thus acting as thermoregulators in fabric and garment form. Typically, these materials are based on a wax type of compound encapsulated within a fibre (Figure 6.3), although this is not the only type of material and technique that can be used. Alongside paraffin waxes, compounds such as hydrated inorganic salts, linear long chain hydrocarbons and polyethylene glycols have been considered in the development of textiles with enhanced thermoregulatory properties, with techniques

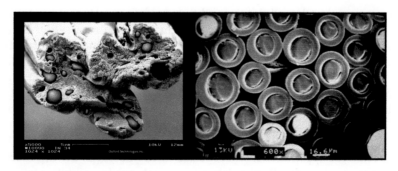

FIGURE 6.3
Outlast phase change material encapsulated in (left) viscose fibre and (right) polyester fibre. (Images courtesy of © Outlast Technologies LLC.)

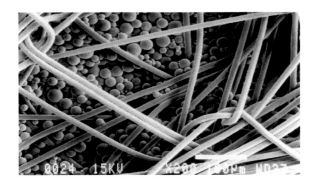

FIGURE 6.4
Fabric coated with Outlast phase change material. (Image courtesy of © Outlast Technologies LLC.)

such as lamination and coating (Figure 6.4) being explored by researchers as alternatives to fibre encapsulation.

The critical factors required for a phase change material to be effective in a textile clothing application are as follows:

1. A melting point between 15°C and 30°C
2. A small difference between melting and solidification points
3. Stability during repeated melting and solidification
4. Excellent thermal conductivity

Additionally, health and safety factors such as low toxicity, resistance to flammability and effects on the environment are key considerations. Ultimately, for the mass market, cost is the deciding factor (Mondal, 2007).

As body and environmental temperatures rise, the compound absorbs the thermal energy and, as it does so, changes state from solid to liquid, drawing heat away from the body and allowing cooling to occur. Conversely, as temperature drops, the compound emits thermal energy and it transforms from liquid to solid state, allowing heating of the surrounding area. Careful engineering of the garment is required to ensure it is the body that benefits from the movement of thermal energy and not the surrounding atmosphere (Lam Po Tang and Stylios, 2005).

Currently, the amount of phase change material that can be incorporated into textile weights common in garments (approx. 150–400 g²) is such that only approximately 15 minutes of effectiveness can be achieved. However, this could be of use in sportswear applications for garments worn immediately after warm-up to ensure muscles are not allowed to cool before activity. There has been some development using this technology in heavyweight garments such as those used in extreme sports in colder climates; gloves

FIGURE 6.5
The hand before the test. (Image courtesy of © Outlast Technologies LLC.)

for snowboarding and skiing have successfully used phase change materials such as Outlast® to impart thermal comfort (Outlast, 2013).

The images in Figures 6.5 through 6.7 show the results of experiments carried out with Outlast phase change materials incorporated into a glove. The images are taken using an infrared camera, which can track thermal emissions; red depicts areas of high thermal emission (warm areas) moving through to yellow and green as the heat gradient decreases. Blue depicts lower thermal emission (colder regions) which progresses to black as the gradient decreases further. The first image (Figure 6.5) shows the hand before the test where the warm regions can clearly be identified.

The second image (Figure 6.6) shows the effects of placing the hand on an ice block for 5 minutes whilst wearing the Outlast glove. There are still warm regions visible showing the hand being 'protected' from the cold by the phase change material.

The final image (Figure 6.7) shows the effects on the hand being placed on the ice block for 5 minutes wearing a non-phase-change material glove. The loss of heat can clearly be seen, which in turn could lead to wearer discomfort.

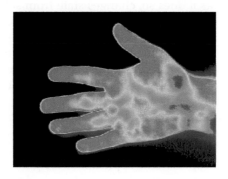

FIGURE 6.6
Hand, wearing a phase change material, placed on ice block for 5 minutes. (Image courtesy of © Outlast Technologies LLC.)

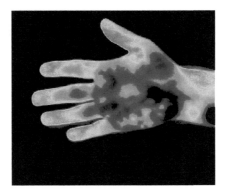

FIGURE 6.7
Hand, wearing a non-phase-change material, placed on ice block for 5 minutes. (Image courtesy of © Outlast Technologies LLC.)

6.3.2.3 Chromic Effects

Materials experiencing a change of colour due to external stimuli are known as chromatic. The specific category of chromatic function is dependent on the stimulus as follows:

Photochromatic – colour change due to light

Electrochromic – colour change due to electrical currents

Piezochromic – colour change due to mechanical deformation

Solvachromic – colour change due to moisture

Thermochromic – colour change due to heat

Thermochromic textiles have been used to some degree in the medical industry in the monitoring of body temperature and could be used in sportswear in a similar way and aid performance monitoring. Similarly solvachromes could aid sportswear performance, being used to monitor the production of perspiration and thus advise the wearer on necessary rehydration.

To date, the use of chromic textiles in apparel has been limited to a niche market in novelty fashion items.

6.4 Specific Applications in Sportswear

6.4.1 Athletics

Professional athletes require perfect conditions in both their own bodies and the external environment to perform at their optimal level (Hassan et al., 2012).

Smart textiles can offer monitoring techniques in the training environment which are noninvasive and therefore do not distract wearers from their performance (Cho, 2010).

However, monitoring of the body's response to exercise is not in demand exclusively by professional sportspeople. Sport as a recreational activity is commonplace, with many participants having a keen interest in monitoring their performance. Many companies have recognised this 'High Street demand' for such monitoring devices and several are now available for use in both the leisure and professional sportswear market.

Companies such as SmartLife have developed the HealthVest. The T-shirt type garment is constructed from a knitted fabric base containing elastane to improve fit and comfort properties. In the front portion of the garment, ECG electrodes and a respiratory monitor are knitted into the fabric structure alongside the circuitry, which is created from silver-coated polyester yarns. The vest can provide real-time monitoring and data output, which can enable athletes to understand and adapt their performance and training regimes (SmartLife, 2013).

In a similar way, NuMetrex has developed the 'sports bra'. The sensors and conductive components are all inherent in the garment, thus minimising any distractions that could be caused by the monitoring equipment. The information is transmitted wirelessly to a watch via a clip-on transmitting device, enabling the wearer to monitor heart rate in real time. Additionally, the garments and watch accessory can be purchased in a variety of colours to cater to the fashion requirements of the wearer (NuMetrex, 2013).

6.4.2 Ski Wear

The sport of skiing involves periods of high activity (skiing downhill) followed by those of relative inactivity (resting at the end of the ski run, or the lift climb to the top of the slope). Ski wear therefore must be thermoregulatory – supplying the wearer with both cooling and insulating properties.

Phase change materials lend themselves perfectly to the shorter ski runs associated with recreational skiing. Where the downhill ski runs and lift climbs are short (less than 10–15 minutes), enough phase change material can be incorporated into the clothing to allow effective thermoregulation to take place. Ski clothing is generally thicker than everyday wear; however, the wearer still requires clothing that is flexible enough for normal body movement, thus restricting the amount of phase change materials that can be incorporated into clothing covering the arms and legs (and torso to some degree).

Areas of the body where heat loss due to conduction is known to be greatest in cold climates are the hands and the feet. In the case of the feet, less apparel flexibility is required; thus the textiles used can be thicker and more phase change materials can be incorporated, increasing the length of the effectiveness of the material. A degree of manual dexterity is still required in skiing; however, more phase change material can be incorporated in the body of the glove to improve insulation without compromising movement.

A critical factor in mountain sports is protection from the wind, which can cause accelerated heat loss and hypothermia if the body is not sufficiently shielded. Coupled with this, it is essential that moisture produced by the body is allowed to escape through the garments to allow cooling when required. This moisture must be allowed to escape and not held either on the skin or within the textile structure as this can again accelerate the cooling effects of the surrounding environment. Shape memory polymers, such as those discussed in Section 6.3.2.1, would serve a useful purpose in this case, with the structure being closed if wind chill is a threat, whilst opening up to allow moisture out if the temperature of the body begins to increase.

CASE STUDY: COLUMBIA OMNI-HEAT™

Omni-Heat (Columbia, 2013) is another smart approach to thermoregulation, using materials that are

- **Reflective** – a series of metallic dots printed onto the fabric reflect heat back toward the body, thus aiding thermal retention and reducing the thickness required of a traditional insulation layer (Rodie, 2010). As the body begins to heat up and humidity increases within the garment, the foil dots are activated as conductors to transmit heat away from the body, thus preventing overheating and excessive perspiration.
- **Moisture wicking** – the main fabric is a polyester/elastane blend which enables moisture wicking via the cross-sectional shape of the polyester fibre.
- **Air permeable** – the main fabric structure allows the free flow of air to assist thermoregulation.

Columbia has used this technology and combined it with a smart application in footwear. The Bugaboot™ contains the preceding technology, plus a flexible carbon fibre heating element, powered by a lithium (mobile phone type) battery, which is charged using a mains AC adaptor (Rodie, 2010). The heating element can be operated via a push button on the side of the boot and can supply heat to the foot for up to 8 hours.

6.5 Future Developments in Smart Materials

The majority of developments in smart textiles centre around the miniaturisation of the electronic components used for the monitoring of the wearer. The miniaturisation of microchips and processors has enabled researchers, such as those at Nottingham Trent University, to incorporate chips into yarn structures. Whilst more traditional incorporation of electronics into textiles has involved a 'face' and a 'back' to the textile (the back being the side of the textile on which all the componentry can be seen). The tiny microchips work in conjunction with the yarns, which act as conductors themselves, meaning that the textile structure is inherently the electronic circuit which supports the smart operation (Cork et al., 2013).

From the perspective of truly new developments in the field of smart textiles, the use of graphene as a substrate holds great promise. Research is currently being undertaken to develop graphene as a novel electronic device. As graphene is only one atom thick, it is extremely flexible and therefore could bring great leaps forward in the incorporation of electronics with textiles and garments. The US Army research lab is currently exploring the options of grapheme-based devices as wearable electronics (Dubey, Nambaru and Ulrich, 2012). As was discussed earlier in the chapter, many sportswear applications of smart textiles began life as military research and this could prove to be an exciting new era for smart textiles in sportswear.

Further developments see researchers working on flexible batteries, which would remove the current need for a pocket in a garment in which to house the battery power source (Kwon et al., 2012). Highly flexible batteries open exciting new possibilities in clothing design and could offer large improvements in clothing comfort, without a compromise in functionality, a critical area for consideration in sportswear.

6.6 Discussion

The market for smart textiles is growing, particularly in those industries, such as sportswear, in which there is a need for the wearers to be monitored and the information collected to be analysed.

The fashion industry has yet to find a real use for such technology, and smart textiles are still frequently used as novelty items. There is a danger that the fashion industry will continue to see such developments as a 'gimmick' using the technologies to support structures such as those with LEDs. However, designers such as Angella Mackey are keen to explore the concepts of fashion meeting function meeting technology and are creating highly functional, fashionable apparel which incorporates electronic illumination

(Vega, 2013). Improvements in the flexibility of the fabrics which incorporate the componentry to enable operation of the smart garment are also allowing designers to create garments more easily. In early developments the wiring of the components was the deciding factor in the garment design (as seen with the ICD+ jacket), which led to smart garments being heavy and bulky. Conductive yarns, componentry and sensors are now all readily available in flexible forms off the shelf, enabling lighter weight garments to be produced, which are more comfortable to wear (Ohmatex, 2013).

The apparel industry still encounters problems with power sources for smart textiles and as such is limited in the progression of the technology until a viable option is found (O'Mahony and Braddock, 2002). Some encouragement should be taken from the mobile phone industry where much work is being done to improve battery life whilst reducing battery size. Companies such as Apple and Samsung are market leaders in the smartphone arena where their products are providing an increase in functionality with no increase in battery size (Apple, 2013).

Alternative power sources should not be disregarded. Traditional rechargeable batteries, which draw their charge by being plugged into the mains supply for a period of time, could be replaced with more sustainable alternatives. Solar power has seen great leaps forward in the last few years, with solar cells decreasing in size and, as discussed in the previous section, increasing in flexibility. The 'charge whilst you wear' concept would be appealing to consumers, who are quick to highlight their frustrations with the traditional methods of recharging batteries.

In order for the smart textile to become a reliable product for the diagnostic and medical aspects of the sports industry, improvements in wireless integrity still need to be made. There is a perception that wireless technologies are still not as reliable as their hard-wired counterparts and, in order for such textiles to gain credibility, absolute reliability of data transmission is critical (Hunter, 2008). Whilst this is based largely on perception rather than hard fact, wireless technologies do have room for improvement.

Finally, the environmental impact should be considered. In a world where a conscious effort is being made to reuse and recycle, is there really a place for textiles which contain added technology? And what shelf life do these garments have? It could be suggested that in terms of performance sportswear, there will always be a need to monitor athletic performance – be that through heart rate, blood pressure, respiratory rate or one of the other functions discussed previously in this chapter. However, as the technologies are developing at such a rate, will this mean garments can be upgraded as the technology develops, or will current garments be deemed obsolete in a couple of years' time? In this case, does this mean that garments now will go to a landfill site or can they be recycled? An argument could be put forward that, as the garments are the sensors and are transmitting the raw data, it is the processing software that will change and, as such, the garment will have a longer life than that initially thought. This issue will only be brought to light as the smart textile market widens.

6.7 Summary

Smart materials have an important part to play in the future of sportswear. There have been many developments since the integration of electronics into a garment in the late 1990s and, although the novelty factor of this garment was quickly dismissed by the consumer, it served as a basis for innovation. Developments in military and medical applications easily translated into sportswear products. Sports performance monitoring was the first to be considered and the development of integrated sensors and piezoelectric fibres enabled professional athletes to monitor their activity in real time, recording their data for analysis and enabling training sessions to be adapted accordingly. This quickly led to mass market adaptation with products such as the NuMetrex sports bra (NuMetrex, 2013).

In addition to performance monitoring, smart textiles have also been developed to enhance comfort, with shape memory and phase change materials being used. These technologies have allowed apparel with exceptional thermoregulatory properties to be established which adapt in line with wearers' individual needs.

However, the concept of fashion versus function cannot be ignored. Advances in performance technologies cannot be successful without similar developments in garment design. Apparel designers are now embracing such developments and the value of the aesthetics offered by smart textile technologies, such as thermo-, photo- and electrochromic dyes or incorporated illuminations, can be seen in fashion. As these technologies are embraced by fashion, the ease of incorporation of performance technologies into garments is developed and the 'gimmick' factor is replaced by true functionality and ease of garment creation.

References

Apple. (2013). Apple [online], http://www.apple.com/uk/ (accessed September 29, 2013).

Cho, G. (2010). *Smart clothing: Technology and applications*. London: CRC Press.

Columbia. (2013). Columbia [online], http://www.columbia.com/ (accessed September 29, 2013).

Cork, C., Dias, T., Acti, T., Ratnayaka, A., Mbisre, E., Anastasopoulos, I. and Piper, A. (2013). The next generation of electronic textiles. *Proceedings of the First International Conference on Digital Technologies for the Textile Industries*.

Coyle, S., Morris, D., Lau, K., Diamond, D. and Moyna, N. (2009). Textile-based wearable sensors for assisting sports performance. *Proceedings of the Sixth International Workshop on Wearable and Implantable Body Sensor Networks*, pp. 307–311.

Devaux, E., Koncar, V., Kim, B., Campagne, C., Roux, C. et al. (2009). Processing and characterization of conductive yarns by coating or bulk treatment for smart textile applications. *Transactions of the Institute of Measurement and Control* 29: 355–367.

Dubey, M., Nambaru, R. and Ulrich, M. (2012). Graphene based nanoelectronics (FY11). US Army Research Laboratory.

Dynastream. (2013). ANT wireless [online], http://www.dynastream.com/ant-wireless (accessed September 29, 2013).

Gopalsamy, C., Park, S., Rajamanickam, R. and Jayaraman, S. (1999). The Wearable Motherboard™: The first generation of adaptive and responsive textile structures (ARTS) for medical applications. *Virtual Reality* 4 (3): 152–168.

Hassan, M., Qashqary, K., Hassan, H. A., Shady, E. and Alansary, M. (2012). Influence of sportswear fabric properties on the health and performance of athletes. *Fibres and Textiles in Eastern Europe* 20 (4): 82–88.

Hu, J. (2007). *Shape memory textiles*. London: Woodhead Publishing.

Hunter, B. (2008). Smart textiles face tough challenges. *Innovation in Textiles* [online], http://www.innovationintextiles.com/smart-textiles-face-tough-challenges/ (accessed September 29, 2013).

Kim, Y. and Lewis, A. (2003). Concepts for energy – Interactive textiles. *MRS Bulletin*, pp. 592–596.

Kwon, Y., Woo, S., Jung, H., Yu, H., Kim, K. et al. (2012). Cable-type flexible lithium ion battery based on hollow multi-helix electrodes. *Advanced Materials* 24: 5192–5197.

Mintel Group. (2013). Sports Goods Retailing. London: Mintel Group.

Lam Po Tang, S. and Stylios, G. (2005). An overview of smart technologies for clothing design and engineering. *International Journal of Clothing Science and Technology* 18 (2): 108–128.

Mokhatari, J. and Nouri, M. (2012). Electrical conductivity and chromic behaviour of poly (3-methylthiophene) – Coated polyester fabrics. *Fibres and Polymers* 13 (2): 139–144.

Mondal, S. (2007). Phase change materials for smart textiles – An overview. *Applied Thermal Engineering* 28: 1536–1550.

Moonen, P., Yakimets, I. and Huskens, J. (2012). Fabrication of transistors on flexible substrates: From mass-printing to high-resolution alternative lithography strategies. *Advanced Materials* 24: 5526–5541.

NuMetrex. (2013). NuMetrex [online], http://www.numetrex.com (accessed September 29, 2013).

Ohmatex. (2013). Ohmatex [online], http://www.ohmatex.dk (accessed November 18, 2013).

O'Mahony, M. and Braddock, S. E. (2002). *Sportstech: Revolutionary fabrics, fashion and design*. London: Thames & Hudson.

Outlast. (2013). Outlast [online], http://www.outlast.com/ (accessed September 29, 2013).

Park, S. and Jayaraman, S. (2003). Smart textiles: Wearable electronic systems. *Materials Research Society Bulletin* 28 (8): 585–589.

Peratech. (2013). Peratech [online], http://www.peratech.com (accessed September 29, 2013).

Rodie, J. (2010). Trifecta for warmth. *TextileWorld* Mar–Apr: 46.

Schoeller. (2013). c-change: The bionic climate membrane [online], http://www.schoeller-textiles.com/en/technologies.html (accessed September 29, 2013).

SmartLife. (2013). Smartlife [online], http://www.smartlifetech.com (accessed September 29, 2013).

Stylios, G. (2013). A review of aspects of wireless, wearable technology. *International Journal of Clothing Science and Technology* 25 (4).

Taieb, A. H., Msahli, S. and Sakli, F. (2009). Design of illuminating textile curtain using solar energy. *Design Journal* 12 (2): 195–217.

Tao, X. (2001). Smart technology for textiles and clothing – Introduction and overview. In *Smart fibres, fabrics and clothing*, Tao, X. (ed.). London: Woodhead Publishing.

Tuck, A. (2000). The ICD+ jacket: Slip into my office please. *The Independent* [online], http://www.independent.co.uk/news/business/analysis-and-features/the-icd-jacket-slip-into-my-office-please-694074.html (accessed September 29, 2013).

Van Langenhove, L. and Hertleer, C. (2004). Smart clothing: A new life. *International Journal of Clothing Science and Technology* 16 (2): 63–72.

Vatansever, D., Siores, E., Hadimani, R. and Shah, T. (2011). Smart woven fabrics in renewable energy generation. In *Advances in modern woven fabrics technology*, Vassiliadis, S. (ed.). Intechopen.com.

Vega. (2013). Vega wearable light [online], http://www.vegalite.com/pages/about (accessed September 29, 2013).

Zheng, J. W., Zhang, Z. B., Wu, T. H. and Zhang, Y. (2007). A wearable mobihealth care system supporting real time diagnosis and alarm. *Medical and Biological Engineering and Computing* 45: 877–885.

7

Applications of Compression Sportswear

Praburaj Venkatraman and David Tyler

CONTENTS

7.1 Introduction 171
7.2 Background and Rationale 172
7.3 Compression and Its Influence on Physiology 173
7.4 Compression for Medical Uses 175
7.5 Evaluation of Compression for Sportswear 178
    7.5.1 Effects of Using Compression Garments................................ 181
7.6 Applications of Compression Garments in Sportswear.................. 183
    7.6.1 Cycling... 183
    7.6.2 Skiing... 184
    7.6.3 Rugby Sport 185
7.7 Market Trends in Compression Sportswear 188
7.8 Contextual Factors Affecting Compression Garment Performance... 192
    7.8.1 Garment Sizing 193
    7.8.2 Body Shapes.. 194
    7.8.3 Sizing and Designing with Stretch Fabrics............................ 195
    7.8.4 Fabric Panels 195
7.9 Summary and Conclusions 196
References.. 200

7.1 Introduction

Sports and exercise involves physical movement of the body (torso, upper and lower limbs) and, in some cases, amateurs and professionals alike endure soft-tissue injury. At the elite level, improved individual performance during a tournament or a game is vital. Many athletes consider that compression of muscles to support and enhance muscle alignment and improve the efficiency of muscle movements is essential. They also strap the injured body part to assist recovery from injury. In recent years, there has been an increase in usage and demand for compression garments for a number of sportswear applications and recreational activities due to their ability to offer functional support to the wearer.

The main aim of this chapter is to present research relating to compression garments and highlight the recent developments relating to specific sports such as cycling, skiing and rugby. The benefits of compression garments were documented in various settings (sports, clinical and nonclinical), although convincing evidence remains elusive. The reported benefits of using compression garments were mainly in enhancing blood circulation, reducing the recurrence of injury, aiding recovery, providing muscle support and reducing muscle soreness. However, the claims made by manufacturers in boosting athletes' performance are debatable and the evidence gathered so far is less than convincing. The research relating to the benefits and limitations of using compression garments for sportswear is critically reviewed in this chapter. Most studies used different settings and small sample sizes and it is necessary to question whether the findings should be extrapolated to a wider group. Contextual factors that influence garment design and development such as body shapes, fabric panels, materials' properties, sizing and comfort are discussed. In addition, market trends in compression sportswear and factors affecting new product development are discussed in the context of designing innovative compression sportswear.

The main purpose of this chapter is to provide an overview of compression garments available for active sports, including rugby, cycling and skiing. The contents of this chapter will equip readers to understand the use of compression garments for sportswear and specific applications and will further their creative thoughts toward developing innovative functional apparel.

7.2 Background and Rationale

Compression modalities in the form of elastic compression bandages have been used in healthcare to control the oedema or swelling of tissues and to aid the return of venous blood from the lower limb to the heart. Several benefits of compression therapy in healthcare have been noted, including: assisting calf muscle to perform its function by restoring damaged valves to function properly, promoting comfort and quality of life of the patient and enhancing the condition of skin. Compression modalities vary the amount of stretch and elasticity, both intrinsically and extrinsically, in the manner in which they are applied (Figure 7.1). Recent developments in fibres, fabrics and finishing technologies enabled researchers to develop compression sportswear that is based on a graded application of pressure to the extremities of the body in order to pump the blood back to the heart faster. Some of the claims of the compression garment developers include improved circulation and performance and reduced recovery times, reduced muscle soreness and enhanced lactic acid removal.

FIGURE 7.1
A pair of compression stockings. (Courtesy of Shutterstock, Inc.)

The rationale of this chapter is to provide an overview of compression garments available for active sports, including rugby, cycling and skiing. It should be noted that compression garments are also widely preferred for aesthetic reasons, particularly scar minimisation, reduction of postsurgery infection risk, pain reduction and slimming support. Compression panties that provide firmness to the tissue and mask any abdominal bulge are also used in hernia patients (Haldane, 2013). It should be noted that whilst compression modalities have demonstrated their effectiveness, the focus of this chapter is on sportswear compression garments.

7.3 Compression and Its Influence on Physiology

The principle of compression garments is analogous to compression therapy treatment for lower limbs that involves applying a known amount of pressure to aid the return of venous blood to the heart.

Compression therapy is based on the Laplace equation, which states that pressure is directly proportional to the amount of tension and number of layers applied and inversely proportional to the limb circumference and width of the material applied. The effectiveness of compression therapy is well documented. However, evidence of the effectiveness of compression garments remains elusive and fragmented. A graduated compression is required to encourage the blood flow from the lower limb toward the heart (Figure 7.2). Unlike compression therapy, which is intended to aid the return of venous blood by applying sustained and gradual pressure to lower limbs, the claims for compression garments are overly optimistic, ranging from improved performance of athletes to accelerated recovery following an injury. It should be noted that garments applying pressure locally on the body can only support

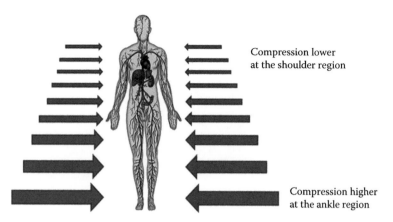

Compression lower
at the shoulder region

Compression higher
at the ankle region

FIGURE 7.2
Vascular system and compression pressure gradient – highest at the lower limb and lowest at the shoulder region. (Courtesy of Shutterstock, Inc.)

physiological processes (increased blood circulation or assisting the recovery of muscle injury) and cannot be guaranteed to enhance and accelerate the performance of athletes, as there are many variables that can affect an athlete's performance (Figure 7.3).

More frequently, elite athletes are determined to achieve personal bests, including breaking world records at each competitive event. Compression has been used since the nineteenth century to treat medical ailments

FIGURE 7.3
Athlete wearing compression sports bra and tights. (Courtesy of Shutterstock, Inc.)

FIGURE 7.4
Typical compression base layer clothing. (Courtesy of Dreamstime.)

(Thomas, 1998; Ramelet, 2002) and has been featured increasingly since the 1980s, when the use of fabrics with elastane gained popularity. This chapter also discusses the benefits and limitations of compression garments used for sportswear. The benefits of compression garments are physiological as well as psychological and evidence from studies is critically evaluated for its usefulness.

A typical compression garment has an intimate and anatomical fit (Figure 7.4), provides support by applying pressure to muscles, increases blood circulation, and reduces blood lactate levels. Athletes require healthy venous blood return. Figure 7.5a and b draws attention to observation that athletes frequently incur hamstring and calf sprain if no garment support is worn.

7.4 Compression for Medical Uses

It is interesting to note that although compression therapy has been widely used in healthcare since the nineteenth century (Thomas, 1998; Ramelet, 2002), the use of bandages for venous diseases can be dated back to 450–350 BC (Van Geest, Franken and Neumann, 2003). Medical practice has found the use of graduated compression favourable, particularly as it works with the muscles to encourage blood flow toward the heart (Moffatt, Martin and Smithdale, 2007). Other notable benefits of compression therapy are

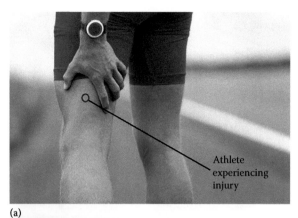

(a)

(b)

FIGURE 7.5
(a) Sports injury with running man. (b) Sports Injury – cramps. (Courtesy of Shutterstock, Inc.)

- Absorption of exudate (fluid) from the wound (Thomas, Fram and Phillips, 2007)
- Reduction of scar size and improvement of scar appearance (Wiernert, 2003)
- Relief of the symptoms associated with venous disease (Moffatt, 2008, p. 339)

Compression therapy is achieved using two methods: by traditional bandaging techniques or by specially manufactured garments such as medical elastic compression stockings (MECS) (Ramelet, 2002; Van Geest et al., 2003; URGO Medical, 2010). Van Geest et al. (2003) explained how these categories can be classified as elastic or inelastic. Although inelastic bandages may be worn for 24 hours due to low resting pressure, elastic compression should be removed during a 24-hour period to avoid pressure sores resulting from constant compression.

An inelastic bandage, also known as a short-stretch bandage, applies light pressure for a short period of time. Due to its inability to conform to the leg shape, much of the pressure applied decreases over time (Ramelet, 2002; Moffatt, 2008). Elastic, or long-stretch bandages, sustain pressure provided for a longer period of time due to the flexibility of the structure (Moffatt, 2008); however, they are more likely to cause discomfort to the wearer (Ramelet, 2002).

MECS are available in a variety of lengths dependent on the requirement of the user. MECS are classified for prescription with the pressure delivered to the ankle varying from 10 to ≥49 mmHg depending on the treatment necessary (Van Geest et al., 2003). Van Geest et al. (2003) reported the classification of the Medical Elastic Compression System (MECS) as light (10–14 mmHg), mild (15–21 mmHg), moderate (23–22 mm Hg), strong (34–46 mmHg) and very strong (≥49 mmHg) based on the pressure applied at ankle. Current classifications of bandages do not solely incorporate those for compression. Made-to-measure MECS are recommended for those patients who have very specific needs in terms of fit (Ramelet, 2002). Additionally, ready-to-wear versions are available in a range of classifications. Ramelet (2002) explained that some patients find MECS hard to wear, particularly the higher classification garments; however, devices are available to deal with this problem and are generally well tolerated.

Although bandages and MECS are the most commonly used forms of compression therapy in medicine, the use of other compression clothing is often associated with the treatment of burns and hypertrophic scarring. Its successful use was investigated in the early 1970s (Wiernert, 2003). Wiernert (2003) reported that compression clothing is available in many forms, including all-in-one body suits and gloves, and is habitually worn throughout the day.

The discussion on the effectiveness of compression therapy for medical ailments is apparent (Weller et al., 2010; Feist, Andrade and Nass, 2011; Miller, 2011). Watkins (2010) highlighted the importance of ensuring that each patient is wearing the correct size compression garment. Miller (2011) pointed out the need for a standardised method for measuring limbs, mainly to ensure that patients are fitted correctly. The lack of fit not only is a cause of discomfort for the wearer but can also result in lack of the desired amount of pressure being applied; thus it could harm rather than treat. Watkins (2010) also explained how a patient's limbs can be measured for postoperative compression garments prior to surgery unless a significant change in body shape or size is predicted.

Feist et al. (2011) and Miller (2011) have reported that in order to achieve complete cure through compression therapy, patient adherence is significant, although discomfort was regarded as one of the main reasons why patients fail to comply with the treatment. Miller (2011) explained that patient awareness is a key factor for a successful treatment. Understanding compression therapy with regard to how long patients must wear bandages or garments and possible problems resulting from removing them prior to this were not highlighted in the majority of cases observed. Furthermore, 100% of the cases observed did not receive any written information relating to compliance. Venkatraman et al. (2005) studied the importance of patient compliance and awareness of compression therapy using a questionnaire in determining the effectiveness of compression modality. However, the continued success of compression therapy is perhaps the main reason as to why sportswear manufacturers began to incorporate similar theory into sporting apparel. The expansion of compression garments in the sportswear market is apparent and growing.

7.5 Evaluation of Compression for Sportswear

Compression garments for sportswear and leisure applications have become widely available, providing increased comfort, fit and muscle support. Voyce, Dafniotis and Towlson (2005) reported that human skin stretches considerably, especially 35% to 45% at knee and elbow regions, and that extensible garments are essential to provide comfort during intense body stretch (Figure 7.5b). Normal body movement expands the skin by 10% to 50% and strenuous movements in sports are facilitated by low resistance from garments and instant recovery. Knitted fabrics are designed in such a way that they possess elastic properties so that the garment offers compression and stretch in both lengthwise and crosswise directions.

In addition, it is important to note that an athlete will stretch in various ways depending on body movement, which is highlighted in Figure 7.6.

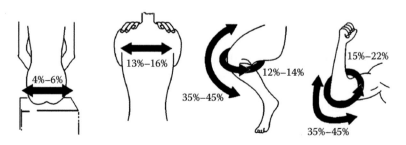

FIGURE 7.6
Key areas of stretch. (Courtesy of Elsevier, 2005.)

Different stretch positions include leg, lower back, thighs, shoulders, abdomen, arms, etc. A compression garment should support numerous different muscles during intense physical activity (Figure 7.7).

A number of brands have promoted compression wearables such as SKINS, 2XU, CEP, Zensah, CW-X and compression-x. Application of compression sportswear in major events has been widespread; for instance, in the 2000 Olympic Games at Sydney, the sports enthusiasts were focused on Fastskin swimsuits, which were both praised and criticised during the games. The skintight compression body suits by Speedo, which aimed to reduce drag whilst allowing full body movement, were worn by almost 85% of the gold medal winners in swimming during the games (Swim-Faster.com, 2012). Craik (2011) reported that the controversy surrounding the suits, which gave the wearers an increased ability to break personal best and world records, led to its ban during 2010. However, this ban was not enforced until after

FIGURE 7.7
Different types of body stretch. (Courtesy of Shutterstock, Inc.)

much development of the suits and the introduction of other models including the Fastskin FSII, Fastskin FS-PRO and, most notably, the 'world's fastest' suit, the LZR Racer Suit (McKeegan, 2008). Speedo's LZR Racer, which is made from an ultralightweight fabric called LZR pulse, has low drag and is both water repellent and fast drying. It was worn by the majority of medal winners at Beijing 2008, including US swimmer Michael Phelps, who tallied up a collection of eight gold medals.

Cipriani, Yu and Lyssanova (2014) investigated the opinion of experienced cyclists on the perceived influence of a 'posture cueing shirt' on comfort and recovery. It was found that the athletes reported increased benefits in riding posture, postride posture, spine discomfort and recovery. Duffield, Cannon and King (2010) reported that muscle recovery after sprinting and exercise over 24 hours showed minimal effect on performance, but lowered levels of muscle soreness. It can be noted that compression garments assist in posture support and post-training muscle recovery and in reducing muscle soreness.

Compression garments sparked media attention not only in swimming at this time but also in other sports including track and field. The all-in-one, head-to-toe Nike Swift Suit aims to provide athletes an advantage, in a way similar to swimsuits, with reduced drag and increasing aerodynamics (Bondy, 2000). American athlete Marion Jones wore a Nike suit to run competitively (Mayes, 2010); however, the trend for head-to-toe suits for running events does not seem to have the prolonged success as with swimming. Similarly, Nike Swift Suits were used in other sport disciplines, including speed skating and cycling, and demonstrated positive effects (Voyce et al., 2005).

The introduction of compression T-shirts has been well received by rugby players (Voyce et al., 2005). The much tighter fit of the shirts, compared to the traditional rugby jersey, meant that not only the players benefitted by the compression physically, but also that other players could not easily grip the tops. McCurry (2004), Shishoo (2005), Cole (2008) and the Mintel Group (2009) stated that public demand for performance sportswear has increased in recent years and all note a rise in compression garments being sold. Walzer (2004) described how compression garments have advanced since the 1990s to include a wider variation of products and colours for all genders, identifying factors that explain the greater demand for such products.

Cortad (2011) reported that although the new garment 'Quicksilver' has the conventional appearance of board shorts, there is a hidden compressive short underneath with taping ergonomically positioned to support muscles. The shorts, which utilise the technology usually seen in other sports, are proving successful for surfing champions wearing them. Quicksilver explosive technology uses a four-way stretch dobby fabric that has less contact with the skin, promotes blood flow, increases lymphatic drainage and supports muscle recovery (Explosive board shorts, 2015). Furthermore, compression sportswear garments are diversifying. The Proskins Co. (2012) has created a range of compression clothing with ingredients such as caffeine and vitamin E incorporated into the fabric to help reduce cellulite. Proskins are marketed for use as day-to-day clothing as well as for sports training. As the compression market continues to grow in popularity in both the professional and consumer markets, the benefits often cited in marketing are continually questioned.

7.5.1 Effects of Using Compression Garments

In the following section, various research reports on the effects of using compression garments during sports and exercise were critically reviewed. As has been noted there is an increasing trend to wear compression sportswear for increasing blood flow and aiding muscles whilst training. However, there is much debate over the effectiveness of wearing such garments for sporting activities. Many compression sportswear companies claim that the garments will improve circulation, improve performance and reduce recovery times (2XU Pty Ltd, 2009; Skins™, 2012).

These claims have led to a plethora of research in the area. But there still needs to be a holistic agreement on the benefits of performance and recovery because of conflicting results emerging from many of the investigations. Brophy Williams et al. (2014) used a strain gauge to assess changes in limb volume among active males while they wearing compression socks, leggings or no compression garment. They reported that both compression leggings and socks are effective in reducing the limb swelling and that further investigation was required to assess whether these changes affect exercise recovery. This study indicated that the compression garments had an effect on blood flow, particularly in the lower limb. Hill, Howatson, van Someren, Leeder et al. (2014) reported a meta-analysis to investigate the effects of compression garments on recovery following exercise. They evaluated 12 studies where assessments were taken at 24, 48 and 72 hours post exercise regime. The research concluded that compression garments were effective in enhancing recovery from muscle damage.

de Glanville and Hamlin (2012) investigated the effect of wearing graduated compression garments during recovery on subsequent 40 km time trial performance. During the study, the participants wore either a graduated full-leg-length compressive garment (76% Meryl elastane, 24% Lycra®) or a similar-looking non-compressive placebo garment (92% polyester, 8% spandex) continuously for 24 hours after performing an initial 40 km time trial in their normal cycling attire. The participants had a second trial with garments following a 24-hour recovery period. A week later, the groups were reversed and tests repeated. The performance time in the second trial was substantially improved with compression garments compared to placebo garments. The researchers concluded that graduated compression garments during recovery were useful and less likely to be harmful for well-trained endurance athletes. Compression garments can influence wearer proprioception (which is an unconscious perception of movement and spatial orientation arising from stimuli within the body itself). Hooper et al. (2015) investigated whether wearing compression garments would enhance proprioception and comfort affecting sports performance, especially on high-level athletes. The athletes, who wore either a compression garment or a non-compression garment, were involved in baseball and golf activities. Researchers reported that comfort and performance can be improved with the use of compression garments in high-level athletes, who are most likely to be influenced by improved propriocep-tive cues, especially while engaging in upper body movements.

Luke and Sanderson (2014) undertook a meta-analysis to determine whether wearing compression clothing affected athletic performance by increasing endurance and aerobic activity as measured using VO_2 max (maximum oxy-gen consumption/uptake) and heart rate. They included only those studies that examined continuous running. Of the four studies considered, only one concluded that wearing compression garments improved performance, and three studies reported that compression garments were highly effective in reducing muscle soreness and oedema. The authors concluded that wearing lower extremity compression socks can be very effective in reducing oedema during or after exercise.

Duffield and Portus (2007) monitored the effects of full-body compression garments. Participants in the study completed a series of distance and accu-racy throwing tests along with sprints in either a control garment or one of three brands of full-body compression garments. However, there were no significant differences between the control condition and the three brands of compression garment. Hill, Howaston, van Someren, Walshe et al. (2014) investigated the efficacy of a lower limb compression garment in accelerat-ing recovery from a marathon run among 24 subjects who were assigned to 'treatment' or 'sham' groups. The researchers reported that lower limb com-pression garments improved subjective perceptions of recovery. However, there were neither improvement in muscular strength nor significant changes in exercise-induced muscle damage and inflammation. There were other studies that demonstrated small benefits to performance (Doan et al., 2003). Sperlich et al. (2010) observed the differences to performance benefits of compression socks, compression tights, whole-body compression gar-ments and control running clothing. All 15 participants completed running tests on a treadmill in each type of clothing. Performance was measured by monitoring lactate concentration and oxygen uptake. There were no perfor-mance benefits observed when wearing compression garments.

Twenty track athletes completed a series of tests to measure sprint times, muscle oscillation and jump power in both loose gym shorts and compression shorts. The jump heights recorded were increased by 2.4 cm when wearing the compression garment. There were some noticeable benefits on performance when compression stockings were worn (Doan et al., 2003). Ali, Creasy and Edge (2011) investigated the effect of varying the level of compression stock-ings (low, medium and high) during a series of countermovement jumps before and after running trials. It was found that the changes in jump height from before to after exercise were much bigger when wearing the low and medium stockings compared to when wearing a control garment. Miyamoto et al. (2011) focused on the effect of compression on torque of the triceps muscle. Triplet torque was monitored both before and after calf raise exercises and there was a smaller reduction of power after exercise when wearing the compression stocking with 30 mmHg at the ankle. However, there was no evidence for the 18 mmHg ankle stockings having any effects.

Other studies also reported some benefits of wearing compression garments among athletes who participated in circuit training that included sprint times, flight times and jump height (Higgins, Naughton and Burgess, 2009), reduced injury and recovery time (Rogers, 2012), and improved recovery following exercise-induced muscle damage (Jakeman, Byrne and Eston, 2010). Compression garments were used as a recovery tool for a 3-day exercise protocol. Cold water bathing and carbohydrate consumption along with postexercise stretching were two other conditions for the investigation. Under these circumstances, cold water bathing was deemed to be more beneficial to recovery than the use of compression garments or carbohydrate consumption and postexercise stretching (Montgomery et al., 2008). It can be noticed from the preceding that the documented effects of compression garments were not consistent across different sporting activities. It should also be highlighted that the research studies were conducted using different protocols and, in most cases, the changes in sports performance were reported by elite athletes. However, it may be significant that most of the studies reporting benefits relate to blood flow in lower limbs as opposed to upper body movements. Thus, compression appears to be effective in reducing oedema, enhancing muscle recovery and reducing muscle soreness.

7.6 Applications of Compression Garments in Sportswear

In this chapter, specific attention is given to three sports: namely, cycling, skiing and rugby. Compression garments have become a staple for athletes in these sports, as support is required for muscles continuously, providing aid in reducing muscle fatigue. Areas of importance to these sports are highlighted in the following sections and in Figure 7.8.

7.6.1 Cycling

A wide range of compression wear is available for professional cycling and is elaborated in Venkatraman et al. (2013). Leg muscles are the main source of power and endurance. During cycling, a cyclist uses the following sets of muscles:

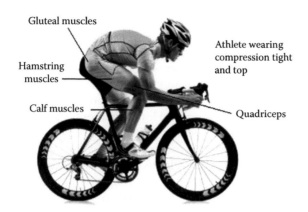

FIGURE 7.8
A cyclist in action wearing a compression garment. (Courtesy of Istockphoto.)

- Gluteal muscles – bottom area pushing pedals at the top of a stroke.
- Quadriceps – large muscles at the front of the thigh that straighten the leg when pushing the pedal down to the ground.
- Calf muscles – ankle movement and calf muscles facilitate return of blood from lower limbs when the heel strikes the ground (Morrison et al., 1998). Hence good ankle movement facilitated by calf muscles only achieves a healthy venous return.
- Hamstrings – located at the back of the thigh, working with calf muscles lifting the pedals up from the bottom of a stroke.

There are support muscles working together with leg muscles, such as upper muscles engaging with the handlebar. Their use is terrain dependent. For example, a hill climb will increase handlebar pressure, so the biceps are used to enhance power. Back and abdominal muscles are also important to stabilise cyclists whilst riding (Yake, 2011). A range of garments are available for professional cycling including sleeve jerseys, sleeveless jerseys, long-sleeve jerseys, jackets, vests, long- and short-sleeve skin suits, bib shorts, tights and knickers. Recently, Venkatraman et al. (2013) critically appraised the specifics of cycling compression garments including requirements of garments/fabrics and fabric panels to support muscle groups. In addition, consumer perspectives on using compression tights for cycling were also described.

7.6.2 Skiing

Compression base layer tights for skiing help to regulate body temperature, support muscles, aid in blood circulation and support in muscle recovery and fatigue. By reducing the amount of lactic acid in the tissue during and after a physical activity, the athlete's recovery is greatly enhanced and prevents soft-tissue injury. Typical base layer garments for skiing include full-sleeve tops, tights and one-piece garments that cover torso, arms and legs (Figure 7.9). A ski base layer compression garment keeps the wearer warm in cold conditions as well as supports muscle movements.

The garments are generally made of synthetic fibre blends or wool fibre blends. The fabric is intended to be lightweight, soft next to the skin and abrasion resistant; some of the features of a typical compression base layer for skiing are highlighted in Figure 7.9. Skiing base layer compression tights are constructed from a lightweight knitted fabric that is brushed inside to provide comfort to the wearer. In addition, the fabric is wind- and waterproof and able to wick moisture away from the skin. The fabric has plenty of stretch as it is composed of 10% to 40% elastomeric filaments and 60% to 90% synthetic filaments (nylon or polyester). The garment is commonly sewn using flat lock seams for additional comfort. To reduce chaffing, seamless garments are preferred. Typical requirements of ski wear are highlighted in Figure 7.10.

7.6.3 Rugby Sport

Rugby is a high-contact sport where one in four players will be injured during a season. The number of injuries in rugby is three times higher than the number of injuries in football. Most injuries are experienced by youth 10–18 and adults aged 25–34. Figure 7.11 shows a typical player wearing a compression garment. In rugby, 57% of most sports injuries occur during matches rather than in training – particularly when a player tackles or is being tackled (South Wales Osteopathic Society, 2009). Hence, most compression garments

FIGURE 7.9
Mathias Elmar Graf (Austria) places third in the men's slalom, on January 21, 2012 in Patscherkofel, Austria. (Courtesy of Shutterstock, Inc.)

FIGURE 7.10
Typical requirements of skiing compression base layer.

will have protective pads in tops (Figure 7.12) and shorts. The range of compression garments for rugby includes sleeve tops, full-sleeve tops, shorts, tights, calf sleeves and socks.

Reported benefits of compression wear in rugby include

- Reduced muscle soreness and swelling
- Reduced muscle oscillation during a vertical jump or fall

FIGURE 7.11
Rugby player wearing compression tops and tights. (Courtesy of Shutterstock, Inc.)

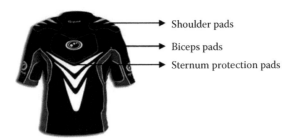

Shoulder pads

Biceps pads

Sternum protection pads

FIGURE 7.12
Typical tops for rugby with impact protection pads. (Source: Optimumsport.com)

- Increased VO_2 max (a physiological index of sports performance)
- Reduced collection of blood lactate levels in the tissue
- Reduced muscle injury or cramps

The use of tapes and compression may assist in muscle recovery and keep the body in a safe position during contact sports such as rugby. In rugby, adults endure more injuries to head, shoulder and lower limb (thigh), whereas children suffer from head/neck injuries followed by injuries to the upper and lower limbs (http://www.injuryresearch.bc.ca).

Approximately half of all injuries occur while a player is tackling (Figure 7.13) or being tackled. Hookers and flankers sustain the most injuries. Forwards are more frequently injured than backs because of their greater

FIGURE 7.13
Rugby player tackling.

involvement in physical collisions and tackles (British Columbia Injury Research and Prevention Unit, 2012). Due to the nature of injuries sustained, clubs may frequently lose players from sports participation; hence they are under pressure to prevent or reduce injuries to players by requiring them to wear protective gear. Table 7.1 highlights some of the popular brands of protective apparel for cycling, rugby and skiing.

It can be observed that popular brands for compression wear make the following specific claims for their products: support for muscles, reduced muscle oscillation, improved recovery, the ability to target specific muscle groups, involving grippers to prevent chaffing, lightweight fabric that is breathable, moisture wicking and thermal insulation. In addition to the compression, comfort and fit play a vital role in designing the garment. In the case of cycling, prominent features include an integrated bib shorts seat pad cushion and provision of mesh shoulder panels. In rugby, to protect from soft injury, pads are integrated in the tops while the tights use quick-dry technology. Finally, skiing garments focus on stability, targeting muscles in the lower back, hamstring, and knee joint and providing ankle support. In addition, most fabrics have antibacterial, antiodour and UV protection finishes applied.

7.7 Market Trends in Compression Sportswear

'Trend Insight' is a feature in *Outdoor Insight* that includes consumer research and retail point-of-sale data from the Leisure Trends Group. Its recent report stated that commonly purchased compression items were tights/capri tights, sport tops/bra tops, shorts, socks and arm/leg sleeves (http://www .leisuretrends.com).

According to a recent report by Leisure Trends Group's Running Retail-TRAK™, which traced sales in the compression apparel category, compression apparel and apparel accessories grew 56% in 2009 and 170% in 2008. Tights were the largest compression category for runners during 2010, followed by arm/leg sleeves. Other compression categories included capri tights, sport tops/bra tops, shorts and socks. Women's apparel styles outsell men's in running speciality and this trend holds true with compression. Woman-specific shorts and capris are larger in terms of overall volume (especially compression sports bras and sports tops in 2011), with leading brands such as Zensah and Nike (http://www.leisuretrends .com).

It can be noted that tights for bottoms (lower limbs) were frequently purchased (32%) followed by arm or leg sleeves (27%). In addition, compression socks were also widely preferred (15%) followed by base layer bottoms (Figure 7.14).

TABLE 7.1

Market Brands: Compression Base Layer

Compression Wear	Popular Brands	Major Claim	Specific Feature
Cycling	Adidas bib shorts	British cycling brand for added comfort and fit	82% polyester and 18% elastane; silicone elastic gripper on hem for perfect fit and mesh strap in shoulder and synthetic chamois padding
	Dhb Vaeon Roubaix padded bib tights	Anatomical fit; lightweight; thermal insulation	Breathable; made of 85% nylon and 15% elastane; brushed inner side
	Castelli Garmin Sanremo Speedsuit	Body paint speed suit; seamless and integrated cushion seat pad	Integrated grippers; no seams to avoid chaffing; flat lock seams; four-way stretch fabric
	Altura	Cool mesh bibs and a ProGel seat pad for cushioning and comfort	Lightweight stretch fabric; breathable, antibacterial; silicone grippers
	Assos tights	F1 Uno_S5 seat pad; an anatomic six-panel cut for added comfort and performance	75% nylon and 25% elastane; shock-absorbent foam cushioning that is bonded onto the elastic; frictionless; skin contact textile
	Giordana	Fusion bib shorts with great fit; supports a range of body motion	Made of Moxie fabric, lightweight, breathable, with micro mesh for ventilation and silicone injected grippers
	Gore bike wear	Oxygen waist shorts; high-end padded cycling tights for the medium- to long-distance road cyclist	Offer high functionality and comfort; reinforced inner leg panels and an Ozon seat pad for comfort, durability and longevity
Rugby	Canterbury	Mercury TCR compression legging reduces muscle oscillation by compression at key zones in lower limb	Made of Coolmax polyester 74.5% and 25.5% Lycra; moisture management; flat seams and antibacterial; reduced seams for added comfort
	Nike Pro	Prodirect compression tops with 'dri-fit' fabric	Moisture management; flat tape at shoulders to reduce friction; mesh panels at underarm and flat seams to prevent chaffing

(Continued)

TABLE 7.1 (CONTINUED)

Market Brands: Compression Base Layer

Compression Wear	Popular Brands	Major Claim	Specific Feature
	Kooga	Tops – EVX protection offers the optimum fit and freedom of movement.	The pads consist of a Lycra body, shoulder, sternum, bicep padding – all comprising high expanded EVA
	2XU	2XU Men's Elite Compression Top for athletes requiring more power and support	Garment helps to reduce muscle fatigue and damage; the graduated fit supports in increasing blood circulation
	There are other popular brands for rugby such as Skins, BSc, Odlo, Under Armour, Red Ram and Helly Hansen		
Skiing	Skins	Skins S400 long-sleeved top offers comfortable and precise fit coupled with engineered gradient compression to create the ideal compression garment for colder activities such as skiing	Warp knit fabric; brushed inner side; thermal insulation; moisture management; anti-bacterial; UV protection
	CWX	Stabilyx tights target lower back/core and knee joint and surrounding muscles for stability during skiing	Made of 80% polyester and 20% elastane; designed to reduce the quick buildup of lactic acid in the muscles, whilst keeping comfortable and dry during exercise by wicking moisture away; body-moulded compression garments improve recovery after periods of sustained exercise

Note: From various sources.

The chart in Figure 7.15 shows that compression garments are preferred when compared to non-compression sportswear. This is true for short-/long-sleeve shirts, sleeveless shirts, shorts, tights, socks and arm or leg sleeves. The retail price for compression garments is 1.25 to 2.5 times higher than that for non-compression wear, but in the case of socks the prices are four time higher than those for normal socks, as highlighted in Figure 7.15. The market trend for compression wear has been steadily increasing since 2008; the apparel market has fared better compared to socks and sleeves for arms/legs. The overall market size in 2010 for all compression wear was $8.4 million as opposed to $3.1 million in 2008. Hence, it is predicted that the market

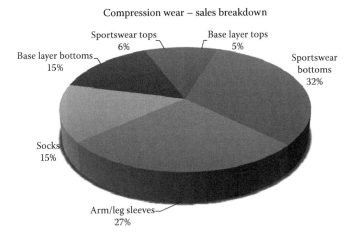

FIGURE 7.14
Compression wear: breakdown of sales (2010). (From Trend Insight, http://www.sportsinsightmag
.com.)

will continue to increase with new developments in fabric and garment technologies for a number of sports.

It can be observed from Figure 7.16 that compression garments in the
form of apparel are frequently purchased compared to socks and arm or leg
sleeves. The market trend is predicted to grow rapidly due to a number of

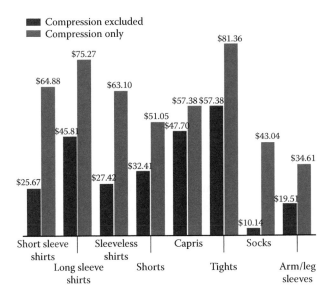

FIGURE 7.15
Average retail selling price – compression garments. (From Trend Insight, http://www
.sportsinsightmag.com.)

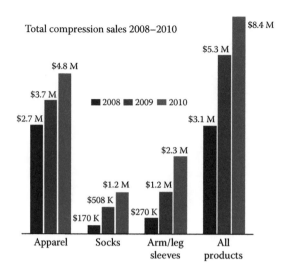

FIGURE 7.16
Compression wear sales. (From Trend Insight, http://www.sportsinsightmag.com.)

factors such as increase in sports participation and athletes being educating about care products such as compression wear. It is also anticipated that, due to rapid developments in the field of smart textiles, a number of new and innovative materials will emerge in the future, creating a plethora of products to choose from for a sporting activity.

7.8 Contextual Factors Affecting Compression Garment Performance

This following section highlights some of the factors that affect the application of compression (Figure 7.17):

- Pressure applied by the garment (support for specific muscle groups)
- Graduated compression (maximum pressure at the extremities and decline in pressure closer to the heart; see Figure 7.2)
- Fabric offering sufficient stretch and recovery at the key points – knees, elbows, buttocks, shoulder and arms

For instance, in cycling the athlete requires endurance and preparation, and there are many factors affecting the overall performance of the wearer. Three core factors have been highlighted: clothing/material worn,

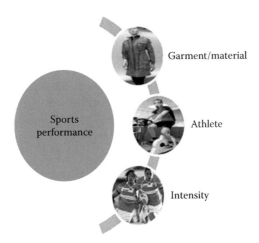

FIGURE 7.17
Factors affecting sports performance.

athletes' capabilities and the intensity of the sport. Compression garments are designed to provide intimate contact with the human skin; hence there is interaction of garment with athlete. This obviously depends on the musculature, microclimate of the fabric and physiological response of the athlete that depends on the intensity of the sport. In addition to these, the three major contextual factors that affect compression garment development are discussed next.

7.8.1 Garment Sizing

Designing compression garments for various body shapes and body types is a complex procedure. It should be noted that most brands develop sizes based on fit athletes or specific target populations. Grading of compression garments for creating various sizes (small, medium, large and extra large) has created a challenge in garment manufacturing by using size charts derived from anthropometric studies (Allsop, 2012). Le Pechoux and Ghosh (2002) reported that consumers were often unhappy because of the large variations between sizes of different brands or in different retail outlets. Loker (2007) reported that sizes are developed on the basis of large anthropometric studies with samples representing the entire target population. Such anthropometric studies have benefitted from 3D body scanners in recent years to reduce the time taken to collate body measurements. However, they are expensive and time consuming to conduct (Le Pechoux and Ghosh, 2002; Yu, 2004a; Loker, 2007; Otieno, 2008). Le Pechoux and Ghosh (2002) stressed that variations between gender, race and generations are apparent through studies on body shape. It could be noted that in the case of the medical compression garments (Jobst stockings), specific measurements of consumers are taken using a body measurement chart to produce a close fit to the body shape and size (Absolutemedical.com, 2014). Certain sports companies such as Giordana (2013) also produce custom-made cyclist compression garments such as bib tights, long- and short-sleeve jerseys, socks, etc. that provide good fit. Some firms also offer custom-made socks and pro-calf tube, hamstring tube and ankle socks (UKsportsproducts.com) and compression leggings and tops that offer optimum pressure distribution (Kurioperformance.com). It should be noted that in the case of stretch garments such as cyclist's shorts that closely fit to body contours, the patterns are shaped for the active position so that they fit well when the rider is on the bike (Ashdown, 2011).

Ideally, a compression garment stays close to the skin with a tight fit and has the ability to stretch and recover based on the activity of the wearer. If the garment has a tight fit in resting position, the increase in blood flow during intense activity is likely to increase the volume of limbs and change the perception of overall garment fit. On the other hand, if the compression garment is baggy, it may not apply the desired compression at the region. Hence, it becomes necessary to understand the pressure required at various zones of the garment, garment sizing allowance during manufacturing and fabric properties during grading of garment sizes.

7.8.2 Body Shapes

Male body shapes are implicitly labelled as endomorph, mesomorph and ectomorph, as illustrated by Figure 7.18. Ectomorph shapes are characterised by being tall, lean builds with little excess body fat. Mesomorphs

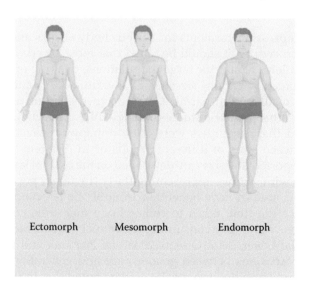

Ectomorph   Mesomorph   Endomorph

FIGURE 7.18
Body types. (Courtesy of Dreamstime.)

are a medium build and have a more athletic frame with broad shoulders and a narrow waist. Endomorph shapes have a wider frame and generally more fat. As previously noted, the most common body shapes in a population are affected by many factors, including race, gender and lifestyle (Le Pechoux and Ghosh, 2002). Furthermore, although these specific body types are widely recognised, many people have variations from these body types. These factors mean that sizing for ready-to-wear garments can be extremely difficult, thus resulting in consumer dissatisfaction. Although custom-made garments produced to specific measurements can ensure a perfect fit, the time and cost to produce such garments is so much that this is not viable for a mass market (Loker, 2007).

7.8.3 Sizing and Designing with Stretch Fabrics

A sizing system involves developing a pattern that relates to a fit model and adding a set of graded patterns based on the assumption of proportional measurement changes larger and smaller than the fit model (Branson and Nam, 2007). Generally, a wear ease is allowed for body movement. However, in the case of close-fitting garments such as compression garments, the wearing ease is negligible as the fabrics stretch because they are made from elastomeric filaments. The patterns developed for stretch fabrics are smaller than patterns created for nonstretch woven fabrics; in other words, patterns are smaller than the body measurements. Therefore, patterns developed for stretch fabrics are often complex and it is necessary to know the fabric behaviour. Yu (2004b) stated that it is incorrect to assume that stretch fabric garment will automatically fit the body contour in all places and provide ease of body movement. Hunter and Fan (2004) identified that fabric properties (bending property, formability and fabric thickness) affect the fit, sizing and seam quality. In addition, the usage of garments following washing also contributes to shrinkage or growth, which can affect close-fitting compression garments. Hardaker and Fozzard (1997) also highlighted the importance of appropriate fabric selection, as an inappropriate fabric may send production back to early stages of product development. The designer should have a good understanding of the fabric behaviour and characteristics, because this fabric knowledge will affect the garment grading, especially in determining the increase or decrease of patterns derived from body measurements and size charts.

7.8.4 Fabric Panels

Compression garments are usually made of fine knitted fabric that contains 75%–80% polyester or nylon and 20%–25% elastomeric filaments, which stretch and recover back to their original shape during physical movements of the body. In a typical flat knit, stretch fabrics can be produced using inlay and body yarn (elastomeric yarn) that imparts stretchability to the fabric.

A recent US patent (Young, 2009) on the manufacture of compression garments for sportswear highlighted the number of panels required to isolate and support specific muscle groups and aid in blood circulation or assist in reducing soft-tissue injury. It also added that panel shapes and seams correspond to various muscle groups for a whole-body compression garment.

In order to support the human anatomy, various fabric panels of a full-body compression wear are proposed that contain 19 different panels (front and back). Recent research (Allsop, 2012), investigating the pressure profile using Tekscan pressure sensors, identified that when a compression garment is made of several panels sewn together, the pressure applied by the garment on a specific muscle group is not uniform. Hence, the lesser the number of sewn fabric panels is, the better the pressure applied will be. This is presented in Figure 7.19, where fabric panels are designed based on the specific muscle groups identified in Figure 7.20 and Table 7.2. It could be noted that factors discussed before that affect compression garment development should be taken into account whilst designing garments.

7.9 Summary and Conclusions

Compression for sportswear is a growing market and, in recent times, greater awareness among athletes has led to increased usage in a wide range of sports. Major international events such as the Olympics and athletic championships also have triggered new technological developments in compression garments. New innovative compression garments for sportswear will continue to be developed due to interest from athletes, enthusiasm from amateurs and demand for casual leisure wear. Compression garments not only appear to support performance but also, more importantly, to portray a professional approach to sports. In addition, these garments also serve to support body posture during training and preparation. Many studies have reported on improved performance characteristics and recovery times of athletes, but there is still yet to be a clear understanding of whether compression garments are beneficial. It is difficult to apply some of the research findings, as most studies have been conducted in a controlled environment, used a small sample or used professional athletes. A series of claims made by sportswear developers can bewilder consumers in their selection of specific sportswear garments.

Furthermore, some studies have highlighted the importance of determining fabric characteristics that contribute to the pressure distributed on the muscles. Good understanding of fabric characteristics is vital when designing garments with stretch fabrics. Practically, evidence relating to the effectiveness of compression sportswear needs to be collated and disseminated to a wider community so that the users will have complete confidence in

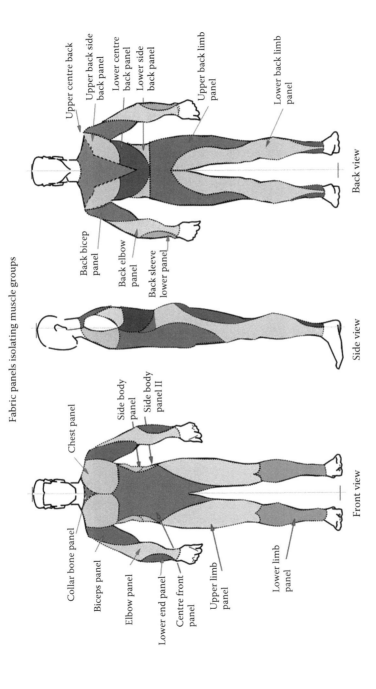

FIGURE 7.19
Various fabric panels to support human musculature.

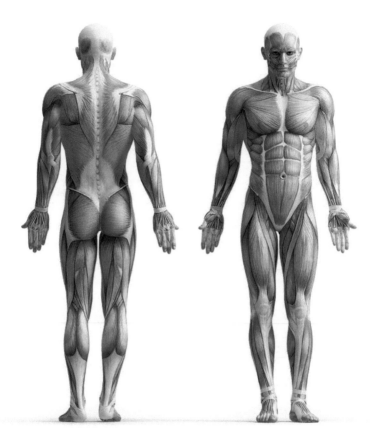

FIGURE 7.20
Human musculature. (Courtesy of istockphoto.)

TABLE 7.2

Fabric Panels for Specific Muscle Groups

Front View	Back View
Collarbone panel	Back bicep panel
Biceps panel	Back elbow panel
Elbow panel	Lower back end panel
Lower end panel (arm)	Lower back limb panel
Centre front panel	Upper back limb panel
Upper limb panel	Lower side back panel
Lower limb panel	Lower centre panel
Side body panel I and II	Upper back side body panel
Chest panel	Upper centre back panel

wearing the compression garments. In addition, there is a dearth of information relating to various body shapes and sizing, as research was generally related to fit athletes.

This chapter has highlighted three key sporting applications where compression garments are widely used: cycling, skiing and rugby. In the case of cycling, compression garments focus on the lower limbs, providing specific support to calf muscles, gluteal muscles, hamstrings and quadriceps. As a rule of thumb, the pressure is applied in a graded way so that the extremities receive slightly more pressure than the regions close to the heart. In the case of skiing, which is a high-intensity sport, the garment is required to provide stretch and recovery, particularly for knee, elbow, pelvic muscles, lower limbs and arms. In addition, the garment should be regulating moisture and keeping the wearer warm by resisting cold wind penetration. Rugby, unlike cycling and skiing, is a high-contact sport, where shoulder/lower limb injuries are often sustained during tackling; hence the key aspect in rugby compression wear is to aid in muscle recovery, prevent soreness and fatigue, reduce blood lactate levels and, more importantly, offer impact protection to shoulder, arms and sternum (chest).

Various contextual factors that influence design and development of compression garments were discussed, including number of fabric panels, body shape and sizing, stretch and recovery and compression applied locally. It should be noted that as the consumer market base is increasing, it is necessary to consider various body shapes and to size the garments accordingly. Stretch fabrics made of knitted construction require good knowledge of fabric behaviour, particularly of stretch and recovery properties following deformation. Market trends for compression garments are positive. There is an increase in usage, particularly in tights that are often sought to facilitate support for lower limbs during intense sport.

A systematic review of evidence relating to compression garments was collated. Although some studies indicated convincing benefits, most research suffered from small sample size and lack of practical significance. Although studies relating to recovery and reduced injury gave positive results, subjects could have been influenced to perform well when they were aware of trials. It should be stated that further work is needed in this area, as evidence is elusive. It can be concluded that users are concerned with garment sizing and comfort; hence further work needs to be carried out in developing a precise measurement system that will evaluate pressure and comfort and predict using a simulation model by utilising fabric behaviour characteristics that would eventually facilitate garment design and development.

The compression sportswear market is predicted to grow significantly in the next decade or so. The aim of this chapter is to help readers understand the reported benefits of compression and to interpret claims of manufacturers – being aware of specific applications of compression wear in cycling, skiing and rugby, factors that affect compression performance and development and recent market trends in compression garments.

References

2XU Pty Ltd. (2009). Compression 2XU: Benefits. [online] http://www.2xushop.co .uk/compression/benefits.html (accessed 7 February 2012).

Absolutemedical.com. (2014). [online]. Available at http://www.absolutemedical .com/customjobstelvarexcompressiongarments.html (accessed 11 July 2014).

Ali, A., Creasy, R. H. and Edge, J. A. (2011). The effect of graduated compression stockings on running performance. *Journal of Strength and Conditioning Research* 25 (5): 1385–1392.

Allsop, C. A. (2012). An evaluation of base layer compression garments for sportswear, MSc dissertation, Manchester Metropolitan University, Manchester, UK.

Ashdown, S. P. (2011). Improving body movement comfort in apparel. In *Improving comfort in clothing*, Song, G. (ed.). Oxford: Woodhead Publications.

Bondy, F. (2000). Garb doesn't suit Olympic athletes well. [online] http://articles .nydailynews.com/2000-07-21/sports/18158625_1_swift-suit-snatchersfull -body-suit (accessed 20 April 2012).

Branson, D. H. and Nam, J. (2007). Materials and sizing. In *Sizing in Clothing: Developing effective sizing systems for ready-to-wear clothing*, Ashdown, S. P. (ed.). Cambridge, UK: Woodhead Publishing.

British Columbia Injury Research and Prevention Unit. (2012). Rugby injuries. http:// www.injuryresearch.bc.ca (accessed on 23 March 2012).

Brophy Williams, N., Gliemann, L., Fell, J., Shing, C., Driller, M., Halson, S. and Askew, C. D. (2014). Haemodynamic changes induced by sports compression garments and changes in posture. Proceedings of the 19th Annual Congress of the European College of Sport Science, Amsterdam, Netherlands, 2–5 July 2014.

Cipriani, D. J., Yu, T. S. and Lyssanova, O. (2014). Perceived influence of a compression, posture-cueing shirt on cyclists' ride experience and post-ride recovery, *Journal of Chiropractic Medicine* 13 (1): 21–27.

Cole, M. D. (2008). Compression apparel brand winning at the 'skins' game. *Apparel Magazine* 50 (3): 58–62.

Cortad, R. (2011). Quiksilver's new board shorts combine compression and taping technologies. [online]. http://www.apparelnews.net/news/manufacturing/10 2011-Quiksilvers-New-Boardshorts-Combine-Compression-and-Taping -Technologies (accessed 2012).

Craik, J. (2011). The Fastskin revolution: From human fish to swimming androids. *Culture Unbound* 3:71–82.

de Glanville, K. and Hamlin, M. (2012). Positive effect of lower body compression garments on subsequent 40 kM cycling time trial performance. *Journal of Strength and Conditioning Research* 26 (2): 480–486.

Doan, B. K., Kwon, Y. H., Newton, R. U., Shim, J., Popper, E. M., Rogers, R. A., Bolt, L. R., Robertson, M. and Kraemer, W. J. (2003). Evaluation of a lower-body compression garment. *Journal of Sports Sciences* 21 (8): 601–610.

Duffield, R. and Portus, M. (2007). Comparison of three types of full-body compression garments on throwing and repeat-sprint performance in cricket players. *British Journal of Sports Medicine* 41 (7): 409–414.

Duffield, R., Cannon, J. and King, M. (2010). The effects of compression garments on recovery of muscle performance following high-intensity sprint and polymetric exercise. *Journal of Science and Medicine in Sport* 13 (1): 136–140.

Explosive board shorts. (2015). http://www.explosive-boardshorts.com/tech.php (accessed online 29 June 2015).

Feist, W. R., Andrade, D. and Nass, L. (2011). Problems with measuring compression device performance in preventing deep vein thrombosis. *Thrombosis Research* 128 (3): 207–209.

Giordana. (2013). Custom cycling apparel. Online portal. http://www.giordana.com (accessed May 2013).

Haldane, J. (2013). Healing hosiery, view on compression. https://www.macom -medical.com (accessed on 3 August 2013).

Hardaker, C. H. M. and Fozzard, G. J. W. (1997). The bra design process – A study of professional practice. *International Journal of Clothing and Science Technology* 9 (4): 311–325.

Higgins, T., Naughton, G. A. and Burgess, D. (2009). Effects of wearing compression garments on physiological and performance measures in a simulated game-specific circuit for netball. *Journal of Science and Medicine in Sport* 21 (1): 223–226.

Hill, J., Howaston, G., van Someren, K., Leeder, J. and Pedler, C. (2014). Compression garments and recovery from exercise induced muscle damage – A meta-analysis. *British Journal of Sports Medicine* 48:1340–1346. doi: 10.1136/bjsports-2013-092456.

Hill, J., Howatson, G., van Someren, K., Walshe, I. and Pedlar, C. (2014). Influence of compression garments on recovery after marathon running. *Journal of Strength and Conditioning Research* 28 (8): 2228–2235.

Hooper, D., Dulkis, L., Secola, P., Holtzum, G., Harper, S., Kalkowski, R., Comstock, B., Szivak, T., Flanagan, S., Looney, D., DuPont, W., Maresh, C., Volek, J., Culley, K. and Kraemer, W. (2015). The roles of an upper body compression garment on athletic performances. *Journal of Strength and Conditioning Research* 21 February, p. 1.

Hunter, L. and Fan, J. (2004). Fabric properties related to clothing appearance and fit. In *Clothing appearance and fit: Science and technology*, Fan, J., Yu, W. and Hunter, L. (eds.). Cambridge, UK: Woodhead Publishing.

Jakeman, J. R., Byrne, C. and Eston, R. G. (2010). Lower limb compression garment improves recovery from exercise-induced muscle damage in young, active females. *European Journal of Applied Physiology* 109 (6): 1137–1144.

Kurioperformance.com. (2015). Custom designed compression garments. [online]. Available at http://www.kurioperformance.com/custom-design.php (accessed 11 June 2015).

Le Pechoux, B. and Ghosh, T. K. (2002). Textile progress: *Apparel Sizing and Fit* 32 (1). Manchester, The Textiles Institute.

Leisure Trend Group (2011). Compression apparel market – A report. http://www .leisuretrends.com (accessed April 2013).

Loker, S. (2007). Mass customization and sizing. In *Sizing in clothing: Developing effective sizing systems for ready-to-wear clothing*. Ashdown, S. P. (ed.). Cambridge, UK: Woodhead Publishing.

Luke, V. and Sanderson, S. (2014). Compression stockings and aerobic exercise: A meta-analysis. *International Journal of Human Movement and Sports Sciences* 2:68–73. doi: 10.13189/saj.2014.020403.

Mayes, R. (2010). The modern olympics and post-modern athletics: A clash in values. *Journal of Philosophy, Science and Law* 10 [online] https://www6.miami.edu/ethics/jpsl/archives/all/Olympics%20and%20Athletics.pdf (accessed 20 April 2012).

McCurry, J. W. (2004). Building on the base layer buzz. *Apparel Magazine* 45 (5): 22–23.

McKeegan, N. (2008). Speedo LZR Racer – The world's fastest swimsuit. [online] http://www.gizmag.com/speedo-lzr-racer-worlds-fastest-swimsuit/8819/ (accessed 2 May 2012).

Miller, J. A. (2011). Use and wear of anti-embolism stockings: A clinical audit of surgical patients. *International Wound Journal* 8 (1): 74–83.

Mintel Group. (2009). Sports clothing and footwear—UK—September 2009. [online] London, Mintel Ltd, http://academic.mintel.com/sinatra/oxygen_academic/search_results/show&/display/id=394601/display/id=479629#hit1 (accessed 2 March 2011).

Miyamoto, N., Hirata, K., Mitsukawa, N., Yanai, T. and Kawakami, Y. (2011). Effect of pressure intensity of graduated elastic compression stocking on muscle fatigue following calf-raise exercise. *Journal of Electromyography and Kinesiology* 21 (2): 249–254.

Moffatt, C. (2008). Variability of pressure provided by sustained compression. *International Wound Journal* 5 (2): 259–265.

Moffatt, C., Martin, R. and Smithdale, R. (2007). *Leg ulcer management*. Oxford, UK: Blackwell Publishing.

Montgomery, P. G., Pyne, D. B., Hopkins, W. G., Dorman, J. C., Cook, K. and Minhan, C. L. (2008). The effect of recovery strategies on physical performance and cumulative fatigue in competitive basketball. *Journal of Sports Sciences* 26 (11): 1135–1145.

Morrison, M., Mofatt, C., Nixon, J. B. and Bale, S. (1998). *A colour guide to nursing management of chronic wounds*, 2nd ed. London: Mosby.

Otieno, R. B. (2008). Improving apparel sizing and fit. In *Advances in apparel production*, Fairhurst, C. (ed.). Cambridge, UK: Woodhead Publishing.

Proskins. (2012). The technology behind Proskins [online] http://www.proskins.co/store/en/content/8-how-proskins-work (accessed 18 April 2012).

Ramelet, A. A. (2002). Compression therapy. *Dermatolgic Surgery* 28 (1): 6–10.

Rogers, L. (2012). Post op compression garments for hockey, football and basketball Players. [online] http://ezinearticles.com/?Post-Op-Compression-Garments-for-Hockey,-Football-and-Basketball-Players&id = 6263104 (accessed 21 February 2012).

Shishoo, R. (2005). *Textiles in sport*. Cambridge, UK: Woodhead Publishing Limited.

Skins (2012). SKINS science. [online]. http://www.skins.net/en-GB/whyskins/skins-science.aspx (accessed 7 February 2012).

South Wales Osteopathic Society. (2009). Rugby injury statistics. http://www.osteopathywales.com/Online portal (accessed 4 January 2012).

Sperlich, B., Haegele, M., Achtzehn, S., Linville, J., Holmberg, H.-C. and Mester, J. (2010). Different types of compression clothing do not increase sub-maximal and maximal endurance performance in well-trained athletes. *Journal of Sports Sciences* 28 (6): 609–614.

Swim-Faster.com. (2012). The Speedo Fastskin fsii swimsuit story. [online]. http://www.swimming-faster.com/ (accessed 20 April 2012).

Thomas, S. (1998). World wide wounds: Compression bandaging in the treatment of venous leg ulcers. [online] http://www.worldwidewounds.com/1997/september/Thomas-Bandaging/bandage-paper.html (accessed 15 September 2011).

Thomas, S., Fram, P. and Phillips, P. (2007). World Wide Wounds: The importance of compression on dressing performance. [online] http://www.worldwide wounds.com/2007/November/Thomas-FramPhillips/Thomas-Fram-Phillips -CompressionWR AP.html (accessed 1 May 2012).

Trend Insight. (2011). Compression apparel tightening grip on market. March/April (http://www.sportsinsightmag.com).

UKsportsproducts.com. (2015). Herzog sport compression. [Online]. Available at http://www.uksportsproducts.com/recovery/herzog-sport-compression/ (accessed 11 June 2015).

URGO Medical. (2010). Two major compression therapy methods: Bandages and medical compression hosiery. [online]. http://www.urgomedical.com/Pathophysiologies /Compression/Veno-lymphatic-compression/Compression-therapy-methods (accessed 1 May 2012).

Van Geest, A. J., Franken, C. P. M. and Neumann, H. A. M. (2003). Medical elastic compression stockings in the treatment of venous insufficiency. In *Current problems in dermatology: Textiles and the skin*, vol. 31, Elsner, P., Hatch, K. and Wigger-Alberti, W. (eds.). Basel, Germany: Karger.

Venkatraman, P. D., El Sawi, A., Afify, S., Anand, S. C., Dean, C. and Nettleton, R. (2005). Pilot study investigating the feasibility of an ulcer specific quality of life questionnaire. *Phlebology* 20 (01): 14–27.

Venkatraman, P. D., Tyler, D. J., Ferguson-Lee, L. and Bourke, A. (2013). Performance of compression garments for cyclists, presented at the Textile Institute's International Conference on Advances in Functional Textiles, 25–26 July 2013, Chancellor's Hotel and Conference Centre, Manchester, UK.

Voyce, J., Dafniotis, P. and Towlson, S. (2005). Elastic textiles. In *Textiles in sport*, Shishoo, R. (ed.). Cambridge, UK: Woodhead Publishing.

Walzer, E. (2004). Focus on function, freshness. *Apparel Magazine* 45 (5): 23–26.

Watkins, W. B. C. (2010). Compression garment sizing challenges, issues and a solution. *Plastic Surgical Nursing* 30 (2): 85–87.

Weller, C. D., Evans, S., Reid, C. M., Wolfe, R. and McNeil, J. (2010). Protocol for a pilot randomised controlled clinical trial to compare the effectiveness of a graduated three layer straight tubular bandaging system when compared to a standard short stretch compression bandaging system in the management of people with venous ulceration: 3VSS2008. *Trials* 11 (26): 1–20.

Wiernert, V. (2003). Compression treatment after burns. In *Current problems in dermatology: Textiles and the skin*, vol. 31, Elsner, P., Hatch, K. and Wigger-Alberti, W. (eds.). Basel, Germany: Karger.

Yake, A. (2011). The best compression garments for sport, http://www.livestrong .com (accessed 5 March 2013).

Young, G. (2009). Compression garment or method of manufacture. US Patent /0113596 A1.

Yu, W. (2004a). 3D body scanning. In *Clothing appearance and fit: Science and technology*, Fan, J., Yu, W. and Hunter, L. (eds.). Cambridge, UK: Woodhead Publishing.

Yu, W. (2004b). Objective evaluation of clothing fit. In *Clothing appearance and fit: Science and technology*, Fan, J., Yu, W. and Hunter, L. (eds.), 72–88. Cambridge, UK: Woodhead Publishing.

8

Impact-Resistant Materials and Their Potential

Praburaj Venkatraman and David Tyler

CONTENTS

8.1 Introduction 205
8.2 Injuries Sustained during Sporting Activities 206
8.3 Commercially Available Impact-Resistant Materials 213
8.4 Rationale for Using Impact-Resistant Materials 216
    8.4.1 Methodology 216
8.5 Impact Attenuation Test 217
8.6 Regulations for Impact Protection 221
8.7 Benchmarking of Impact-Resistant Materials 222
8.8 Garment Design and New Product Development 222
    8.8.1 Design Issues 223
8.9 Summary and Conclusions 226
Acknowledgement 227
References 227

8.1 Introduction

The incidence of sports injuries in high-contact sports such as rugby, football, ice hockey and baseball is high. These sports have always been played with intense competition such that injuries become inevitable. However, the severity of injuries can be reduced with appropriate training and when protective clothing is worn, which has the ability to absorb the shocks or dissipate the energy over a wider area such that it lessens the impact. The protection gear in rugby includes headgear, shirt pads, shin guards and mouth guards. High-performance impact protection materials are embedded in the garment, particularly in the shoulder, biceps and chest region. These materials need to be soft, thin, flexible and durable. Commercial products are usually available in various thicknesses and some can be moulded to fit various contours of the body.

In recent years, there has been an increase in the use of such impact-resistant materials in sportswear, but some users have expressed concerns about the way protection is incorporated in the products. Some garments have pads inserted into fabric pouches which can move during an impact and are unlikely to remain in position and protect the wearer during a slip or fall or collision. A few garments possess bulky pads, which inhibit the breathability and restrict the free movement of the athlete. However, most garment manufacturers and sportswear brands claim that padding reduces the risk of injuries sustained in sports. On the other hand, researchers suggest that such padding gives limited protection against fracture and dislocation of joints. In this chapter, a wide range of impact-resistant materials used for sportswear are appraised. Injuries sustained in high-contact sports such as rugby are identified. Regulations that affect the designing of garments with impact-resistant materials are discussed along with illustrations of various commercially available garments with padding. An impact attenuation test developed to ascertain the energy absorption characteristics of various commercial materials is described. Various products are compared for their performance and benchmarked against natural materials such as leather. In addition, factors affecting garment manufacturing using these impact-resistant materials are presented.

8.2 Injuries Sustained during Sporting Activities

From major competitive team sports like football, baseball, cricket and basketball, to individual sports like cycling, bowling, golf and swimming, the body is at risk from the elements surrounding it when involved in a physical activity. Maron et al. (1995) examined 25 cardiac deaths in young adults from sport injuries and found that the cause was impact from a projectile (e.g. a baseball or hockey puck). They added that, among the sports players, 18 did not wear any protection on the chest zone and the remaining wore some form of protection such as padding. Of the seven players who wore protective gear, four were playing hockey, and the other three were playing football. In 2010, the National Electronic Injury Surveillance System (NEISS) reported over half a million injuries relating to basketball. Another two million injuries were associated with bicycling, football and other sports. The highest numbers of sports-related injuries came from bicycling, basketball, baseball and running. While each sport is played differently, they all take the same toll on the body. Each part of the body is affected by physical contact, collisions and excessive exertion. Some of the most common injuries to the body from various sports are listed in Table 8.1, obtained from NEISS, which collects injury data associated with consumer products from US hospital emergency departments across the country.

Impact injuries sustained by sportsmen and -women are increasing due to the competitive nature of sports such as rugby, football, baseball, etc. and

TABLE 8.1

Sport and Types of Injury

Sport	Type of Injury	Estimated Number of Injuries
Basketball	Cut hands, sprained ankles, broken legs, eye and forehead injuries	529,837
Bicycling	Feet caught in spokes, head injuries from falls, slipping while carrying bicycles, collisions with cars	490,434
Football	Fractured wrists, chipped teeth, neck strains, head lacerations, dislocated hips and jammed fingers	460,210
ATVs, mopeds, minibikes	Riders of ATVs were frequently injured when they were thrown from vehicles; there were also fractured wrists, dislocated hands, shoulder sprains, head cuts and lumbar strains	275,123
Baseball, softball	Head injuries from bats and balls; ankle injuries from running bases or sliding into them	274,867
Exercise, exercise equipment	Twisted ankles and cut chins from tripping on treadmills; head injuries from falling backward from exercise balls; ankle sprains from jumping rope	269,249
Soccer	Twisted ankles or knees after falls, fractured arms during games	186,544
Swimming	Head injuries from hitting the bottom of pools and leg injuries from accidentally falling into pools	164,607
Skiing, snowboarding	Head injuries from falling, cut legs and faces, sprained knees or shoulders	96,119

Source: NEISS, 2010.

they adversely affect their careers. The National Sports Medicine Institute (Nsmi.org.uk, 2009) stated that sports players experience injuries caused by impact or contact with objects, surfaces or other people. Injuries caused by impact and contact occur in common sports such as football and rugby and in more dangerous sports such as motor racing, boxing and skiing. Often, contact with other people can cause an athlete to become off balance, or change direction quickly; this causes damage to the connective tissue, and powerful direct contact may also cause a joint to become displaced. Impact injuries often involve spinal injuries, ligament and tendon damage, fractures and head injuries. Rugby, for instance, has the highest risk per player per hour of all the major sports; 30% of injuries occur in the shoulder region, closely followed by injuries in the knee area (Funk, 2012). Marshall et al. (2002) reported that rugby is a high-contact sport and players are likely to sustain a range of injuries. Table 8.2 illustrates various injuries that are found to occur in rugby during tackling.

During training, players are known to sustain injuries. Researchers Gabbett and Godbolt (2010) investigated the incidence of training injuries among professional rugby players and found that injuries were particularly high in the

TABLE 8.2

Description of Tackle Types in Rugby

Arm	Tackler impedes/stops ball carrier using the upper limbs	
Collision	Tackler deliberately impedes/stops the ball carrier without using the arms	
Shoulder	Tackler impedes/stops the ball carrier with his/her shoulder as the first point of contact followed by use of the arm(s)	

Source: Fuller, C. et al., 2010, BMJ Publishing Group Ltd.

Note: Other types of tackler injuries include the following: *jersey* – holds the ball carrier's jersey; *lift* – the tackler raises the ball carrier's hips above the ball carrier's head; *smother* – the tackler uses the chest and wraps both arms around the ball carrier; and *tap* – the tackler trips the ball carrier with his/her hand on the lower limb below the knee.

thigh region. Hematoma (a type of injury in legs where blood clots, blood flow is restricted and blood vessels in the injured portion break) and muscular strain were associated with rugby players. International Rugby Board (IRB) regulations allow players to wear shoulder pads, provided the pads are made of soft and thin material, which may be incorporated in an undergarment or jersey provided the pads cover the collarbone and shoulder. The padding material may not exceed 45 kg/m^3, providing maximum coverage to the shoulder region (IRB, 2012). In addition, players may wear shin guards, ankle support and head gear that conform to IRB regulation 12.0 (IRB, 2012). Shoulder injuries result in sprains, strains, fractures and dislocations (Brooks et al., 2005). Funk (2012) stated that 35% of all injuries in the shoulder region are recurrent ones and the player has a likelihood of sustaining an injury on the other shoulder.

A recent report stated that one in four rugby players will be injured during a season. Rugby injuries are three times higher than football injuries. Most injuries are experienced by youths of 10–18 years and adults aged 25–34 years. In rugby, 57% of most sport injuries occur during matches (rather than in training) and particularly when a player tackles or is tackled (South Wales Osteopathic Society, 2009). Concussion is an injury to the brain or spinal cord due to jarring from a blow, fall or impact of a collision. In the Australian Football League, concussion is estimated to occur at a rate of approximately seven incidents per team per season (Khurana and Kaye, 2011). The Rugby Football Union (RFU, 2002) stated that despite the increased use of shoulder pads from 20% to 36% between 1999 and 2002 there was an increase in injuries from 12% to 13%. Gerrard (1998) noted that shoulder pads do not protect against fracture, dislocation or rotator cuff tears. It has been suggested that shoulder pads may protect from lacerations and reduce bruising and haemotoma of the soft tissue surrounding the shoulder, but do not prevent major injuries that result from direct blows to the top of the shoulder or falling onto an outstretched hand. Recently, Harris and Spears (2010) conducted an investigation on four commercially produced shoulder pads – PVA foam (Kooga, Canterbury, Gilbert and Terminator). They examined material properties by dropping hard and soft objects onto materials protecting a force plate, recording peak impact forces at predetermined heights and measuring their force-deformation behaviour. Best performing pads were thicker; all pads were able to attenuate force for lower loads, but at higher impact loads, offered little protection. Pain, Tsui and Cove (2008) reported in vivo effectiveness of Kooga shoulder pads using Tekscan sensors that measured impact intensity on actual tackles on six male rugby players. The researchers reported that pads enabled peak forces to be reduced by 35% (impact with an object) and 40% (for all tackles).

Figure 8.1a illustrates that adults endure more injuries to head, shoulder and lower limb (thigh), whereas children (Figure 8.1b) suffer from head/neck injuries followed by the upper and lower limbs.

Approximately half of all injuries occur while a player is tackling (Figure 8.2) or being tackled. Hookers and flankers sustain the most injuries. Forwards are more frequently injured than backs because of their greater involvement in physical collisions and tackles (British Columbia Injury Research and Prevention Unit, 2012). A rugby league team consists of 13 players (six forwards and seven backs); each team has sets of six tackles to advance the ball downfield. Due to the nature of injuries sustained, players have enforced absences from sports participation. Hence, teams are under pressure to prevent or reduce injuries by requiring players to wear protective gear. Figure 8.3 illustrates a shoulder tackle in rugby known to be associated with higher risks of injury. Rugby is a highly intensive team sport and players move fast in

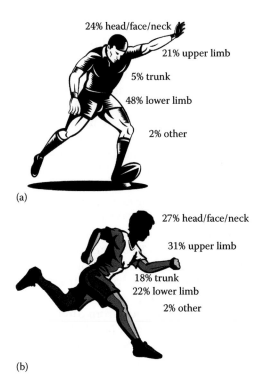

FIGURE 8.1
(a) Proportion of injury sustained – adults. (b) Proportion of injury sustained – children. (From Dreamstime.)

FIGURE 8.2
Rugby player tackling. (From Shutterstock.)

FIGURE 8.3
Shoulder impact – a close-up view of a rugby player being tackled. (From Dreamstime.)

the field. This sport is popular among men; however, women are getting more involved at the college/school level. As discussed earlier, 40% of injuries are muscular strains or contusions and 30% are sprains, followed by dislocations, fractures, lacerations and overuse injuries. Players may also choose to wear protective clothing (in the form of head gear, padded vests, shorts, shin guards, mouth guards and support sleeves) to reduce the risk of sprains and cramps. By 2012 the International Rugby Board had approved 59 brands that supply a wide range of shoulder padding vests to various teams (IRB, 2012).

This chapter reflects on the recent exploration of impact-resistant materials for sportswear using an experimental approach to gain understanding of material properties using an impact attenuation test (which captures peak forces over time). The chapter focuses on the principles of sportswear design when using impact-resistant materials. Figure 8.4 illustrates a recent rugby match between England and Italy, where the English player is being tackled and is uprooted from the ground. The speed at which the collision occurs results in the player falling on the pitch, injuring his head, chest, arms and shoulders.

The force of impact during tackling by rugby players is considered to be 1 to 1.3 times of the body weight (Trewarth and Stokes, 2003) with most of the force acting on the shoulder/collarbone zone. Figure 8.5 illustrates the typical rugby injuries where the direction of the force acts on the tackler (shoulder region) during tackling. A try score is when a player touches the ball in the opposition goal area between the try line and dead ball line. It is worth five points – the maximum points in the rugby union.

FIGURE 8.4
A picture from Mogliano Veneto (Italy): Six Nations 2010, Italy A versus England Red Saxons. Italy's player is tackling hard. (From Dreamstime.)

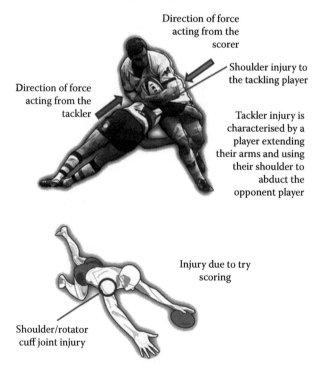

FIGURE 8.5
Rugby player tackling injuries. (Image courtesy of Dahan, 2013.)

A range of protective materials have been evaluated Coir/EVA as nonwoven impact protectors (Maklewska, Krucińska and Mayers, 2005), polypropylene and flax fibre laminate (Van de Velde and Kiekens, 2001); cellular textile materials (Tao and Yu, 2002) as sports protectors for helmets and 3D spacer fabrics from Dow Corning with various thicknesses and levels of protection (Dow Corning, 2011). Maklewska et al. (2004) compared the impact strength of nonwoven fabric pads intended for protective clothing and sportswear; Roger Co. reported two customisable products composed of Poron XRD material. The two products were extreme impact pad and B-guard. The extreme pad was recommended for knee and elbow pads for shin and thigh protection. It was also reported that the product had a fabric backing that allowed moisture wicking air channels to enhance comfort (*WSA*, 2011).

Durá, Garciá and Solaz (2002) investigated the behaviour of dynamic rigidity on shock-absorbing materials (used for shoes) using a viscoelastic linear model. Lam, Tao and Yu (2004) studied various thermoplastic cellular textile composites with knitted and nonwoven fabrics which were sandwiched between layers of thermoplastic matrix. Researchers reported the effect of impact energy of interlock knitted fabric in a matrix – ultrahigh molecular weight polyethylene/low-density polyethylene on three levels of impact 24, 44 and 119 J. The fabrics were in the form of domed-grids (cellular composites) with domes rising up to 15 mm thickness. The fabrics did not collapse with 24 J impacts, but at higher energies the fabric deformed and cells closed. Also, these cellular composites had little recovery soon after impact.

8.3 Commercially Available Impact-Resistant Materials

In this section various materials used for impact protection are described. Dilatants are polymer-based materials commonly used for energy absorption; these are shear thickening materials where viscosity is affected by shear applied strain. At low rate of deformation the material remains soft and pliable, but at elevated rates of deformation it undergoes substantial increase in its viscosity. Hence, at higher deformation the material becomes stiff or rigid (Palmer and Green, 2010). Generally, the viscosity of material decreases as the shear strain increases, but dilatant materials are shear thickening materials where viscosity increases as shear strain increases – hence the name, *non-Newtonian fluid*.

Poron XRD is an open cell urethane foam. When at rest above the glass transition temperature of the urethane molecules, it has softness and flexibility. When impacted quickly, the glass transition temperature of the material drops so that the urethane molecules stiffen to protect the wearer from damage.

D3O comprises a polymer composite which contains a chemically engineered dilatant, an energy absorber. This basic material has been adapted and enhanced to meet specific performance standards and applications. The material is soft and flexible in its normal state; however, when impacted by force it locks itself and disperses energy and returns to its normal state (D3O.com, 2013).

Ethylene vinyl acetate (EVA) foams are described as a specific type of cross-linked closed cell polyethylene foam. They are designed to be soft, with a rubber-like texture and with good shape recovery after deformation.

Deflexion was created by Dow Corning and is made of silicon that has been polymerised into flexible silicone sheets. The company claimed that during a hard impact it acts like a bullet, where all the molecules gather around that area and instantly turn into rock-solid form. They would disperse and absorb the impact, like a bulletproof vest. The idea is to create a kind of armour that can shift around like clothing in a breathable fashion (Dow Corning, 2012).

Sorbothane – a synthetic viscoelastic polymer (thermoset polyurethane material) – is used to absorb shock and is widely used in shoe insoles. It possesses shock absorption, vibration insulation and damping characteristics (http://www.sorbothane.com). Cinats, Reid and Haddow (1987) studied therapeutic implications of sorbothane in orthotic insoles which could absorb the energy from foot strike. The sorbothane had some clinical significance; however, properties of materials may change when bonded to other substances in production of insoles for shoes.

GPhlex is a similar material but not yet in widespread commercial use. The EVA foam was derived from a Canterbury rugby shirt, and the leather was a sample of unfinished material obtained from cow skin.

Spacer fabric is a three-dimensional knitted structure that allows cushioning and shock absorbency with excellent recovery properties. Spacer fabrics are complex 3D constructions made of two separate fabric layers connected vertically with pile yarns or fabric layers. The conventional spacer fabrics composed of two surface layers bound with pile yarns are generally manufactured using weaving and knitting technologies (Abounaim et al., 2010).

Figure 8.6 illustrates a typical rugby top with flexible pads inserted in the shoulder, biceps and sternum region. The pads of varying thickness are included to assist in unrestricted movement. Table 8.3 highlights various rugby tops and impact-resistant materials used.

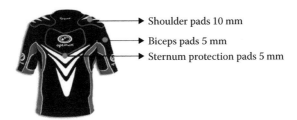

Shoulder pads 10 mm
Biceps pads 5 mm
Sternum protection pads 5 mm

FIGURE 8.6
A typical rugby shirt (optimum) with impact-resistant material insertion.

TABLE 8.3

Leading Brands Providing Shoulder Pads for Rugby

Brand	Product	Impact-Resistant Materials
Canterbury		IRB approved garment with moulded EVA foam to shoulder, biceps and chest to provide enhanced protection against collision.
Optimum		10 mm EVA foam shoulder pads, 5 mm pads for biceps and sternum.
Kooga– EVX V		EVX protection offers the optimum in fit and freedom of movement. The pads consist of a Lycra body, shoulder, sternum, and bicep padding, all composed of high expanded EVA. These pads have been IRB approved.

8.4 Rationale for Using Impact-Resistant Materials

International Rugby Board regulations (IRB, 2012) do not outline the level of protection required of shoulder pads, but stipulate three conditions to be met before the pads can be used in matches. The padding material cannot exceed 45 kg/m³ in density and must provide maximum coverage to the stenocla-vicular, acromioclavicular (AC) and glenohumeral joints. Pads must also be subjected to an IRB-prescribed hammer and anvil test, which involves drop-ping a rigid, flat striking surface (5 ± 0.02 kg) onto a pad resting on a steel anvil.

8.4.1 Methodology

In order to ascertain material properties of impact-resistant materials a range of methods were consulted and, of the many, two methods were suitable. Industrial bump caps (BS EN 812:1997/A1, 2001) and specification for head protectors for cricketers (BS 7928, 1998) involved a striker falling on a surface, with the protective product experiencing the impact. The current research reported here focuses on material properties because measurements of decel-eration are more appropriate to head protection. Consequently, our experi-mental equipment detects the forces experienced by a transducer attached to an anvil located under the protective material. This method is in line with the IRB-prescribed hammer and anvil test, which involves a flat striking sur-face (5 ± 0.02 kg) falling on to a 'pad' resting on a steel anvil (Pain et al., 2008). In this study a range of impact-resistant materials were evaluated and this is presented in Table 8.4.

TABLE 8.4

Commercial Impact-Resistant Materials with Potential for Use in Rugby Tops

Commercial Material	Notes
D3O	Dilatant material
Poron XRD	Open-cell urethane foam
EVA foam	Ethyl vinyl acetate foam
Deflexion S-range	Three-dimensional spacer fabric with silicone
Deflexion TP-range	Dilatant material
Spacer fabrics	Three-dimensional knitted structure
Sorbothane	Synthetic viscoelastic polymer
Leather	Natural benchmarking material

8.5 Impact Attenuation Test

The impact attenuation test has a striker, a steel ball, falling on to a flat anvil on which the protective material is placed. The pressure sensors are located below the sample material and the forces transmitted through the material by the impactor are recorded in the form of a load-versus-time data set. By varying the diameter of the ball, different impact profiles can be created. The mass/height of fall parameters determines the impact energy. For research purposes, impacts of 5, 10 and 15 J are used. An illustration of the test equipment is shown in Figure 8.7.

Impact forces are experienced by the test material, and the forces reaching the support plate are recorded using a load washer. The duration of impact across varying thickness of materials can be measured.

The impact test made use of custom-built equipment from a testing company, INSPEC, UK. A typical set of results recorded for a thin (3 mm) material is illustrated in Figure 8.8. There was insufficient energy taken out by the material, so the ball made a few bounces before coming to rest, and these movements are apparent in the test data.

A selection of materials was tested to compare their abilities to protect against impact. As a control, unfinished leather was tested so that the commercial products could be benchmarked. Leather is a natural material that

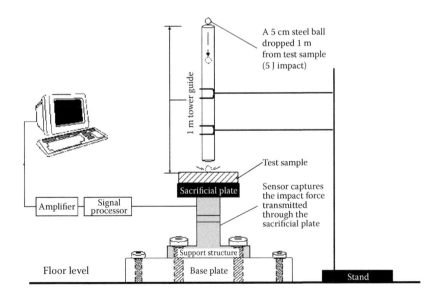

FIGURE 8.7
Impact attenuation test rig.

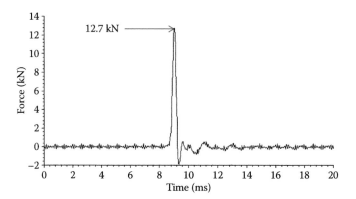

FIGURE 8.8
Impact forces experienced with a protection using 3 mm Poron XRD.

has been used to provide wearers with some protection. Figure 8.9 presents peak force variations for a range of materials with different thicknesses. The materials of 2 to 3 mm experienced high peak forces. As thicknesses increase, the commercial products designed to absorb energy and protect against impacts reduce peak forces more effectively than the leather sample.

The EVA foam samples were taken from commercial garments designed for rugby players. Neither the 5 mm (used for arm protection) nor the 10 mm (used for shoulder protection) sample compared favourably with leather. The maximum thickness for shoulder protection permitted by the International Rugby Board is 10 mm. In general, the branded commercial materials performed better than leather, although less so below 5 mm.

The impact protective pads have been randomly selected from commercial products: Gilbert, Canterbury and Kooga. The illustration (Figure 8.10a to e) shows the tops with protective pads. The pads were also compared

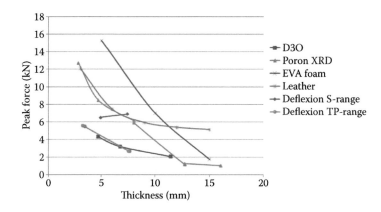

FIGURE 8.9
Findings from impact attenuation tests.

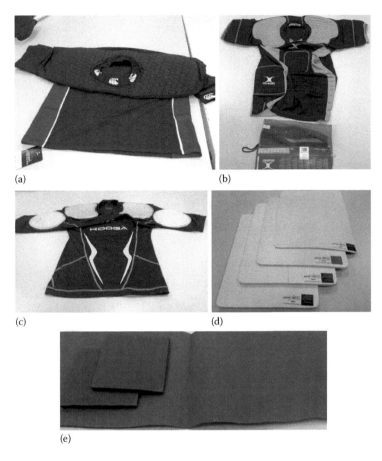

FIGURE 8.10
(a) Canterbury rugby tops. (b) Gilbert rugby tops. (c) Kooga rugby tops. (d) Poron XRD sample pads. (e) GPhlex sample pads.

with leather, D3O and another proprietary material intended to be used in impact protection. Figure 8.11 shows that EVA foam experienced highest peak forces of 15 kN at 5 mm, whilst Poron XRD and leather performed similarly at <5 mm; however, as the thickness increased, the Poron XRD sample outperformed leather. As a general trend, most materials less than 5 mm experienced high peak forces, but D3O was exceptional compared to most materials that were benchmarked. The peak forces experienced by D3O at less than 5 mm were low compared to other materials (GPhlex, Poron XRD, EVA form). As the thickness increased, the peak forces experienced by D3O were comparable to those experienced by GPhlex.

Figure 8.12 illustrates the broadening of the impact forces resulting from the increasing thickness of a protective material. For clothing applications, this factor is of major importance. It can be noticed from Figure 8.12 that

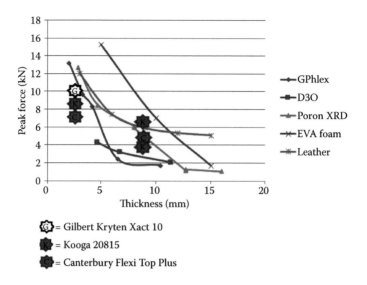

FIGURE 8.11
Peak forces acting on commercially available rugby tops.

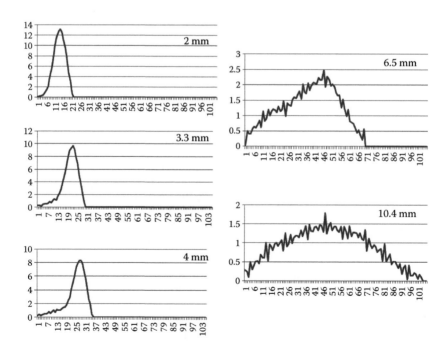

FIGURE 8.12
Reduction and broadening of impact forces with increasing thickness.

the peak forces transmitted through the material reduce as the thickness increases from 2 to 10.4 mm. This can be attributed to the impact energy being dissipated over a longer time interval within the material. The thicker material provides a cushioning effect such that the impact energy is absorbed and diminished by the material subsequently.

The duration of impact increased from 21 to 101 milliseconds; in other words, the material extended the time of the impact with the consequence of reductions in peak values of force. The thickness of material contributes to the reduction of peak forces by extending the duration of impact. This is beneficial in the sense that the material will extend the duration of impact as well as reduce the impact force (absorbing energy) through the material. Earlier research (Venkatraman and Tyler, 2011) revealed that the reduction in peak force with thickness was entirely predictable, as samples of 10 mm thickness or more were effective in protecting against 5 J impacts, and the impacting sphere produced no surface damage. However, at 5 mm thickness, the material experienced high impact forces.

8.6 Regulations for Impact Protection

The International Rugby Board, which was founded in 1886 and has its headquarters in Dublin, Ireland, is the official governing and law-making body for rugby. In its regulation 12.0 it outlines the provisions of clothing in protecting the athletes and includes shin guards, fingerless mitts, shoulder pads, head gear and chest pads as part of the clothing (IRB, 2012).

The impact-resistant materials used in clothing should not be affected when exposed to water, dirt, perspiration or toiletries and during washing. They should also not cause adverse reaction to skin and chafe or abrade the player or any other player in contact. The impact-resistant pads should be homogenous – of same texture, hardness and density. The sandwiching of layers of pads is not permitted. The edges of the pads should be smooth and with no rigid projections which may obstruct players and cause discomfort. Figure 8.13 illustrates the potential padding zone in a typical top.

Shoulder padding should have a maximum zone of coverage including sternoclavicular (SC), acromioclavicular (AC) and glenohumeral (GH) joints. The impact-resistant materials will protect the shoulder and collarbone and extend from the neck to a maximum of 2 cm down the upper arm. An allowance for padding to cover the SC joint is made to a maximum depth of 60 mm and at the back of the neck to a maximum depth of 70 mm. The impact-resistant materials will have a maximum thickness of 10 mm + 2 mm tolerance band plus an additional allowance of 1 mm on each side for fabric.

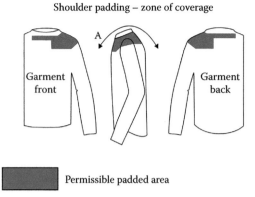

FIGURE 8.13
Shoulder protection in rugby tops recommended by IRB.

8.7 Benchmarking of Impact-Resistant Materials

The materials widely used in the garments have been evaluated with well-known materials such as leather, foam (EVA and urethane), knitted spacer fabric and dilatant materials. The impact forces experienced depend on the thickness of materials. Except for D3O, most materials in the 5 mm range were of limited help withstanding high-impact forces (Figure 8.9). It can be inferred (Figure 8.12) that the thickness of material contributed to reduction of peak forces as well as extending the duration of impact. This is beneficial in the sense that the material will extend the duration of impact as well as reduce the impact force (absorbing energy) through the material. It should also be stated that higher thickness shoulder pads are limited to 10 and 5 mm in the arm and chest region. The impact attenuation test enabled an assessment of the peak forces experienced from a standardised impact when protected by various materials. This analysis also highlights that there is scope for further development, particularly in absorbing shock or impact at lower thicknesses (5 to 8 mm). As a rule of thumb, the bulky inserts restrict athlete's movement and are uncomfortable to wear.

8.8 Garment Design and New Product Development

Rugby tops (long and short sleeves) are made of warp-knitted fabric and are designed for a close fit and impact. There are a number of garment design issues, which are outlined in Figure 8.14. Sportswear products require various performance characteristics such as durability, comfort, identity, recognition

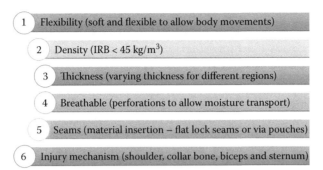

FIGURE 8.14
Design issues for rugby tops.

and functionality. However, these are dependent on the type of sport, level of physical activity, team or individual sport, intensity of sport, indoor or outdoor sport, frequency, age and other special functions (El Moghazy, 2009). In the context of rugby, with particular focus on protection from impact using pads or materials, six factors were selected for discussion: mechanism of injury, flexibility, bulkiness, breathability, thickness and ability to sew these pads onto the clothing. As discussed earlier, bulky inserts are not well received by players as they restrict free movements as well as offer poor comfort. Designing functional protective gear for sports is challenging and demanding since protection is sometimes achieved at the expense of comfort (thick, stiff, heavy, multilayered and nonbreathable pads).

8.8.1 Design Issues

- Generally, most of the pads, or foams, are available in thicknesses above 5 mm, so it becomes a challenge to incorporate these thick pads in the garment.
- Some pads for shoulders and sleeves lack any moulding of the shape. Pads used in some samples are large and garments do not fit well on the body when pads tend to hold the fabric out.
- Pads used in some products were smaller (shoulder region), which offered less protection to the wearer.
- Seams used in these garments were often overlocked. Some products had body panels sewn together with flat lock seams, which enhanced wearer comfort. Other products had pads that were sandwiched between fabrics in the form of pouches.
- Thicker foams or pads have perforations to allow flexibility and moisture transport. However, at other areas the pads block air/moisture movement, allowing low breathability and hence poor comfort.

- Pads were stiffer and allowed very little flexibility to conform to various contours of the shoulder region. Unlike heavy and bulky shoulder pads used for American baseball, the pads used for rugby vests required flexibility.

A recent epidemiological study by Crichton, Jones and Funk (2011) investigated videos of 24 elite rugby players sustaining injury and reported that there were three common injury mechanisms in rugby resulting in serious shoulder injuries: 'try scorer', 'direct impact' and 'tackler'. It is necessary to understand the nature of the injury and design shoulder padding mechanisms which provide maximum protection to these common injury patterns:

1. Try scorer – outstretched arm when scoring a try
2. Direct impact – direct blow to the arm or shoulder when held by the side in neutral or slight adduction (moving of a body part toward the central axis of the body)
3. Tackler – a levering force on the glenohumeral joint (GHJ) due to movement of the outstretched arm (Figure 8.5)

Knowledge of the mechanisms involved in rugby shoulder injury is useful in understanding the pathological injuries and aiding the development of injury prevention methods such as padded vests.

It could be noticed that a typical shoulder padded vest (Figure 8.15) could provide only a limited coverage to the shoulder region in the context of the

FIGURE 8.15
Typical rugby top with padding. (Courtesy of Kooga.)

findings from Crichton et al. (2011). These design issues should be considered while producing garments for shoulder protection for rugby. The shoulder pads should be flexible to cover the regions as shown in Figure 8.16: the sternoclavicular, acromioclavicular and glenohumeral joints. The rugby tops shown in Figure 8.15 may provide limited protection to those regions. Pain et al. (2008) also reported that when tackled using a shoulder pad, the reduction in force was noticed only in the acromioclavicular joint, whilst forces in other areas of the shoulder region were not reduced. In other words, the shoulder experienced considerable impact during tackling. The chapter emphasised the importance of six areas that affect the garment design: mechanism of injury, flexibility, bulkiness, breathability, thickness and ability to sew these pads onto the clothing. The industry intends to explore material that is flexible, lightweight, breathable and thin and that allows easy movement of the body. This is based on the principle that the material is able to extend the duration of an impact and to broaden the area affected by an impact such that the wearer does not experience peak forces during a fall or tackling.

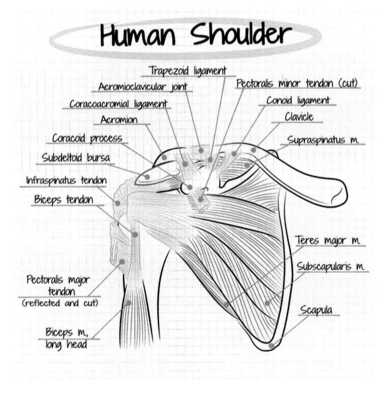

FIGURE 8.16
General anatomy of the shoulder. (Courtesy of Sfischka|Dreamstime.com.)

8.9 Summary and Conclusions

Rugby is a fast-paced contact sport that requires precise protection to the shoulder region, particularly when players fall with outstretched arms (when scoring a try, tackling an opponent or experiencing a direct impact onto the ground). Many studies have reported on the performance of protective materials which have been critically reviewed.

Impact-resistant materials for sportswear, particularly those that are used for rugby, have been discussed in this chapter. Ideally, these materials absorb the impact and provide resistance to shocks from a fall or collision. In recent years the use of such protective pads has increased tremendously such that it becomes difficult for the user to make choices. In addition, it has become a challenge to garment manufacturers to incorporate the pads in the garment. This chapter is concerned with material properties, and the experimental work has focused on the forces experienced by a transducer attached to an anvil under the protective material. A random selection of rugby vests was studied and the results presented. The chapter also highlighted the potential injuries frequently incurred by athletes during training and playing.

A selection of materials was tested to benchmark their capability to protect against impact. These were compared with unfinished leather: a material offering wearers some protection and used here as a control. Figure 8.12 presents peak force variations with a range of materials with different thicknesses. The materials of 2–3 mm were associated with high peak forces. As thicknesses increased, the commercial products designed to absorb energy and protect against impacts reduced peak forces more effectively than the leather samples. A close examination of the sensor signals revealed that the duration of impact was extended and the forces experienced under the material were correspondingly reduced. Figure 8.12 documents the change for one particular sample from 21 ms to 101 ms as the thickness increased from 2.0 to 10.4 mm. The impact attenuation test records all the forces transmitted through the material and does not illuminate how those forces are distributed at the zone of impact. The areal distribution of forces at the point of impact can be potentially explored using pressure sensors.

Six factors affecting the design and development of performance sportswear are represented in Figure 8.14. The analysis of a selection of commercially available and IRB-approved rugby shirts (with protective pads) using the six factors approach has documented a number of issues for designers and product developers to consider. This analysis provides a framework for enhanced design and for the formulation of design principles for protective sportswear. Mechanisms of injury are discussed to highlight the fact that sportswear development is user centred and it becomes essential to understand the frequency of injuries and various parts of the body being affected during a sporting activity.

Acknowledgement

The authors would like to thank Moran Filson-Dahan for kindly offering CAD drawings for the chapter (Figure 8.5).

References

Abounaim, M. D., Hoffmann, G., Diestel, O. and Cherif, C. (2010). Thermoplastic composite from innovative flat knitted 3D multi-layer spacer fabric using hybrid yarn and the study of 2D mechanical properties. *Composites Science and Technology* 70 (2): 363–370.

British Columbia Injury Research and Prevention Unit. (2012). Rugby injuries. [Online]. Available at http://www.injuryresearch.bc.ca (accessed 23 March 2012).

Brooks, J. H. M, Fuller, C. W., Kemp, S. P. T. et al. (2005). Epidemiology of injuries in English professional rugby union: Part 1: Match injuries. *British Journal of Sports Medicine* 39:757–766.

BS 7928. (1998). Specification for head protectors for cricketers. BS 7928:1998.

Cinats, J., Reid, D. C. and Haddow J. B. (1987). A biomechanical evaluation of sorbothane. *Clinical Orthopaedics and Related Research* 222:281–288.

Crichton, J., Jones, D. and Funk, L. (2011). Mechanisms of shoulder injury in elite rugby players. FOSC. [Online]. Available at http://www.shoulderdoc.ac.uk (accessed 23 March 2012).

D3O.com. (2013). Impact protection | D3O lab. [Online]. Available at http://www .d3o.com/materials/impact-protection/ (accessed 16 November 2013).

Dow Corning. (2011). Dow Corning Corporation expands impact protection materials range. Innovation in textiles, technical textiles. [Online]. Available at http://www.innovationintextiles.com (accessed 25 July 2011).

Dow Corning. (2012). Deflexion Technology, TechTextil India Conference. Mumbai, October 11, 2011. [Online]. Available at http://www.dowcorning.com (accessed 22 March 2012).

Durá, J. V., Garciá, A. C. and Solaz, J. (2002). Testing shock absorbing materials: The application of visco-elastic linear model. *Sports Engineering* 5:9–14.

El Moghazy, Y. E. (2009). *Engineering textiles: Integrating design and manufacture of textile products*. Cambridge, UK: Woodhead Publishing.

EN812:1997. (2001). Industrial bump caps. BS EN 812:1997 updated to EN 812:1997/A1:2001.

Fuller, C., Ashton, T., Brooks, J., Cancea, R., Hall, J. and Kemp, S. (2010). Injury risks associated with tackling in rugby union. *British Journal of Sports Medicine* 44 (3): 159–167.

Funk, L. (2012). The rugby shoulder. [Online]. Available at http://www.shoulderdoc .co.uk (accessed 22 March 2012).

Gabbett, T. J. and Godbolt, R. J. B. (2010). Training injuries in professional rugby league. *Journal of Strength and Conditioning Research* 7:1948–1953.

Gerrard, D. F. (1998). The use of padding in rugby union. *Sports Medicine* 25:329–332.

Harris, D. A. and Spears, I. R. (2010). The effect of shoulder padding on peak impact attenuation. *British Journal of Sports Medicine* 44:200–203. doi: 200 10.1136/bjsm.2008.047449.

IRB. (2012). International Rugby Board: Laws of the game. Dublin, Ireland. [Online]. Available at http://www.irb.com (accessed 22 March 2012).

Khurana, V. G. and Kaye, A. H. (2011). An overview of concussion in sport. *Journal of Clinical Neuroscience* 19 (1):1–11. Epub 5 December 2011.

Lam, S. W., Tao, X. M. and Yu, T. X. (2004). Comparison of different thermoplastic cellular textile composites on their energy absorption capacity. *Composites Science and Technology* 64:2177–2184.

Maklewska, E., Tarnowski W., Krucińska, I. and Demus, J. (2004). New measuring stand for estimating a material's ability to damp the energy of impact strokes. *Fibres and Textiles in Eastern Europe* 12 (4): 48–52.

Maklewska, E., Krucińska, I. and Mayers, G. E. (2005). Estimating the shock absorbing ability of protector materials by use of pressure films. *Fibres and Textiles in Eastern Europe* 13 (4): 52–55.

Maron, B. J., Poliac, L. C., Kaplan, J. A. and Mueller, F. O. (1995). Blunt impact to the chest leading to sudden death from cardiac arrest during sports activities. *New England Journal of Medicine* 333 (6): 337–342.

Marshall, S. W., Waller, A. E., Dick, R. W. et al. (2002). An ecologic study of protective equipment and injury in two contact sports. *International Journal of Epidemiology* 32:587–592.

NEISS. (2010). National Electronic Injury Surveillance System. US consumer product safety commission, NEISS data 2010. [Online]. Available at http://www.cpsc.gov.

Nsmi.org.uk. (2009). Sports medicine information. [Online]. Available at http://www.nsmi.org.uk/ (accessed 26 July 2011).

Pain, M. G., Tsui, F. and Cove, S. (2008). In vivo determination of the effect of shoulder pads on tackling forces in rugby. *Journal of Sports Sciences* 26:855–862.

Palmer R. M. and Green P. C. (2010). Self-supporting composite of solid foamed synthetic polyurethane elastic matrix, polyborosiloxane dilatent distributed through matrix, and fluid/gas distributed throughout; compression set; impact protection. US Patent 7794827 B2.

RFU. (2002). Rugby Football Union. Governance committee players safety. A statement of serious and very serious injuries reported to the RFU by clubs and schools for the five seasons (1997–2002). Twickenham, UK, Rugby Football Union, pp. 1–8.

South Wales Osteopathic Society. (2009). Rugby injury statistics. [Online]. Available at http://www.osteopathywales.com/ (accessed 4 January 2012).

Tao, X. M. and Yu, T. X. (2002). Lessening impact. *Industrial Fabric Products Review* 87 (8): 20–21.

Trewarth, G. and Stokes, K. (2003). Impact forces during rugby tackles [abstract]. *Proceedings of the International Conference on the Science and Practice of Rugby*, 5–7 November 2003, Brisbane, Australia, 25.

Van de Velde, K. and Kiekens, P. (2001). Thermoplastic pultrusion of natural fibre reinforced composites. *Composites and Structures* 54:355–360.

Van de Velde K., Kiekens, P., Lauwereys, Y. et al. (1998). Textile reinforced composites for sports protectors. *Textile Asia* 29:48–51.

Venkatraman, P. D. and Tyler, D. (2011). A critical review of impact-resistant materials used in sportswear clothing. Paper presented at International Conference in Advances in Textiles, Machinery, Nonwovens and Technical Textiles, ATNT 2011, held at Kumaraguru College of Technology, Coimbatore, India, 15–16 December 2011.

WSA. (2011). The international magazine for performance and sports materials. Textile Trades Publishing, UK.

9

Seamless Knitting and Its Application

Kathryn Brownbridge

CONTENTS

9.1   Introduction...231
    9.1.1   Circular Seamless Knitting 232
    9.1.2   Complete Garment Knitting 233
9.2   Implications of Knitted Fabrics on Garment Fit; Knit Structures.......233
    9.2.1   Application of Body Measurement 235
    9.2.2   Creating Garment Shape through Stitch Manipulation...........235
9.3   Seamless Garment Fit..236
9.4   Seamless Knits and Their Application within Sportswear 238
    9.4.1   Garment Comfort..238
    9.4.2   Garment Aesthetics 240
9.5   Specific Fit-Related Opportunities 241
9.6   Current Limitations within the Industry...................................242
9.7   Summary...243
References...243

9.1 Introduction

This chapter will focus on two different types of weft knit technologies that are associated with seamless garments: circular machinery and flat bed machinery.

Circular knitting traditionally produces a tube of fabric and is commonly used to produce hosiery. When used for garments, the tubular structure is often cut down one side and opened out to create a flat length of fabric (Brackenbury, 1992), which is cut into the component parts of a garment. These are then sewn together. This technique is similar to the process used for woven garments and clearly creates a garment with seams. However, a more recent application is to utilise the tubular structure to create partially formed garment parts; namely, the tubular structure becomes the body and sleeves, creating a garment with fewer seams.

Flat bed knitting machinery was originally used only to produce lengths of knitted fabrics. However, the need to decrease labour and costs and save cutting waste has driven the development of many machine improvements that have increased capabilities including the ability to construct garment panels. Perhaps the most innovative development has been the ability to construct tubular garments that can be constructed and assembled in one process, creating a garment that has a completely seamless structure.

9.1.1 Circular Seamless Knitting

Circular weft knitting machinery is traditionally used for hosiery. Developments came from the Italian machine builder Lonati in 1988, which were aiming to extend the capabilities of the machinery to knit seamless underwear. This led to the creation of a new company, Santoni, which was promoted as a specialist producer of machines with the capability to knit what was claimed to be seamless garments. It is, however, important to stress that, although the body and sleeves can be knitted without seams, these component parts must be constructed once they have been knitted. Therefore, it is perhaps a misnomer to call garments knitted on circular machinery 'seamless' (Semnani, 2011; Power, 2012). Circular machinery is highly efficient and has a higher productivity rate than flat bed machinery (Magnus et al., 2009). Santoni is still generally acknowledged to be the leader in seamless knitting produced on circular knitting machinery.

Circular machinery is capable of producing fine gauge knitwear but it is difficult to create patterns and sophisticated garment shaping (Spencer, 2001; Power, 2008). It is also difficult to change the diameter of the seamless knitted tube (Semnani, 2011), a considerable limitation in terms of garment design and the ability to fit the complex shape of the body. Selective use of the circular needle bed can, however, enable the designer to knit garment shapes other than a basic tubular construction (Figure 9.1).

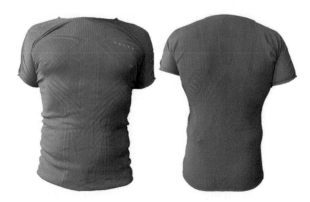

FIGURE 9.1
Circular knitted garment.

9.1.2 Complete Garment Knitting

Previous to the development that enabled complete garment knitting, fully fashioned garments were knitted on straight bar machines as shaped panel pieces and then constructed using a separate postproduction process (Spencer, 2001). Since the 1940s it had been the aim of many knitters to develop methods that enable complete garments to be produced on flat bed machinery. It is easy to understand why this capability would be beneficial, as it not only eradicates the need for time-consuming postproduction processes but it also minimises waste. A considerable number of different methods were patented but none were commercially viable until 1995, when Shima Seiki launched WholeGarment® technology at the International Textile Machinery Association (ITMA) tradeshow (Gibbons, 1995). The German company Stoll also made important developments that enhanced commercial viability (Gibbons, 1996). Shima Seiki and Stoll are at the forefront of commercial application. Although Shima Seiki has patented its machinery as WholeGarment and Stoll has coined the term 'Knit and Wear', *complete garment technology* is a term commonly used to describe flat bed machinery with this capability.

9.2 Implications of Knitted Fabrics on Garment Fit; Knit Structures

The structure and properties of a knitted fabric influence the way a knitted garment fits the body. Weft-knitted fabric is extensible across the width and the length and diagonally. Its extensible structure enables it to mould itself to a 3D shape through deformation (Power, 2004). These fabric properties impact the way a knitted garment relates to the shape of the body, allowing it to stretch and move on the body much more comfortably than a woven fabric would. Because of this it is possible to wear tight, streamlined knitted garments without restricting comfort even when the fibres are nonelastic.

Having an understanding of different knit structures and their properties helps designers and product developers create garments that fit. It is possible to categorise three basic groups of knitted structures:

- Plain structure consists of loops which are all identical and intermeshed in the same direction. These structures are all face loops on one side and all reverse loops on the other.
- A rib structure is made of alternating face and reverse loops and the number of needles used to create a rib can vary. A two by two rib, for instance, is created by alternating two face and reverse loops at a time.
- A purl structure consists of alternate courses of back and face loops (Figure 9.2).

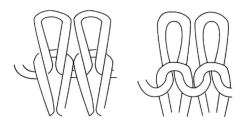

FIGURE 9.2
Face and back loops.

Each different knit structure has a range of differing properties which will impact garment shape and, ultimately, the way the garment fits the body. A knitted rib collapses across the width when relaxed, decreasing the dimensions of the width and increasing the thickness of the fabric. When tension is exerted across the rib courses, the fabric is reported to extend up to 120% (Brackenbury, 1992). Knitted ribs will therefore reduce the width of the garment if it is integrally incorporated into a plain knitted garment without compromising comfort. Ribs have traditionally been used to create shape within a garment – often to provide a snug fit on the cuffs and welt (bottom edge of the body panel). However, designers can also insert bands of rib in other areas of the garment to create shape; this is particularly effective at the waist where the body narrows. This practice exploits the rib structure's propensity to expand and contract.

A plain knitted fabric extends more across the courses than the wales. It is therefore common practice for the courses to be positioned horizontally on the garment to exploit this increased extensibility where it is needed most. A purl knit, which is less commonly used within commercial garments, contracts longitudinally. It therefore has more lengthwise elasticity than a rib or plain knitted structure (Raz, 1993). Experienced knitwear designers will have a good understanding of how to use knit structures. Evaluation of the properties of knit structures, however, tends to be made by the technicians using their experience and judgement (Brownbridge, 2010). Figure 9.3 shows the three basic knit structures.

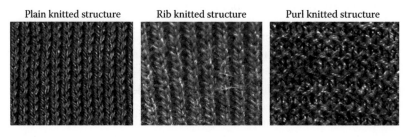

FIGURE 9.3
Knit structures.

9.2.1 Application of Body Measurement

To create knitted garments that fit the body, it is important to have a method of applying body measurement within product development processes. For woven garments it is the pattern-cutting process that translates body measurement from size charts into flat patterns and then through to flat fabric pieces. These are then constructed to make three-dimensional garments. For knitted garments the pattern-making process uses a calculation to translate key body dimensions such as chest, shoulder and waist into stitch numbers. This calculation is based on the stitch density, which is the number of loops in a given area of knitted cloth.

As knitted structures differ greatly from woven ones, the key body measurements used to achieve fit are also different. As a consequence of their extensible qualities, knitted garments tend to conform to simple flat shapes without darts, seams or other methods used when developing 3D shapes for more stable woven garments. In women's knitwear, a bust girth measurement is not commonly used and garments are unlikely to include bust shaping. The fit of these simple garment shapes clearly relies on the knitted fabric's ability to stretch and mould to the body. To achieve woven garments that allow for comfort and movement of the body, an extra allowance is added to body measurement – namely, ease allowance (Gill, 2009). Extensible knitted garments do not always include ease allowance, and garment dimension can conform to body measurement and still produce a comfortable fit. The degree of ease needed to achieve fit on knitted garments can vary from a negative amount to a positive amount at differing body regions and will depend on the style and type of fit that the designer is aiming to achieve. Historically, garment measurement rather than body measurement is used to guide the process of defining garment dimension. Knitwear manufacturers therefore work from previously established garment measurements, often generated through trial and error.

9.2.2 Creating Garment Shape through Stitch Manipulation

When knitting on flat bed machinery, shape can be created through adding or reducing stitches, a technique known as fashioning. To increase the width, the knitted loops are transferred onto a needle outside the selvedge edge; narrowing is done by transferring a loop to a needle inside the one to which the loop was previously attached. This needle will then cease knitting. It is also possible to transfer a number of loops to create a loss or gain in fabric dimension within the body of the fabric. The smallest dimension that can be gained or reduced will only ever equal the stitch dimension. It is not possible to create the required dimensions with the same precision as that when garments are cut to shape.

The angle created by the fashioning process can be controlled by the number of courses that are knitted between each fashioning course (fashioning frequency). It is also possible to transfer more than one loop at a time to create a more acute angle. There are methods of integrally shaping in order to create 3D contours within the garment which serve the same functions as a dart, used in woven construction (Black, 2001). However, this method has not been widely used in traditional industrial knitting. It has been argued that in order to develop knitted garments that have a more sophisticated approach to achieving fit, it is necessary to use shaping in order to achieve 3D shapes that relate to the shape of the body (Guy, 2001; Haffenden, 2009; Brownbridge, 2012).

9.3 Seamless Garment Fit

Machinery that has the ability to construct garments three dimensionally, such as complete garment technology, implies the possibility of creating a product that can conform to the three-dimensional shape of the body. The process of shaping complete garments, however, relies on the capabilities of the machinery and its operator. There are difficulties associated with knitting tubular garments on flat bed technology. Shaping the garment is limited by the number of needles available on which to knit and transfer. The body and sleeve tubes must commence knitting on specific needles with an exact number of available needles between each tube. These needles are introduced to create underarm widening. The three tubes must meet at the correct location to merge. The sleeve tubes can then be gradually moved across and narrowed to eventually create the neck dimension. This is a complex operation and is not easy for technicians to master (Brownbridge, 2012). There are, however knitwear teams who are successfully exploiting the ability to knit three dimensionally and create garments that have 3D shaping.

Figure 9.4 shows an example of a garment where an integral shaping line has been incorporated on the back and the front of the garment, which creates a 3D draped shape. Further shaping is created on the lower half of the body to enhance the 3D garment shape. Figure 9.5 shows a close-up of the widening technique where every three courses become one. The image reveals that it is only the courses on the left side of the shaping line that are lost. The courses on the right-hand side remain constant. Figure 9.6 shows that more narrowing occurs running into the pointelle pattern.

The process used to create circular knitted seamless garments is far more limited in terms of stitch manipulation as a method of shaping. It is therefore a more limited knitting technique and it is perhaps more difficult to create shape innovation.

FIGURE 9.4
Seamless draped top.

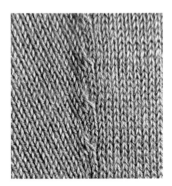

FIGURE 9.5
Detail showing integral shaping.

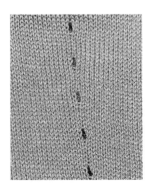

FIGURE 9.6
Integral shaping incorporated into pointelle detail.

The garment pressing process which, in the fully fashioned industry generally uses a wooden frame custom made to specific dimensions in order to maintain and set garment dimension (known as boarding), will also affect the stitch density. Fully fashioned practices are still used for complete garment production and this has been found to cause problems as the structure of the complete garments tends to be far less formulaic in shape than traditional fully fashioned styles. The seamless structure of complete garments also means they are less stable and more difficult to press flat. These differences have been found to render the use of frames obsolete. This makes the finishing of complete garments difficult.

9.4 Seamless Knits and Their Application within Sportswear

Seamless knitted garments have been available to the consumer for a number of years and some forward-thinking knitwear manufacturers have embraced the opportunities this method of knitting offers in terms of technical sportswear development. The following sections outline how the use of seamless technology can create functional features that meet specific needs of the sportswear consumer in terms of comfort and sportswear aesthetics.

9.4.1 Garment Comfort

Comfort is a garment attribute that the consumer seeks and is particularly important for sportswear as it can affect performance. The evaluation of comfort from a consumer's perspective is subjective and complex as it takes into account a number of various factors. Studies relating to comfort tend to list these factors separately as follows:

Sensorial comfort in relation to garments includes the subjective evaluation that the wearer makes about how the garment feels against the skin. It has considerable bearing on the comfort properties of seamless garments because clearly the lack of seams eliminates any potential for seam abrasion caused by rubbing against sensitive areas of skin. This is clearly advantageous for sports garments, particularly those worn next to the skin, such as base layers and swimwear (Magnus et al., 2009).

Ergonomic comfort relates to the ability of the wearer to move easily within a garment. In relation to seamless garments it has been noted that the customer should experience heightened ergonomic comfort as the seamless structure is claimed to be able to stretch and mould to the body more effectively than a knitted garment with seams (Brownbridge, 2012).

Thermophysiological comfort is when a person is satisfied with the thermal environment. To sustain thermal comfort the production and loss of heat from the body must balance. In addition to clothing, a number of factors can impact the thermal environment, including humidity, air movement, air temperature and activity. Thermophysiological encompasses heat and moisture transported through the body and impacts a person's ability to regulate body temperature. Wicking is a process that transports moisture away from the body. Knitted fabrics have a capacity to absorb water. This will vary depending on fibre content and knit structure. Knitted fabrics also have insulation properties because the knitted structure traps air, which acts as an insulating layer, and differing knit structures have different thermal insulation properties. Fibres vary in their ability to insulate and therefore can be used strategically within garments to target areas of the body that need more insulation than others. This is where the ability to use a variety of knit structures within one garment is advantageous (Abreu et al., 2011). Figure 9.7 shows a garment that incorporates a variety of knit structures mapping different areas of the body, and Figure 9.8 shows a detail of the knit structures on the sleeve of the garment.

Psychological comfort relates much more to how someone feels in a garment and will be more influenced by factors such as styling, colour, fashion and aesthetics. When relating psychological comfort to sportswear, garments that are perceived to look and feel as if they will provide high performance may also provide the wearer with higher levels of psychological comfort. Psychological factors may also relate to how effectively the garment provides thermal, ergonomic and sensorial comfort to the wearer in order to create a sense of well-being and psychological comfort.

FIGURE 9.7
Circular knitted garment incorporating a variety of knit structures.

FIGURE 9.8
Detail of differing knit structures on a sleeve.

9.4.2 Garment Aesthetics

Through the use of both circular and flat bed seamless machinery it is possible to use a number of different knit structures and patterning details within one garment. In addition it is possible to vary the types of yarn on specific areas of the garment. This enables designers to create patterns, textures, surface design and styling features. The use of elastomeric (stretch) yarn can reduce the diameter of the tubular structure, without the loss of comfort to the wearer. This can introduce body shaping, structure to the garment, which will have an impact on the aesthetic and could be used by an informed designer to create aesthetically pleasing garment shaping (Magnus et al., 2009).

The ability to map the body with the garment and create zoning areas that function differently creates a sports aesthetic, which suggests innovation and high performance. These garments draw attention to muscle groups and use body contour panels, a technique that not only has functional value but also is becoming accepted as signalling that this garment is designed as a high-performance sports garment for a high-performing sports body. The performance aspects of seamless garments have therefore also created the aesthetic aspects. Not only does the garment aid performance but it also looks like it aids performance and, arguably, this will increase psychological comfort for the wearer.

Complete garment technology has the ability to shape the garment three dimensionally and this can create an innovative aesthetic. Unbroken patterns, striping and pointelle structures can be knitted around the whole circumference of the garment, which is particularly effective when used on a yoke. The stitch manipulation techniques (fashioning) used to shape complete garments form a visible patterning within the knitted fabric. This can be aesthetically pleasing and can draw attention to innovative shaping techniques and, consequently, in aesthetic terms, makes the garment look like a product of sophisticated engineering.

9.5 Specific Fit-Related Opportunities

Body mapping can also be described as comfort mapping and is a technique that is gaining popularity amongst sportswear developers. It is hoped that, through in-depth analysis of functional requirements of differing parts of the body in relation to temperature, moisture control and the consideration of skin sensory needs, the need to wear multiple layers will no longer be necessary. Different knit structures and a variety of fibres with a range of abilities to insulate, ventilate and control moisture are placed where they are needed on the body. The chest area, for instance, may be matched with a wind-chill panel, whereas the underarm area will have wicking and breathable panels. In terms of fit this type of development creates opportunities to link anthropometric (the study of body measurement) research with technical garment development. Santoni has developed two methods that are claimed to help regulate temperature control by targeting specific areas of the body: (1) the use of mesh-like knit structures to allow venting within regions of the garment, such as the underarm, to help the body lose unwanted heat and (2) the ability to create thermal pockets which are effectively double layers of fabric strategically placed to trap heat at areas of the body that are more vulnerable to heat loss.

   If the specific areas of the garment must map to specific areas of the body in order to achieve this end, it must be presumed that development should include a sophisticated understanding not just of the functions of the body but also of body size, shape and proportion. In terms of the development of sizing systems for such garments, it would seem logical that this type of body mapping approach would demand a more sophisticated sizing system, which would take into account the variety of heights, body shapes and sizes within a population in order for the mapping to function correctly.

   The use of panels to compress specific muscle groups is another area of development which creates opportunities for further study in relation to anthropometrics. These panels are knitted with extremely elastic yarns. They are said to have a number of benefits to the sportsman, including reduction of muscle strain and the time it takes for muscles to repair themselves. There are also claims that compression panels can increase performance by improving oxygenation to working muscles. In terms of fit, any compression panel incorporated within a sports garment must be located on the correct muscle group when the garment is worn. Therefore, for garments that are mass produced, the addition of compression panels that must conform to very specific regions of the body introduces additional fit complexities when creating sizing systems. The knitwear industry have been found to lack the skills and data needed to achieve this type of specific fit (Brownbridge, 2012).

It therefore may be difficult for some consumers to get the full benefit from these highly technical garments because they are not made for their body dimensions.

9.6 Current Limitations within the Industry

Currently the seamless technology (circular and flat bed) that is available to knitwear producers is expensive and requires a high level of design and technical skills in order to fully exploit its advantages. This is a relatively new type of manufacturing process and requires a considerable amount of research and development work. In the current economic climate, it is perhaps a more logical choice for knitwear producers to reduce both costs and risk-taking decisions rather than to invest in the development of new and innovative products. Consequently, although a growing number of sportswear companies specialise in seamless knitted garments, it is perhaps a market that is yet to be fully exploited.

In terms of fit, there is evidence to suggest that, although knitwear is stretchy and therefore there is an assumption it will fit a wide variety of body shapes and sizes, there is still a need for fit improvements.

Complete garment development forces practitioners to consider all the component parts of a garment simultaneously, which challenges traditionally trained personnel. Difficulties can occur when shaping the armhole for a set-in sleeve and the desired effect is not always achieved. Various factors have been blamed for this, including fashioning restrictions, a lack of previous knowledge to draw on and no available garment templates or patterns on which to base methods (Troynikov, 2008). The lack of templates is related to the fact that the technology is still relatively new. It is therefore likely that some of these restrictions will be alleviated as those who are driving development become more skilled and knowledgeable. The restrictions in shaping may also have originated from the fully fashioned industry, where there have always been inconsistencies and inaccuracies in the methods used (Eckert, 2001). There is also a lack of attention paid to the application of anthropometric data within the knitwear industry, which could well have a detrimental effect on the development of new methods for seamless garments.

There is, however, a further restriction that relates to the acquisition of appropriate skills and knowledge and that is the propensity within the knitwear industry to protect intellectual property with patents. The machine builders themselves have actively been developing new methods of knitting garments. In 2002 an interview with the president of Shima Seiki suggested that only 1% of machine capability was at the time being exploited (Mowbray, 2002).

Whilst this statement is making claims for a vast, untapped potential, it also acknowledges a skills and knowledge gap within the industry. Shima's response to this need for new knitting techniques was to produce 2500 new sample garments a year. Through this approach, it is claimed that a whole new range of knitting techniques have been created and, further, potential for design innovation (Mowbray, 2002). However, information is slow to be released and typically in formats that protect what is considered to be proprietary information. This slows down creative innovation and contributes to an aura of exclusivity around seamless garments.

9.7 Summary

Two different types of seamless garments have been identified: those knitted on circular machinery and those knitted on flatbed machinery. Each of these methods of knitting creates slightly different opportunities and limitations. Currently there are a number of specialist sportswear producers who are using both types of seamless circular garments to create highly technical sports garments using specific knit structures to map the body for warmth, comfort and performance. The ability to knit a number of different knit structures has been exploited by product developers to create highly specialised performance garments. Two innovative techniques have been discussed: the use of pressure panels and body mapping. Both of these techniques, however, demand a really sophisticated understanding of how to relate garment dimensions to the human body. Currently there is little evidence to suggest the industry is producing sizing systems that can cater to a variety of sizes within a population. Therefore, although innovative practice is evident, there is still a skills and knowledge issue which limits the potential to innovate performance sportswear. In addition, skills issues have been found to limit the ability of product development teams to achieve the fit results they desire in terms of shaping the garment, and the protection of intellectual property limits the spread of knowledge. There is therefore a need for companies to invest in research and development in order to improve knowledge and understanding and truly exploit the potential capabilities of seamless knitting technologies.

References

Abreu, M. J., Catarino, A., Cardoso, C. and Martin, E. (2011). Effects of sportswear design on thermal comfort, *Proceedings for the Autec Conference*, Mulhouse, France.
Black, S. (2001). *Knitwear in fashion*. London: Thames and Hudson Ltd.

Brackenbury, T. (1992). *Knitted clothing technology*. London: Blackwell Science Ltd.

Brownbridge, K. (2010). Anthropometrics and the complete knitted garment. *86th Textile Institute World Conference*, Manchester, UK, November.

————. (2012). The development of a conceptual model for anthropometric practices and applications regarding complete garment technologies for the UK women's knitwear industries. Thesis in partial fulfilment of a PhD, Manchester Metropolitan University.

Eckert, C. (2001). The communication bottleneck in knitwear design: Analysis and computing solutions. *Computer Supported Cooperative Work* 10: 29–74.

Gibbons, J. (ed.). (1995). Future technology towards 21st century manufacture. *Knitting International* 102 (1222): 60–61.

————. (1996). Stoll in Dublin. *Knitting International* 103 (1232): 26.

Gill, S. (2009). Determination of functional ease allowances using anthropometric measurement for application in pattern construction. Thesis in partial fulfilment of a PhD, Manchester Metropolitan University.

Guy, K. (2001). A design perspective on shaping possibilities with new technology V-bed knitting machines (part fulfillment of PhD), Nottingham Trent University.

Haffenden, V. (2009). Knit to fit: Applying technology to larger sized women's body shapes. *11th Annual IFFTI Conference*. London College of Fashion, 2–3 April, 2009, UK P 1-122.

Magnus, E., Broega, A. C. and Catarino, A. (2009). Interactions between apparel design and seamless technology. *Proceedings of First International Conference on Design, Engineering and Management for Innovation*, IDEM109 Porto, Portugal, 14–15 September.

Mowbray, J. (2002). The quest for ultimate knitwear. *Knitting International* 109 (1289): 22–23.

Power, J. (2004). Knitting shells in the third dimension. *Journal of Textile and Apparel Technology and Management* 3 (4): 1–13.

————. (2008). Developments in apparel knitting technology. In *Advances in apparel production*, Fairhurst, C. (ed.). Cambridge, UK: Woodhead Publishing Ltd.

————. (2012). Sustainable development in knitting. *International Journal of Business and Globalisation* 9 (1): 1–11.

Raz, S. (1993). *Flat knitting technology*. Westhausen: Universal Maschinenfabrik.

Semnani, D. (2011). Advances in circular knitting. In *Advances in knitting technology*, Au, K. F. (ed.). London: Woodhead Publishing.

Spencer, D. J. (2001). *Knitting technology: A comprehensive handbook and practical guide*, 3rd ed. Cambridge, UK: Woodhead Publishing Ltd.

Troynikov, O. (2008). Smart body – Ergonomic seamless sportswear design and development. *The Body – Connections with Fashion*, Melbourne, IFFTI, pp. 1–20.

10

Garment Fit and Consumer Perception of Sportswear

Simeon Gill and Jennifer Prendergast

CONTENTS

10.1 Factors Affecting Fit 245
    10.1.1 Garment Fit and the Pattern 246
    10.1.2 Fabrics 246
    10.1.3 Function 247
10.2 Sensorial Comfort 249
    10.2.1 Psychological Considerations of Wearer Perceptions 249
    10.2.2 Pressure Comfort 250
    10.2.3 Pressure Comfort and Gender 251
10.3 Performance Expectations 252
10.4 Ease Levels within the Garment 254
10.5 Summary 254
10.6 Future Developments 257
References 257

10.1 Factors Affecting Fit

Clothing fit is important to wearers on an everyday level and is frequently covered in clothing texts (Brown and Rice, 2001; Le Pechoux and Ghosh, 2002; Fan, Yu and Hunter, 2004), often with reference to the variables identified by Erwin, Kinchen and Peters (1979). Whilst these highlight important attributes of fit aesthetics, the fit of sportswear clothing can have a direct bearing on performance and potentially the outcome of the sports event. Fit can be considered in terms of function and comfort perception, but from a product development perspective fit can and should be engineered with consideration of how each variable interacts. Importantly, developments of 3D product development reinforce the geometry and numeric nature of product development and support the assertion for numerically understanding the garment and its fit (Gill, 2011). We would propose that wearer perception plays only a small part in engineering fit and that the following variables require careful and integrated consideration: fabric, function, sensorial comfort, performance expectations and ease determination, especially regarding their numeric application in terms of the pattern.

10.1.1 Garment Fit and the Pattern

Whilst garment fit assessment is often a reflective tool assessed after the development of the product (Le Pechoux and Ghosh, 2002; Ashdown et al., 2004; Bye and McKinney, 2010), we would argue that the tools exist for considering fit prior to the development of the product, though this requires enumeration of key concepts of fit (Gill, 2011). Pattern development using many of the available methods (Beazley and Bond, 2003; Aldrich, 2008; Armstrong, 2010) requires the plotting of a geometric coordinate system using a combination of human measurement and proportional rules, which describes the 3D garment shape in two dimensions (Gill and Chadwick, 2009; Gill, 2011). The constraint of the pattern as a geometric object requires that all changes be applied as point repositioning described as X, Y movements from one location to another. Whilst fit analysis does not always enumerate the analysis of the five key variables – grain, line, set, balance and ease (Erwin et al., 1979) – or comfort perceptions, it is through alterations to pattern geometry that these changes informed through fit assessment will be applied. The numerical nature of fit is further evidenced through the application of tension maps within virtual fit software (Ashdown and Dunne, 2006), though these have little in historical fit practices to contextualise what is observed through virtual fit interfaces.

10.1.2 Fabrics

Fabrics as the structure from which garments are made have a direct bearing on how fit is perceived. Often research focuses mainly on the thermophysiological aspects of the fabric as it insulates and controls temperature; this is clearly important in sportswear as exertion causes temperature increases which, if not correctly controlled, will lead to overheating and loss of performance (Fourt and Hollies, 1970; Gaul and Mekjavic, 1987). Thermophysiological factors can also be measured related directly to temperature on a scale which can be easily understood, unlike other factors related to comfort, which are taken to be more subjective and have proven more difficult to quantify in any meaningful and applicable manner. The restrictions to function in clothing related to fabric characteristics have been investigated by Huck (1988) and requirements for elasticity in fabrics for function by Kirk and Ibrahim (1966). However, both focus on specific garment types and neither looks specifically at sportswear. Kirk and Ibrahim (1966) and Watkins (1995) suggest that fabric characteristics including stretch and slip are important considerations in terms of retaining function when wearing garments, though these are difficult to assess in existing practices. The predominant use of knitted fabrics in sportswear, which, due to their structure, have high extensibility (see Chapter 4) provides less restriction to function; similarly, the inclusion of elastomeric yarns within knitted and woven structures enables them to perform very much as a second skin enabling the body to move and not inhibiting function/performance. However, it must be argued that there is an optimum relationship in terms of body pattern relationship, considering garment dimensions, function and fabrics, which should be fully understood.

The engineering of fabric to enhance performance and comfort can be seen in the garments of X-BIONIC, a Swiss company specialising in engineered sportswear (X-BIONIC.co.uk, 2014). Garments are made using different fabrics, and even fabric structures, for different parts of the body to try to maximise performance for the wearers and satisfy their comfort requirements. This shows the importance of engineering approaches when creating performance and sportswear garments.

10.1.3 Function

The function of the wearer within the garment is important and any impingement on function will have a direct bearing on fit perception and, potentially, performance. Movement requirements are often directly related to sporting disciplines, though little reported work in clothing exists that looks at key movements by discipline. Watkins (1977) conducted research looking at movement during ice hockey to recognise how to create better protective equipment with clear considerations of the types of postures and the movements made during play. Functional considerations have also been outlined in research looking at considerations of anatomy in human function (Watkins, 1995; Ashdown, 2011) and directly related to functional considerations of ease (Gill, 2009; Choi and Ashdown, 2011). It is evident that discipline-specific considerations of clothing function are required to ensure that design is engineered to allow for maximum performance.

The function of a person in clothing and, most especially a highly functioning athlete, must take account of the structure of the body (Watkins, 1995; Gill, 2009; Ashdown, 2011). This will not only include skeletal structure, muscle and fat deposition, but also the movements specific to the sport (Watkins, 1977) and the anatomy that underpins this. The anatomy considerations include the skeletal joints, their degrees of freedom and permissible movement, especially as athletes can be found to be more flexible than the general population (Marrin and Bampouras, 2007). One possible focus of this may be through functional change to the body surface directly related to the anatomy and having an effect on the engineering of fit. More recently, a number of studies have looked at measurement of functional changes to the body surface, comparing changes between a static control posture and a number of dynamic postures representing key movements (Gill, 2009; Choi and Ashdown, 2011). There are clear opportunities to look at dynamic changes using evolving technologies to capture real movement and how this affects the body surface lengths and the ease or stretch requirements of a garment and fabric. However, little exists which looks fundamentally at clothing performance in a dynamic environment, though some considerations of how this may be done are shown by Watkins (2011), who observed movement of a participant wearing a skinsuit with grid patterns marked on it. There is also discussion of analysis of the active body in clothing by Ashdown (2011), who provides a number of case studies of existing research and raises the important consideration of how the garment may restrict function through tethering at key body locations.

The body is recognised to be composed of different anatomical structures and joints (Watkins, 1995; Gill, 2009; Ashdown, 2011) with certain areas of the body having higher movement potential or degrees of freedom due to the type of joint and the musculature controlling movement. The hip and shoulder areas permit a large range of movement (Kapandji, 2003, 2007) and this requires careful consideration in the selection of the fabric and its structure when this area will be covered. Looking specifically at the upper body, levels of change through representative movements can cause quite large changes to the body surface (Gill, 2009). Whilst considerable change may be found within certain locations on the body, the anatomy will be a controlling factor over whether high positive or negative changes will occur. In the research by Gill (2009), 24 males were measured in a static control posture and five representative movement postures (Figure 10.1) to determine the levels of body surface change in measurements which had a clear context in the garment, as a means to inform functional ease requirements in the garment pattern. In some cases large changes were observed as either negative amounts like the across front with a decrease of 22% or positive amounts of up to 27% in the across back (Figure 10.2). Whilst movements used do not represent the absolute limits of function, these can be related to movements undertaken within sports. A key consideration raised by this work is how to establish and apply the results of cumulative change where, throughout a measurement's length, it both decreases and increases in different sections. These changes will be influenced by the sport, flexibility and build of the athlete and the types

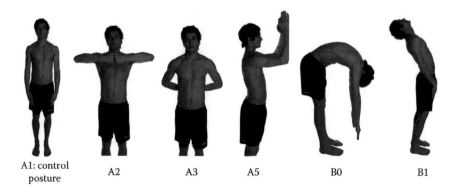

A1: control posture   A2   A3   A5   B0   B1

FIGURE 10.1
Control and movement postures in assessment of upper body surface change.

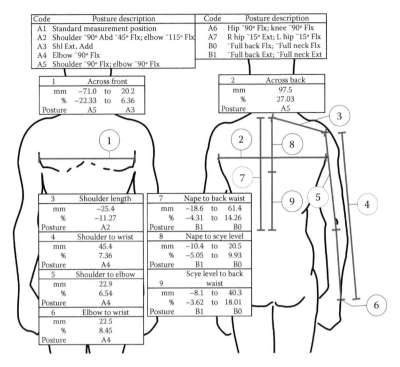

FIGURE 10.2
Functional surface changes by measurement and posture.

of garments worn. Therefore, it is imperative that movements are related to disciplines as with the work of Watkins (1977), but also have a clear context in the pattern as with the study by Gill (2009).

The types of studies discussed focused on the undressed body and further study is required to consider how the sporting body interacts with clothing. Image capture techniques like body scanning and motion capture provide opportunities to analyse garments in sporting environments and can be used with suitable software and researcher skills to analyse performance of clothing through surface appearance examples, and considerations for this area are discussed in Chapter 13.

10.2 Sensorial Comfort

10.2.1 Psychological Considerations of Wearer Perceptions

Sportswear in terms of 'activewear' is becoming increasingly popular in mainstream fashion. No longer is it committed to the professional athlete within the sporting area; consumers are now purchasing sportswear items for everyday wear. Therefore, comfort is paramount, suggesting that the complexities associated with the wearers' perceptions during an activity must be addressed. Many of our comfort perceptions are based on experiences from childhood and can be heavily influenced by our peers (Daters, 1990). Sportswear that does not fit correctly has the potential for the wearer to underperform during an activity, which can result in injury (particularly an activity that requires simultaneous movements, such as running or swimming). However, physical exertion will vary between people depending on fitness levels and their unique physiology; therefore, the levels of sensorial comfort will depend widely upon these factors (Bartels, 2011). Kilinic-Balci (2009) presented a comprehensive overview of the factors that affect a person's ability to judge comfort, in that they differ significantly because of the person's complex genetic disposition combined with physiological and psychological experiences, including learned fit experience. This is known as 'subjective perception' (Kilinic-Balci, 2009; Troynikov, Ashayeri and Fuss, 2011) whereby the verbal interpretation of a tactile sensation from one person to another will vary. This is significant, particularly where key demographic characteristics such as age, race and gender relate directly to comfort. In the research undertaken by Sontag (1985) in temperature-controlled conditions, mature female participants were asked to wear trial garments on different parts of the body to ascertain comfort levels; the garment that was worn closest to the body was deemed as having a high emotional comfort value due to the fact that this was concealed and not visible to others. This suggests that comfort may have a dual purpose in consumer perceptions: in relation to where the garment is positioned on the body and the environment in which the garment is worn (Woodward, 2007). Comfort in clothing is not only an issue for mature consumers; in a larger study of consumers from a wide ranging demographic undertaken by Kaplan and Okur (2008) based on clothing comfort, there were similar findings to Sontag's research. The participants also concluded that the personal concealment of garments and outward appearance were important and highly regarded factors in clothing comfort. These factors are highly important to sportswear comfort, particularly in activities that require garments to be layered, such as skiing, climbing and other outdoor pursuits – in particular, lower body movements that require varied ease levels to obtain maximum comfort levels (Ashdown and DeLong, 1995).

10.2.2 Pressure Comfort

Pressure comfort plays an important role in sensorial comfort, particularly in relation to physiology. The body itself controls heat management in respect of a cooling system through perspiring; however, extreme temperatures and each consumer's unique physiological DNA should be considered (Choudhury, Majumdar and Datt, 2011). A study undertaken by Sweeney and Branson (1990), whereby a numerical ranking system was assigned to specific comfort level descriptors, was used to test mental reactions to water pressure on an isolated area of the body. In the study, participants wore a base layer that was subjected to different intensities of water pressure under controlled conditions. This resulted in the participants having similar, more closely related reactions based on the ranking system. However, pressure from both physiological and external forces such as sweat, rain and wind varies during activity, suggesting that a garment needs to be multipurpose and respond to sudden and frequent changes that occur during active exercise. Pressure comfort is affected by external factors, such as extreme climates, ultimately affecting the air quality, which can have a dramatic effect on performance during sporting activities (Yan et al., 2012). The body's internal organs are therefore also under pressure as they try to cope with the external factors, and whilst perspiring will help to cool a person down, muscle fatigue may also be prevalent, which can lead to injury (Jarvinen et al., 2007). Pressure comfort studies (Kamalha et al., 2013; Liu et al., 2013) indicate that weight, height and individual physiology do not have a significant effect on the casual sportsperson who participates in a noncompetitive environment. However, many professional sportspeople compete with similar physiologies, particularly weight. This is evidenced in the case of professional gymnasts, whereby they train under very similar conditions, which are often strict both physically and mentally; therefore, controlled pressure comfort during activity is paramount for sporting achievement (Yan et al., 2012).

It can be concluded from the research that pressure comfort for sportswear and casual wear differs significantly. The professional athlete requires garments that aid both muscle function and physical performance in order to focus directly on the activity; the casual sportswear consumer may not. However, in both cases pressure comfort in relation to garment body contact is important (Senthilkumar, Kumar and Anbumani, 2012).

10.2.3 Pressure Comfort and Gender

Pressure comfort in relation to gender also differs significantly whether the consumer is a professional sportsperson or not. Ashdown and DeLong (1995) clearly identified that small incremental differences in garment ease within women's clothing designed for the lower body can be detected by the wearer, demonstrating the body's sensitivity to discomfort. Therefore, the need for gender-specific, activity-orientated sportswear with performance qualities is important as it allows the wearer to concentrate on performance rather than clothing discomfort.

Ho (2010) addressed the shape and fit of indoor cycling wear, whereby male and female participants were required to discuss the fit and finish garments. The male participants required a 'relaxed' fit and the female participants required a 'baggy' fit on the upper garment. These requirements suggest that pressure comfort and physiology play important roles in the comfort factors during exercise. The function of comfort in this example is that the garment has to accommodate the differences between chest (men's), bust (women's) and waist (both genders); in essence, both descriptions of the garments indicate that they are similar in shape but perform different functions in relation to pressure comfort, which is the same function. However, as the study focuses on indoor cycling, which is performed in climate-controlled conditions, there are significant differences in outdoor cycling, where the climate can be unpredictable. A study performed in similar environmental conditions by Liu et al. (2013) focused on male subjects and 'upper body pressure', in which the participants were required to wear a dry, seamless T-shirt during intermittent active and inactive periods. The authors stated that the areas least affected by pressure comfort were the torso, arms and back. In contrast, gymnasts wearing a leotard of a similar construction experienced body pressure within those areas (Yan et al., 2012). The study implies that total body measurements between the nape of the neck and the front and back rises should be considered in order to support the functions of the activity (Yan et al., 2012).

10.3 Performance Expectations

Clothing comfort is also derived from our homogenous cultural, historical and physiological makeup. However, the interpretation of clothing comfort will vary because of the dual-purpose elements of sports clothing (i.e. the casual wearer and the professional athlete will have different performance needs, as will the requirements of the male and female consumers) (Fan, 2004).

Garment silhouettes have significantly affected the fit and performance of sportswear for both men and women. Historically, men's garments influenced women's sportswear clothing; moreover, women had to wear men's clothing because that was the only sportswear available. The adoption of looser fitting men's clothing was found to be the case when Ledbury (2009) investigated women's clothing worn during outdoor pursuits. However, women wanted a clear distinction from the masculine silhouette and the comfort issues that were associated with poorly fitted garments cut to a male body shape. Women required the same performance qualities, particularly in the case of contact sports such as basketball, which allow them to focus on the activity rather than be distracted by the clothing misfit (Klomsten, Skaalvik and Espnes, 2004).

Clothing comfort, specifically for women during exercise, is influenced by their physical shape, as well as weight, body measurements and BMI (body mass index); overall, the body experiences the same amount of pressure during activity. However, many women do not have equally proportioned upper and lower body measurements, which indicates that comfort for the whole body may not be achievable during exercise without consideration of proportional differences within athletic and individual bodies (Liu et al., 2013). Table 10.1 highlights the four areas in relation to clothing type, comfort

TABLE 10.1

Performance Expectations in Sports Clothing

Sports Discipline	Clothing Type	Comfort Impact	Performance Expectations	Gender Considerations
Cycling (road)	Tight fit; double layer; bib shorts and jersey (long/short sleeves); socks; boots	Seat/groin/shoulders, arms	Breathable; sweat wicking; flexibility	Male and female outfits differ
Football/soccer	Loose or tight fit; shirt (long/short sleeves); shorts; socks; boots		Breathable; sweat wicking	
Gymnastics	Tight fit, one layer; leotard, sleeveless/long sleeved	All areas	Flexibility; dynamics; breathable; stable	Male and female outfits may differ
Track running	Tight fit, all-in one; vest, cropped top; pants or leggings or shorts; training shoes	Legs, arms, hips, knees, feet	Aerodynamic; breathable; sweat wicking	
Skiing (alpine)	Tight fit, all-in one; boots; goggles; gloves; helmet	Upper body, back, arms, hips, thigh, knees, shins, feet	Air-permeable sweat wicking; muscle warmth; breathability; flexible; reduced muscle fatigue	
Snowboarding	Semifitted; several layers; jackets; pants; thermal boots; goggles; gloves; board; helmet	Upper body, hips, knees, arms, legs, core, back	Aerodynamic; maximum flexibility	
Swimming	Single layer; all-in one costume; trunks; sleeveless; cap; goggles	Arms, shoulders, legs, core	Buoyancy; speed; endurance; reduction of drag	Male and female outfits differ

impact, performance expectations and gender considerations. Each area has an important bearing in relation to fit and the consumer.

10.4 Ease Levels within the Garment

Ease in garment development relates to the specific amount added or subtracted from the dimensions of the pattern in addition to the dimensions of the wearer (Beazley, 1997; Gill, 2011). Whilst ease in the context of the pattern may be applied as a singular figure, its determination is impacted by a number of factors, including garment styling and the interrelated variables of function, comfort, fabric characteristics and oversize (Gill, 2011). Whilst styling may be dependent on the designer, the other variables could be determined quantitatively through different methodological techniques. Functional ease considerations have been investigated and numerical levels suggested (Kirk and Ibrahim, 1966; Gill, 2009; Choi and Ashdown, 2011). Whilst the other suggested variables (comfort, oversize and fabric characteristics) have been considered, the findings from existing studies have not been enumerated in a manner applicable directly to pattern development. Without determining an ease value as a positive or negative amount in addition to the body, it will often be subjectively determined by the product developer in reference to his or her experiences and then adjusted subjectively during the product development process. Generally, the application of ease differs between pattern construction methods (Gill, 2010) and is not always explicitly stated during the pattern construction process (Gill and Chadwick, 2009). However, whilst methods to determine numeration of the interrelated variables of ease may be difficult, some successful investigations of ease in wear have been undertaken (Ashdown and DeLong, 1995). These have been able to suggest sensitivity to ease levels, though this avenue of research has had little exploration, possibly due to its complexities. With regard to stretch fabrics, ease can also be considered regarding proximity of the garment to the wearer's body (Watkins, 2006, 2011). Though proximity ratings have been suggested, these have no attributed numeric values which could be adopted by the pattern technician and incorporated into ease considerations.

10.5 Summary

Whilst research continues to develop the means to construct garments with engineered fit, there still remain some key areas of focus to connect the analysis of fit with the attainment of fit during the product development

CASE STUDY: PUMA UK – PROTECTIVE
LAYERS THAT PROVIDE COMFORT

Designers at Puma, a leading sportswear brand, developed a prototype football kit for the UEFA Euro 2008 football tournament without full consideration of comfort and performance for the football players. In collaboration with Manchester Metropolitan University, research was undertaken to improve the wearer comfort experience of the football kit.

The prototype sport shirt was designed with two mesh layers in order to make the product aesthetically pleasing as well as providing ventilation during performance. However, the designers were concerned that during physical activity the layers, when in contact with perspiration, would result in the garment clinging to the body and ultimately affecting the players' performance as well as reducing the product's aesthetic appeal. Therefore, as part of the product development process, it was essential to test the product in order to make improvements prior to the tournament.

The study undertaken between Puma and the academics at Manchester Metropolitan University was to determine how well the football shirt would perform at three stages – prematch, during play and postmatch – in relation to the following factors:

- Analysis of injuries during performance
- Warming up and cooling down exercises
- Base layer protectors
- Hydration levels
- Interaction between body and layer during contact

The researchers were then able to categorise the factors into two main areas: comfort and comfort control. The approach involved both a qualitative and quantitative approach in the form of a focus group and laboratory tests.

During the focus group, the researchers were able to collect extensive data in relation to the interaction between the wearer and product performance, particularly during the 'warming up' and 'cooling down' phases. From these findings, they were able to recommend the use of thermochromatic dyes that changed colour when they interacted with the changes in body temperature, suggesting that this would counteract some of the negative psychological impact the players may experience from the physiological changes encountered during both phases. However, Puma wanted to maintain the aesthetic qualities of the garment throughout performance and it was felt that this would add further discomfort to the players in relation to team identity.

During phase two of the study, the academics undertook quantitative research to test football shirts within a controlled laboratory setting. The laboratory was prepared to simulate environments similar to the locations where the Euro 2008 tournaments were to be played in Austria and Switzerland. Both of these countries encountered similar weather patterns and, during the tournament from 7 to 29 June, the average temperate ranged between 13°C and 24°C, with an average wind speed of between 6 and 9 km/h and an average wind gust speed of between 43 and 48 km/h.

The Puma shirt was tested against two other shirts of their main sportswear competitors. It was then proposed that there would be four main areas associated with the predicted weather conditions in order to provide comfort factors prematch and postmatch and during performance:

- Ventilation with dry fabrics
- Ventilation with sweaty fabrics
- Insulation with dry fabrics
- Insulation with sweaty fabrics

Within the controlled laboratory environment, an industrial fan was used to simulate wind and a treadmill used to simulate running on the pitch as well as a thermistor to measure skin temperature. Each shirt was tested four times (twice with wind, twice without) and skin temperature was measured on the upper arm, upper back, chest and abdomen. The garments were weighed before and after exercise.

Through the interviews and laboratory tests, the data collected allowed the researchers to conclude that the area which required the most ventilation was the upper back, which is consistent with several studies (Sweeney and Branson, 1990; Okubo, Saeki and Yamamoto, 2008; Hassan et al., 2012) in relation to sweat production during exercise. The Puma shirt and one other were similar in ventilation and insulation as both were made of similar mesh fabrics. However, the Puma shirt was superior in relation to reducing sweat mark visibility during the study. In addition to this, there was a limited negative effect on the aesthetic qualities.

process. As consumers develop more sophisticated requirements for sportswear clothing in relation to function and fit, the gap between the functional requirements of the professional sportsperson and the consumer is closing rapidly as demonstrated through technological advancements in product development. Three-dimensional technology provides the platform for garment simulation essential to custom fit sportswear, which has been used extensively in the development of professional sportswear apparel. Consumers now expect sportswear garments to fit exactly to the shapes of their bodies without restricting movements and with a great deal of comfort. In addition to this, their unique physiology needs to be addressed during the pattern development process, allowing the wearer to perform complex movements such as running, walking and jumping. Within this, fabric plays an important role to enhance the performance of the wearer as well as to offer some protection.

In addressing the psychology of fit, sports apparel in many cases has a personal association to the professional sportsperson and the consumer. This is increasingly important in team cohesion, which can improve performance. However, as fashion often dictates fit, therein lie many problems in how to ensure that fit, function and consumer needs are met.

10.6 Future Developments

Considering fit in sportswear, future developments in terms of greater theoretical underpinnings are necessary. These require a detailed understanding of the body, its anatomy and how it functions in a discipline specific environment. These considerations will likely be influenced by gender and should be sensitive to the culture of both the country and the sport as an influence over subjective perceptions. Technology like body scanning can help provide the detailed measurement and visual analysis which can better inform clothing engineering, though full benefits will not be realised until these data can be suitably linked to more subjective evaluations related to fit perceptions of individual wearers.

References

Aldrich, W. (2008). *Metric pattern cutting for womenswear*. Oxford, UK: Wiley-Blackwell.
Armstrong, H. (2010). *Patternmaking for fashion design*, 5th ed. Upper Saddle River, NJ: Pearson/Prentice Hall.

Ashdown, S. (2011). Improving body movement comfort in apparel. In *Improving comfort in clothing*, 1st ed., G. Song (ed.), 278–298. London: Woodhead Publishing.

Ashdown, S. and DeLong, M. (1995). Perception testing of apparel ease variation. *Applied Ergonomics* 26 (1): 47–54.

Ashdown, S. and Dunne, L. (2006). A study of automated custom fit: Readiness of the technology for the apparel industry. *Clothing and Textiles Research Journal* 24 (2): 121–136.

Ashdown, S. P., Loker, S., Schoenfelder, K. A. and Lyman-Clarke, L. (2004). Using 3D scans for fit analysis. *Journal of Textile and Apparel, Technology and Management* 4 (1). Retrieved from http://www.tx.ncsu.edu/jtatm/volume4issue1/articles/Loker/Loker_full_103_04.pdf.

Bartels, V. (2011). Improving comfort on sports and leisure wear. In *Improving comfort in clothing*, 1st ed., G. Song (ed.), 385–411. London: Woodhead Publishing Ltd.

Beazley, A. (1997). Size and fit: Procedures in undertaking a survey of body measurements. *Journal of Fashion Marketing and Management* 2 (1): 55–85.

Beazley, A. and Bond, T. (2003). *Computer-aided pattern design and product development*. Malden, MA: Blackwell Pub.

Bishop, P., Balilonis, G., Davis, J. and Zhang, Y. (2013). Ergonomics and comfort in protective and sport clothing: A brief review. *Journal of Ergonomics S* 2: 2.

Brown, P. and Rice, J. (2001). *Ready-to-wear apparel analysis*. Upper Saddle River, NJ: Prentice Hall.

Bye, E. and McKinney, E. (2010). Fit analysis using live and 3D scan models. *International Journal of Clothing Science and Technology* 22 (2/3): 88–100.

Choi, S. and Ashdown, S. (2011). 3D body scan analysis of dimensional change in lower body measurements for active body positions. *Textile Research Journal* 81 (1): 81–93.

Choudhury, A., Majumdar, P. and Datt, C. (2011). Chapter factors affecting comfort: Human physiology and the role of clothing. In *Improving comfort in clothing*, 1st ed., G. Song (ed.), 3–60. London: Woodhead Publishing Ltd.

Das, A. and Alagirusamy, R. (2010). *Science in clothing comfort*. New Delhi: Woodhead Publishing India Pvt Ltd.

Daters, C. (1990). Importance of clothing and self-esteem among adolescents. *Clothing and Textiles Research Journal* 8 (3): 45–50.

Dolan, E., O'Connor, H., McGoldrick, A., O'Loughlin, G., Lyons, D. and Warrington, G. (2011). Nutritional, lifestyle, and weight control practices of professional jockeys. *Journal of Sports Sciences* 29 (8): 791–799.

Erwin, M., Kinchen, L. and Peters, K. (1979). *Clothing for moderns*, 6th ed. New York: Macmillan Publishing.

Fan, J. (2004). Perception of body appearance and its relation to clothing. In *Clothing appearance and fit: Science and technology*, 1st ed., J. Fan, W. Yu and L. Hunter (ed.), 1–14. London: Woodhead Publishing Ltd.

Fan, J., Yu, W. and Hunter, L. (2004). *Clothing appearance and fit*. Boca Raton, FL: CRC Press.

Fourt, L. and Hollies, N. (1970). *Clothing comfort and function*. New York: Marcel Dekker.

Fowler, D. (1999). The attributes sought in sportswear apparel: A ranking. *Journal of Marketing Theory and Practice* 7 (4): 81–88.

Gaul, C. and Mekjavic, I. (1987). Helicopter pilot suits for offshore application. A survey of thermal comfort and ergonomic design. *Applied Ergonomics* 18 (2): 153–158.

Gill, S. (2009). Determination of functional ease allowances using anthropometric measurement for application in pattern construction (unpublished PhD thesis), PhD, Manchester Metropolitan University.

———. (2010). Determination of ease allowances included in pattern construction methods. In *Textile Institute 100th World Conference*, Manchester, UK.

———. (2011). Improving garment fit and function through ease quantification. *Journal of Fashion Marketing and Management* 15 (2): 228–241.

Gill, S. and Chadwick, N. (2009). Determination of ease allowances included in pattern construction methods. *International Journal of Fashion Design, Technology and Education* 2 (1): 23–31.

Hassan, M., Qashqary, K., Hassan, H., Shady, E. and Alansary, M. (2012). Influence of sportswear fabric properties on the health and performance of athletes. *Fibres & Textiles in Eastern Europe* 93: 82–88.

Ho, Y. (2010). Indoor cycling wear: A needs assessment. MSc thesis, Oregon State University.

Huck, J. (1988). Protective clothing systems: A technique for evaluating restriction of wearer mobility. *Applied Ergonomics* 19 (3): 185–190.

Jarvinen, T. A., Jarvinen, T. L., Kaariainen, M., Aarimaa, V., Vaittinen, S., Kalimo, H. and Jarvinen, M. (2007). Muscle injuries: Optimising recovery. *Best Practice & Research Clinical Rheumatology* 21 (2): 317–331.

Jeong, Y., Hong, K. and Kim, S. (2006). 3D pattern construction and its application to tight-fitting garments for comfortable pressure sensation. *Fibers and Polymers* 7 (2): 195–202.

Jin, Z., Luo, X., Yan, Y. and Tao, J. (2010). Pressure distribution and comfort pressure range of men's seamless underwear. *Journal of Textile Research* 10: 22.

Kamalha, E., Zeng, Y., Mwasiagi, J. and Kyatuheire, S. (2013). The comfort dimension; A review of perception in clothing. *Journal of Sensory Studies* 28 (6): 423–444.

Kamijo, M., Uemae, T., Kwon, E., Horiba, Y., Yoshida, H. and Shimizu, Y. (2009). Comfort evaluation of T-shirt type underwear made of spun silk yarn. *International Conference on Biometrics and Kansei Engineering*, pp. 100–105.

Kapandji, I. A. (2002). The Physiology of the Joints: Volume 2 Lower Limb (5th ed.). Edinburgh: Churchill Livingstone.

Kapandji, I. A. (2007). The Physiology of the Joints: Volume 1 Upper Limb (6th ed.). London: Churchill Livingstone.

Kaplan, S. and Okur, A. (2008). The meaning and importance of clothing comfort: A case study for Turkey. *Journal of Sensory Studies* 23 (5): 688–706.

Kilinic-Balci, F. (2009). How consumers perceive comfort in apparel. In *Improving comfort in clothing*, 1st ed., G. Song (ed.), 97–113. London: Woodhead Publishing Ltd.

Kirk, W. and Ibrahim, S. (1966). Fundamental relationship of fabric extensibility to anthropometric requirements and garment performance. *Textile Research Journal* 36 (1): 37–47.

Klomsten, A., Skaalvik, E. and Espnes, G. (2004). Physical self-concept and sports: Do gender differences still exist? *Sex Roles* 50 (1–2): 119–127.

Le Pechoux, B. and Ghosh, T. (2002). Apparel sizing and fit. *Textile Progress* 32 (1): 1–60.

Ledbury, J. (2009). Beyond function: A design development case study. In *IFFTI 11th Annual Conference*. London.

Liu, H., Chen, D., Wei, Q. and Pan, R. (2013). An investigation into the bust girth range of pressure comfort garment based on elastic sports vest. *Journal of the Textile Institute* 104 (2): 223–230.

Lloyd, E. (1994). ABC of sports medicine. Temperature and performance. I: Cold. *British Medical Journal* 309 (6953): 531.

Marrin, K. and Bampouras, T. (2007). Anthropometric and physiological characteristics of elite female water polo players. In *Kinanthropometry X*, 1st ed., M. Marfell-Jones and T. Olds (ed.), 151–164. Glasgow: Routledge.

Okubo, M., Saeki, N. and Yamamoto, T. (2008). Development of functional sportswear for controlling moisture and odor prepared by atmospheric pressure non-thermal plasma graft polymerization induced by RF glow discharge. *Journal of Electrostatics* 66 (7): 381–387.

Senthilkumar, M., Kumar, L. and Anbumani, N. (2012). Design and development of a pressure sensing device for analysing the pressure comfort of elastic garments. *Fibres & Textiles in Eastern Europe* 20 (1): 90.

Sontag, M. (1985). Comfort dimensions of actual and ideal insulative clothing for older women. *Clothing and Textiles Research Journal* 4 (1): 9–17.

Sweeney, M. and Branson, D. (1990). Sensorial comfort part I: A psychophysical method for assessing moisture sensation in clothing. *Textile Research Journal* 60 (7): 371–377.

Troynikov, O., Ashayeri, E. and Fuss, F. (2011). Tribological evaluation of sportswear with negative fit worn next to skin. *Proceedings of the Institution of Mechanical Engineers, Part J: Journal of Engineering Tribology*, doi: 1350650111425876.

Watkins, S. (1977). The design of protective equipment for ice hockey. *Home Economics Research Journal* 5 (3): 154–166.

———. (1995). *Clothing the portable environment*, 2nd ed. Ames: Iowa State University Press.

Watkins, P. (2006). Custom fit: Is it fit for the customer? In *8th Annual IFFTI Conference*. North Carolina, USA.

———. (2011). Garment pattern design and comfort. In *Improving comfort in clothing*, 1st ed. London: Woodhead Publishing.

Woodward, S. (2007). *Why women wear what they wear*. Oxford, UK: Berg.

X-BIONIC.co.uk. (2014). *Welcome to X-BIONIC® United Kingdom*. [online]. Available at http://www.x-bionic.co.uk (accessed 8 February 2014).

Yan, Y., Gao, J., He, W., Jin, Z. and Tao, J. (2012). Study on the pressure comfort of women's seamless gymnastics clothing. *Computational Intelligence and Design* 2: 502–505.

Haruko, M., Hiroko, M., Tamaki, M. and Kazuo, U. (1991). A study of clothing pressure developed by the brassiere. *The Japan Research Association for Textile-end uses* 32 (9): 416–423.

11

Evaluating the Performance of Fabrics for Sportswear

Praburaj Venkatraman

CONTENTS

11.1 Introduction 262
11.2 Physical Measurements for Woven and Knitted Fabrics 263
    11.2.1 Fabric Area Density 264
    11.2.2 Fabric Thickness 265
    11.2.3 Fabric Bulk Density 266
    11.2.4 Fabric Construction 267
    11.2.5 Fabric Cover Factor 267
    11.2.6 Fabric Count 268
11.3 Evaluating Durability 269
    11.3.1 Abrasion Resistance 269
    11.3.2 Fabric Pilling 271
11.4 Fabric Handle in Sportswear 274
    11.4.1 Fabric Stiffness 274
    11.4.2 Fabric Drape 275
    11.4.3 Fabric Stretch and Recovery 276
11.5 Measurement of Fabric Comfort 277
    11.5.1 Wicking and Its Effect on Fabric Comfort 278
    11.5.2 Moisture Management Tester 278
    11.5.3 Influence of Moisture Transfer in Functional Clothing
        (Permetest) 280
11.6 Fabric Specifications and Interpreting Results 282
11.7 Summary 285
References 286

11.1 Introduction

Evaluation of fabrics for their performance is mandatory, particularly during design and development of sportswear and performance apparel. By investigating the properties of fabrics through textile testing it is possible to determine their suitability for their intended application. Specific fibre types can be blended to meet consumer requirements; for instance, in sportswear stretch plays an important role in adapting to various body movements (Roy, 2014). Stretch can also be controlled in specific directions, which allows support to specific muscles. The X-BIONIC® bib tight (X-BIONIC, 2014) has the ability to apply pressure to specific muscles as well as to transport moisture during intensive sessions of the sport. Such a product design would not exist if properties of materials were not explored in the first instance. ISPO

TABLE 11.1

Various Sports and Their Requirements

Popular Sport	Fabrics Used	Frequently Used Product Name	Specific Properties
Cycling	Brushed knitted fabrics with stretch	Cycling tights	Three-way stretch Insulation during winter Soft next to the skin
Skiing	Warp knitted fabric	Base layer thermal	Thermal resistance Stretch and recovery Moisture permeable
	Single weft knit, napped technical back	Midlayer fleece	Thermal insulation
	Breathable coated woven fabrics (high density)	Outer shell jackets	Waterproof Breathable Durable
Running	Warp/weft knitted fabric with bidirectional stretch	Compression tights	Stretch and recovery Comfort
	Knitted fabric	Tops and leggings	
	Woven fabric	Jogging shorts	Lightweight Quick drying
Football	Warp and weft knit	Tights	Stretch and recovery Moisture permeable
	Knitted fabric	Tops	Comfortable/moisture management
	Woven fabric	Trousers	Durable
Swimming	Woven and knit fabrics with elastomeric filaments	Female/male swimsuits	Stretch and recovery
		Training jackets and trousers	Quick drying

FIGURE 11.1
Different types of individual sports.

TEXTRENDS 2015/2016 categorises fabrics used in sports apparel into eight applications: base layer, second layer, outer layer, membrane and coatings, fibres for insulation, trims, accessories and soft equipment used in sports.

Evaluation of sportswear involves a series of tests to ascertain the performance of the garment and accessories and it varies based on the application and requirements (Table 11.1). The most important aspect in sportswear is moisture management and thermal balance.

In this chapter, specific test methods relating to fabric durability, fabric handle, stretch and recovery, moisture transport, water vapour permeability and thermal resistance are discussed. The test methods are referred to in British standards, which provide definition of various parameters. Examples are included to enable the reader to interpret the test results and explanation is provided on how these parameters affect garment performance and research relating to these parameters (Figure 11.1).

11.2 Physical Measurements for Woven and Knitted Fabrics

Fabrics' physical parameters play an important role in determining their characteristics and are widely used by professionals as 'specifications' whilst making decisions such as suitability for a particular end use or to communicate across the fashion supply chain. In this section, the following parameters will be discussed:

- Fabric area density or fabric weight
- Fabric thickness
- Fabric bulk density
- Fabric construction
- Fabric count
- Fabric cover factor

11.2.1 Fabric Area Density

Fabric weight is mass per unit area, denoted in grams per square metre (g/m^2). It is an important fabric property that is often used in determining the fabric cost and quality. For woven fabrics it is calculated by preparing a fabric specimen to a dimension of 100 × 100 mm and weighing it on a scale. The resultant value is multiplied by 100 to calculate it for a metre. In industry, the fabric area density is also calculated using a gsm (grams per square metre) cutter (Figure 11.2), which is circular in shape.

For knitted fabrics,

$$\text{area density} = (s \times l \times T) \div 100$$

where
  s = stitch density per square centimetre
  l = stitch length – length of yarn in a knitted loop
  T = yarn count in tex units (Anand, 2000)

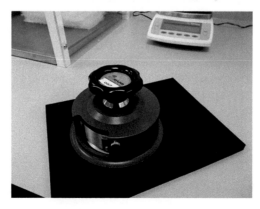

FIGURE 11.2
Fabric gsm cutter. (Image courtesy of MMU Textile Lab.)

TABLE 11.2

Fabric Weight for Fabrics Used in Sportswear

Fabric Type	Grams/Square Metre (g/m²)	Function/End Uses
Very lightweight fabric	18	Base layer, lightweight, moisture management, smooth fabric worn next to skin
Lightweight fabric	<100	Midlayer fabric used for tops, trousers, trainers, etc.
Medium weight fabric	130–180	Soft shell fabrics used in outer layer jackets; these are either laminated or coated with finishes
Heavyweight fabrics	250+	Hard shell heavyweight fabrics used for jackets, trousers and high-performance technical materials used for outdoor applications

Stitch density refers to the total number of loops and is obtained by determining the wales per centimetre (wpc) and courses per centimetre (cpc) in a measured area and multiplying both the values, which is 315 cm². For example, wpc = 15 and cpc = 21 for single jersey fabric, and the stitch density is obtained multiplying these values. A typical stitch length is 2.55 and yarn count 16 tex. The area density for a knitted fabric is obtained using the formula on the previous page, which is 128.5 g/m². Fabric weight affects the fabric stiffness, especially the bending rigidity, particularly in the warp direction (Mandal and Abraham, 2009). Table 11.2 classifies area density of fabrics based on the application.

11.2.2 Fabric Thickness

Fabric thickness is the distance between the upper and lower surface of the fabric and is measured using a thickness gauge or tester (Figure 11.3). The test sample is placed between two reference plates which exert a known pressure on the sample. The distance between plates is recorded in millimetres (BS EN ISO 5084:1997). The fabric thickness affects garment production, especially in adjusting the sewing machine settings. This could be selection of a needle or fabric feed system. Selection of needle depends on the stitch density (seams per inch) required. In sportswear, stitch density of base layer apparel is finer than for a jacket. For knitwear, a ball-pointed tip is preferred because it prevents fabric damage (laddering effect). For instance, consider the case of knitted stretch fabrics with fine thickness that slip during sewing, resulting in fabric being gathered or staggered; in this case, a differential feed system at the top and bottom will be used in the sewing machine. One end will feed the fabric quickly and the other feeds slowly, resulting in a good-quality seam. Fabric thickness also affects the overall performance of a garment, especially the abrasion resistance of fabrics; the higher the fabric thickness, the higher

FIGURE 11.3
Fabric thickness tester. (Image courtesy of MMU Textile Lab.)

TABLE 11.3

Fabric Thickness

Type	Thickness (mm)
Thin	<0.20
Medium	0.23–0.46
Thick	>0.47

Source: Collier, B. J. and Epps, H. H. (1999). *Textile testing and analysis.* Upper Saddle River, NJ: Pearson, Prentice-Hall Inc.

the resistance to abrasive action (Özdil, Kayseri and Mengüç, 2012). Table 11.3 generally classifies the thickness of fabrics.

11.2.3 Fabric Bulk Density

Fabric bulk density takes into account the fabric weight and thickness. It represents the bulkiness of the fabric relative to its thickness. It is an important factor in determining the garment comfort. A thick fabric with an average weight is more comfortable in cold conditions or outdoor sports; on the other hand, a thin fabric of the same weight will be ideal in warm conditions. It is generally expressed in grams per cubic centimetre (g/cm^3). Bulk density is calculated using the following equation:

$$\text{Bulk density} = \frac{\text{fabric thickness in cm} \times \text{fabric weight in } g/cm^2}{10,000}$$

11.2.4 Fabric Construction

The repeat of the design is presented by shading the box that represents the warp interlacing over the weft yarn. This is called 'fabric design' construction (Figures 11.4 and 11.5).

11.2.5 Fabric Cover Factor

Cover factor for woven fabrics indicates the extent to which a fabric area is covered by one set of yarns. In woven fabrics, cover factor is determined in warp and weft directions. In knitted fabrics, cover factor is also termed as tightness factor. It is generally denoted by K. It is calculated using the formula for woven fabrics

$$K = \frac{\text{threads per cm } \sqrt{\text{tex}}}{10}$$

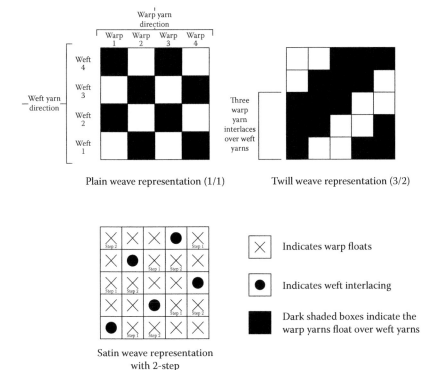

Plain weave representation (1/1)

Twill weave representation (3/2)

Satin weave representation
with 2-step

FIGURE 11.4
Plain weave, twill and satin weave.

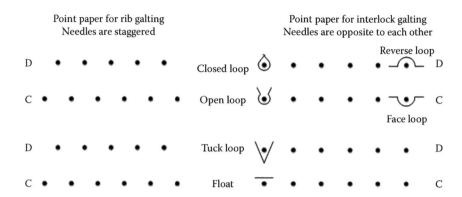

FIGURE 11.5
Point paper for weft knitted fabric representation. (From Taylor, 1999.)

Tightness factor for knitted fabrics is

$$K = \frac{\sqrt{tex}}{stitch \ length \ in \ mm}$$

11.2.6 Fabric Count

A fabric counter or magnifying glass is used to determine fabric count. Fabric count is assessed by counting the number of courses and wales in knitted fabrics (Figure 11.6a) and counting the number of warp or weft yarns in woven fabric (Figure 11.6b). The fabric count plays an important role in determining the closeness of the weave or knit that affects various properties such as porosity, permeability and durability of the fabric.

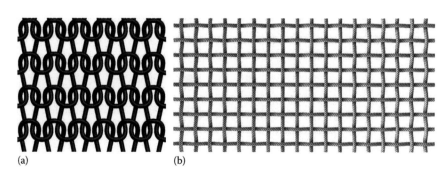

(a)   (b)

FIGURE 11.6
(a) Knitted fabric structure. (b) Woven fabric structure. (From Dreamstime.)

11.3 Evaluating Durability

Durability is one of the important parameters when selecting a fabric for a particular end use. Sports activity involves repeated body movements and sportswear can abrade in several ways – for instance, fabric rubbing with another layer of fabric; fabric abrading in particular areas including crotch, knee and underarms; fabric rubbing against another object due to tripping or falling or garment abrasion while laundering. Abrasion can also occur between yarns and fibres when fabric is stretched repeatedly. The effect of fabric abrasion depends on various elements: fibre type and its properties, yarn quality and its structure and fabric construction.

Factors that affect abrasion resistance of fabrics include:

- Presence of longer fibres in the yarn offer better resistance to abrasion than short fibres.
- Increased fibre diameter enhances resistance to a certain extent.
- Optimum yarn twist offers good resistance to abrasion.
- Increasing yarn linear density increases resistance with a constant fabric density.
- Uniform yarn crimp in the fabric (warp and weft) enables even wear across the fabric.
- Warp or weft floats are highly susceptible to abrasion.
- Fabric weight and its relative thickness affect the abrasion resistance.
- Abrasion also depends on fabric count (ends and picks per inch): the more threads per inch, the lesser the wear.
- Type of yarns also affects abrasion resistance, for instance, air jet spun yarn has high resistance to abrasion, ring spun yarn possesses moderate resistance and open-end spun yarn possesses low resistance to abrasion.

11.3.1 Abrasion Resistance

The usual method to evaluate fabric abrasion is a Martindale abrasion tester, where the instrument (BS EN ISO 12947-1:1998) subjects a specimen to a uniform rubbing motion (Lissajou's figure), which is repeated until two threads are broken (woven fabric). Various methods used to determine the end of the test are the evaluation of change in before- and after-sample weight, change in colour using a colour change chart (ISO 105 A02) or examining whether the sample distortion is being rubbed away and completion of a specified number of cycles (Cohen et al., 2010). However, this section discusses the Taber abrasion tester (BS EN ISO 5470-1:1999) that works on the principle of a rotary platform, which tests flat abrasion and is intended for heavier woven and knitted fabrics as the abrading action is severe (Figure 11.7). In this method, the instrument uses a flat abrasive action in which the fabric is placed on a rotary platform and is abraded by two abrasive wheels. Six samples of dimension 114 ± 1 mm diameter with a central hole of diameter 6.35 mm are chosen and conditioned. The load applied can be varied depending upon the type of abrasion required – for instance, from very gentle (2.5 N) to harsh action (9.8 N). Similarly, the rotary wheels can be either rubber and abrasive grain or a vitrified version depending on the required abrasive action. The sample weight is measured prior to test in milligrams. To determine the average rate of loss in mass, the loss in mass for every 100 cycles should be recorded. The end of the test is determined by change in colour of the abraded portion, change in mass or change in surface distortion.

Laminated or coated or heavy multilayered fabrics that are intended for jackets, trousers, backpacks and footwear (e.g. Cordura® Naturalle) are often subjected to this test to evaluate their resistance to abrasion. Three heavy-weight woven fabrics that are intended for outer wear for jackets were randomly selected: fabric 1 (282 g/m²), fabric 2 (375 g/m²) and fabric 3 (245 g/m²). Fabric 1 is laminated at its back side; fabrics 2 and 3 are coated fabrics. The standard load of 7.35 N was applied and an abradant H18 (nonresilient) vitrified surface that applied medium abrasive action.

It could be observed from Table 11.4 that in the case of fabric 1, the specimen endured surface distortion between 300 and 500 cycles, and rate of loss of mass is higher especially at 500 cycles. At 500 cycles, the top surface of the fabric was lost, leaving the coating exposed. In the case of fabric 2, the rate of loss of mass varied, the fabric surface was distorted and loss of

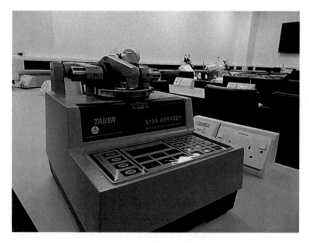

FIGURE 11.7
Taber abrasion tester. (Image courtesy of MMU Textile Lab.)

TABLE 11.4

Percentage Change in Mass

Abrasion Resistance		Fabric 1 (Yellow)	Fabric 2 (Green)	Fabric 3 (Camo)
Initial weight (mg)		2530	3260	2080
Average rate of loss in mass	100 cycles	4.74% (2410 mg)	1.84% (3200 mg)	2.40% (2030 mg)
	200 cycles	5.13% (2400 mg)	3.37% (3150 mg)	2.88% (2020 mg)
	300 cycles	6.32% (2370 mg)	5.52% (3080 mg)	5.05% (1980 mg)
	400 cycles	7.50% (2340 mg)	6.44% (3050 mg)	5.76% (1960 mg)
	500 cycles	7.90% (2330 mg)	8.58% (2980 mg)	6.25% (1950 mg)
	Average	6.32%	5.15%	4.46%

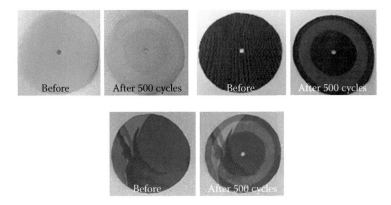

FIGURE 11.8
Taber abrasion test results.

colour was also noted. However, the fabric structure remained unaltered. The thickness of fabric 3 was less compared to remaining samples, and fabric surface was distorted with change in colour and threadbare was also noted at 500 cycles. The samples are illustrated in Figure 11.8, which shows the material at the start and at the end of 500 cycles.

11.3.2 Fabric Pilling

Pilling is a fault commonly observed in knitted woollen goods or fabrics made from soft twist yarns. Pilling occurs when rubbing action in wear causes loose fibres from the surface of yarn coils/loops to form pills on the fabric surface due to brushing up with a surface (washing or daily wear). Formation of fuzz on the fabric surface can be due to

- Brushing of free fibre ends from yarn structure
- Formation of fibre loops into fibre fuzz (Figure 11.9)

FIGURE 11.9
Typical knit fabric with surface fuzz.

In fabrics with synthetic fibre content the pills are stronger and remain on the fabric surface, causing unsightly appearance to the product.

In this method (BS EN ISO 12945-1:2001), fabrics are subjected to constant rotation in a pill box (Figure 11.10) to tease the fibres to form pill and grading it against the original specimen. The method is suitable to a wide range of fabrics. The test specimen is conditioned to a temperature of 20°C ± 5°C

FIGURE 11.10
Pilling box used to measure pilling resistance of fabrics. (Image courtesy of MMU Textile Lab.)

and a relative humidity of 65% ± 2% for at least 6 hours. There are five specimens, each 125 × 125 mm: two for machine direction and two for cross direction of the fabric; an additional specimen is required for assessment. A sample specimen is sewn to form a tube and is mounted on a polyurethane tube as shown in Figure 11.11. At the end of the set number of revolutions (18,000) the test specimen is assessed visually using the pictures as shown in Figure 11.12. Pilling is graded visually using a rating scale where 5 indicates no change and 1 indicates dense fuzz (Table 11.5).

FIGURE 11.11
Preparation of sample for pilling test.

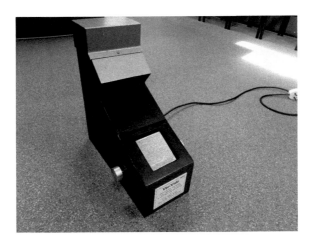

FIGURE 11.12
Pilling assessment photographs. (Image courtesy of MMU Textile Lab.)

TABLE 11.5

Visual Assessment: Fabric Pilling

Rating	Description	Notes
5	No change	No visible change
4	Slight change	Slight surface fuzzing
3	Moderate change	Exhibits fuzzing and/or pills
2	Significant change	Distinct fuzzing and/or pilling
1	Severe change	Dense fuzzing and/or pilling covering specimen

11.4 Fabric Handle in Sportswear

In the fashion industry, a number of personnel handle fabrics to test their suitability to an end use, especially during manufacturing. Fabric handle is an individual's response to touch when a fabric is handled. This could also refer to how a fabric drapes. A number of subjective attributes have been used to refer to fabric handle – for instance, smooth, rough, stiff, soft, crisp, silky, etc. Subjective grading is often not consistent, as what a person perceives as appealing may not be appealing for another person. In this section, various methods to determine the fabric handle are discussed which can be utilised to characterise the fabrics for performance and determine relevance to a particular end use. These are fabric stiffness, fabric drape and crease recovery. Fabric stiffness refers to the ability of the fabric to bend on its own weight. Stiffness refers to resistance to bending. In casual sports tops, the most widely preferred fabric is a single jersey, which is not stiff compared to a tightly woven twill fabric.

11.4.1 Fabric Stiffness

In this method (BS 3356:1990), bending length is measured, in which a rectangular strip of material (200 × 25 mm) is slowly slid on a horizontal grip

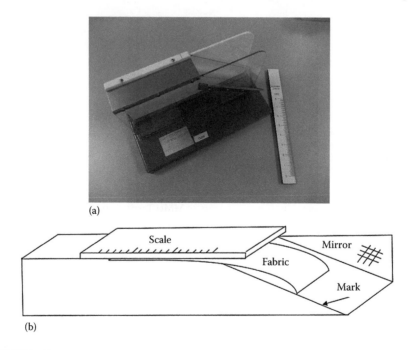

(a)

(b)

FIGURE 11.13
(a) Shirley stiffness tester. (b) Close-up view. (Images courtesy of MMU Textile Lab.)

as shown in the diagram (Figure 11.13a and b) until the edge of the fabric touches the marked line. The length of material protruding beyond the edge is recorded. The test is repeated separately in warp and weft directions over a number of times ($n = 5$) and the average bending length is reported. Generally, a stiffer fabric possesses a greater length compared to a limp fabric. Care should be taken to ensure the test sample is conditioned and the experiment conducted in a standard atmosphere.

Flexural rigidity (G) is calculated using the equation $G = 0.1\ MC^3$, which determines the resistance of fabrics to bending by external forces. M is mass per unit area in grams per square centimetre and C is bending length in centimetres. Flexural rigidity is reported in milligrams per centimetre.

11.4.2 Fabric Drape

Drape is a characteristic of a material to freely fall or hang over a three-dimensional form. This parameter is important to determine fabric handle. In this method (BS 5058:1973), a fabric specimen of 30 cm diameter for medium fabrics (24 cm diameter for limp fabrics; 36 cm diameter for stiff fabrics) is placed on a circular disc (Figure 11.14) and the specimen is allowed to drape on its own weight. Using a light source placed beneath the specimen, a shadow of the draped specimen is cast on a paper ring (Figure 11.14a). The outline of the fabric shadow is traced on the paper (of known mass, W_0) and the traced paper ring is weighed (W_1). Drape coefficient is calculated as the percentage of the total area of the paper ring obtained by vertically projecting the shadow of the draped specimen.

A fabric with a drape coefficient closer to 100% is stiffer, whilst a fabric with drape coefficient closer to zero is pliable and more drapable. In woven fabric, grain affects drape in garments. Figure 11.14b illustrates a

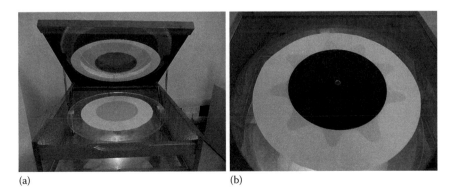

(a)   (b)

FIGURE 11.14
(a) Drape meter. (b) Shadow of the draped specimen. (Images courtesy of MMU Textile Lab.)

plain single jersey fabric whose drape coefficient was 20%, indicating a limp fabric:

$$\text{Drape coefficient (\%)} = \frac{W_1}{W_0} \times 100$$

where W_1 is weight of paper in the shadow region (g) and W_0 is weight of paper ring (g).

The experiment is repeated for six specimens and the average is reported and is carried out in standard atmospheric conditions (BS 1051) (i.e. a relative humidity of 65% ± 2% and a temperature of 20°C ± 2°C). Limp fabrics are difficult to handle during the garment manufacture, mainly during fabric laying-up, spreading and stitching. Stiff fabrics are difficult to form body form shapes, particularly shoulder, armhole, etc.

11.4.3 Fabric Stretch and Recovery

The importance of stretch and recovery was highlighted in Chapter 7. During wearing and removal of knitwear or certain sportswear, stretch is an important parameter. BS 4294:1968 is commonly used to determine the stretch and recovery of woven and knit fabrics using the apparatus Fryma fabric extensiometer. It consists of two jaws (75 mm wide) – a fixed and a movable jaw capable of holding the specimen without slippage (see Figure 11.15). The apparatus is capable of applying a load of 6 kg. The conditioned test specimen is prepared to 200 × 75 ± 1 mm in both warp and weft directions for woven fabrics and course and weft directions for knit fabrics. The number of test specimens is usually five in both directions.

For most knitted fabrics, the distance between the inner edges of two clamps is set at 75 mm (L_1) and 200 mm for woven fabrics. A reference 'mark' is made on the fabric at the inner edges of the clamp using a marker. The load (e.g. 6 kg for knits) is applied slowly for 10 seconds and immediately reduced within 7.5 seconds; the test specimen returns to the original position. Immediately, the load is reapplied for 1 minute and the stretch of the fabrics is recorded (L_2). Now the load is reduced and clamps are brought to the original position. The test specimen is removed from the apparatus and allowed to remain flat for 1 minute and the distance between the two reference marks is now recorded (L_3). Using the preceding values, the stretch properties of the fabric are calculated:

$$\text{Mean extension (\%)} = 100 \times \frac{L_2 - L_1}{L_1};$$

$$\text{Mean residual extension (\%)} = 100 \times \frac{L_3 - L_1}{L_1}$$

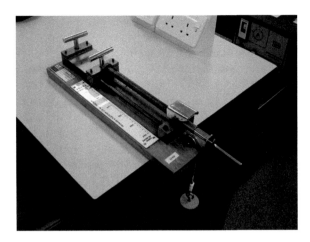

FIGURE 11.15
Fryma extensiometer. (Image courtesy of MMU Textile Lab.)

On a general note, fabrics that offer stretch are expected to recover within 3% of the original extension (Saville, 1999). The stretch and recovery properties of fabrics are essential whilst designing garments for sportswear, particularly those that can stretch either in one direction (warp or weft) or in both the directions. In addition, it is very important to comprehend the amount of stretch required for different body movements. Voyce, Dafniotis and Towlson (2005) explained how a person's skin stretches considerably with some key areas of stretch including 35% to 45% at knees and elbows and 13% to 16% at the shoulder back; with sporting activities increasing such numbers, stretch of sportswear apparel is a key element for comfort. Normal body movement expands the skin by 10% to 50% and strenuous movements in sports will require least resistance from garment and instant recovery. Hence, it is vital that stretch and recovery values are taken into account while creating fabric panels for garment manufacture, particularly those that are intended for form-fitting gear for swimming, cycling, etc. For a compression top made of warp knit structure with fibre composition of 63% nylon, 23% polyester and 14% elastane, the typical stretch and recovery values are stretch in course direction – 95%; wale direction – 94%; recovery – 96% in course direction and 95% in wale direction (Allsop, 2012). This indicates that the fabric can provide uniform stretch in both directions, which is ideal for compression sportswear.

11.5 Measurement of Fabric Comfort

Athletes perceive comfort as an important factor while making a purchase or selecting a garment for training or an event. During intensive sport, the core temperature of the human body changes and perspires to balance the excessive heat (Pocock and Richards, 2009). The most important factor while designing sportswear is moisture management and wicking. Performance of fabrics is affected by a combination of fibre properties and composition, yarn formation and fabric structure (see Chapter 3).

In sportswear, particularly athletic apparel, fabrics chosen should have quick-drying and good wicking properties to handle excess sweat produced by the body. Zhou, Yu and Qin (2014) highlighted that fibres with greater surface area had better wicking. They reported the comfort properties of six interlock knitted fabrics made from various blends of chemically and physically modified polyethylene terephthalate (PET) fibre with trefoil cross section and cotton fibres with trefoil cross section and cotton fibres. Modified polyester fibre had enhanced moisture management, particularly 35% PET and 65% cotton blend knitted fabric.

11.5.1 Wicking and Its Effect on Fabric Comfort

Wicking is the movement of liquid by capillary action, provided that the liquid wets the assembly of fibres so that it can move from its source to some distance against the gravitational forces by occupying the available capillary spaces. The smallest capillary possesses greater capillary forces, hence the wickability. Wicking depends upon the surface properties of fibres, surface area, density, thickness and the capillary path through the fabric. Moreover, the rate of wicking is different along warp (wale) than weft (course) direction.

11.5.2 Moisture Management Tester

The moisture management tester (MMT) was developed by Hong Kong Polytechnic University and SDL Atlas Textile Testing Solutions to determine the dynamic liquid transport properties of knitted and woven fabrics. The test method depends on the change in the contact electrical resistance of fabric during moisture transport (Yao et al., 2008). The apparatus is used to measure the liquid transport in multiple directions. MMT consists of upper and lower concentric moisture sensors (Figure 11.16). The predefined amount of saline solution (simulated sweat) is introduced onto the upper side of the fabric, which allows the saline to spread onto the fabric outer surface, through the fabric and lower surface of the fabric. The following six moisture management properties are measured:

1. Wetting time for top (WT_t) and bottom (WT_b) surfaces in seconds
2. Absorption rate for top (AR_t) and bottom (AR_b) surfaces of the fabric in percentage per second
3. Maximum wetted radius (mm) for top (MWR_t) and bottom (MWR_b) surfaces of the fabric in millimetres

FIGURE 11.16
Moisture management tester (MMT). (Image courtesy of MMU Textile Lab.)

4. Spreading speed for top surface (SS_t) and bottom surface (SS_b) in millimetres per second
5. Accumulated one-way transport capacity in percentage
6. Overall moisture management capacity (OMMC)

MMT is connected to a personal computer and software records the moisture transport through the fabric and produces various illustration including water content versus time, water location versus time (Figure 11.17) and fingerprint of moisture management properties (Figure 11.18). McQueen et al. (2013) reported a development of a protocol to assess the fabric suitability for testing moisture management properties using MMT.

MMT grades fabric for six indices and classifies fabrics as waterproof fabric, water-repellent fabric, slow-absorbing and slow-drying fabric, fast-absorbing and slow-drying fabric, fast absorbing and quick-drying fabric, water-penetration fabric and moisture-management fabric (See fingerprint.). The moisture management equipment has been recognised by the American Association for Textile Chemists and Colorists (AATCC) in its test method AATCC 195:2009.

MMT is suitable to determine the liquid transport in multiple directions, and fingerprint (see Figure 11.18) is a useful and instant method to identify the moisture management of fabrics. For instance, in the sample as shown in Figure 11.18, the top wetting time is *very good* (grade 4), absorption rate is *good* (grade 3), top surface wetting radius is *very good* and spreading speed was between grades 3 and 4, indicating good wicking capacity. It could be noted that the top surface wicks and transports moisture better than the bottom surface. The one-way transport of liquid was also *excellent*. The fingerprint also provides the overall moisture management as *very good* (grade 4). It could be observed that the fabric is suitable for moisture management, where

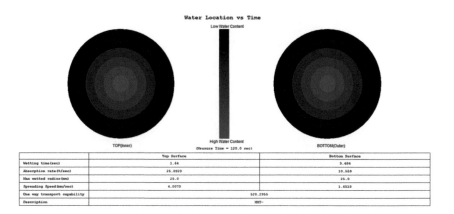

FIGURE 11.17
MMT output – water location versus time. (Image courtesy of MMU Textile Lab.)

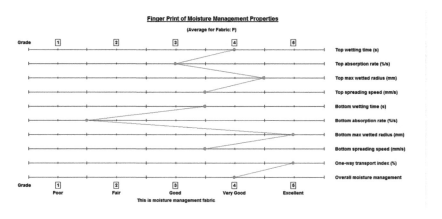

FIGURE 11.18
Fingerprint for moisture management properties. (Image courtesy of MMU Textile Lab.)

the top surface of the fabric wicks moisture slightly more than the bottom surface of the fabric. The surface of the fabric that wicks away the moisture quickly would be in contact with the skin and helps to prevent stickiness due to excess perspiration during intense activity.

11.5.3 Influence of Moisture Transfer in Functional Clothing (Permetest)

Permetest was developed by Sensora Instrument, Czech Republic, which measures water vapour resistance, relative water vapour permeability and thermal resistance of woven, knitted and nonwoven fabrics (Hes, 2014). The equipment works on ISO 11092 standard that measures the thermal and water vapour resistance under steady-state conditions (sweating guarded hot plate test). Bogusławska-Bączek and Hes (2013) recently presented the working principle of Permetest as a skin model that provides a reliable measurement of the water vapour permeability of fabrics in dry and wet states. Permetest is often called a 'skin model', where the instrument measures the water vapour permeability of fabrics. This instrument simulates the wet and dry human skin conditions in terms of thermal feeling and serves for the determination of water vapour permeability of fabrics (Figure 11.19). The instrument works on the principle of heat flux sensing. When water flows into the measuring head, some amount of heat is lost. This instrument measures the heat loss from the measuring head due to the evaporation of water in bare conditions and while being covered by the fabric. The relative water vapour permeability of the fabric sample is calculated by the ratio of heat loss from the measuring head with and without fabric. Das et al. (2009) reported the water vapour permeability of blended fabrics using Permetest.

The results are presented in the equipment as well as via software connected to a PC. The image in Figure 11.20 illustrates a typical output from the Permetest. The instrument also allows recording the test and calculating simple statistical test, mean, median and standard deviation. The relative water vapour permeability (RWVP) is determined using the following formula:

$$\text{RWVP} = \frac{\text{Heat lost with the fabric placed on measuring head } (u_1)}{\text{Heat lost with bare measuring head } (u_o)} \times 100$$

Table 11.6 highlights the results of widely used sports fabrics (base layer, fleece, coated fabric and outer layer) including dry thermal resistance, relative water vapour permeability and water vapour resistance.

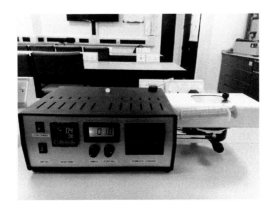

FIGURE 11.19
Permetest. (Image courtesy of MMU Textile Lab.)

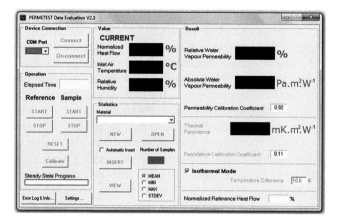

FIGURE 11.20
Typical output from Permetest.

TABLE 11.6

Outcomes from Permetest

Type of Material	Water Vapour Resistance (m² Pa W⁻¹)	Relative Water Vapour Permeability (%)	Dry Thermal Resistance (km²W⁻¹)
Base layer knit 1 (skins)	3.7	72.8	2.8
Base layer knit 1 (SUD)	4.0	75.3	2.8
Fleece	7.3	57.1	7.9
Coated fabric (grey woven fabric)	8.5	54.0	Not applicable as the fabric is very thin
Outer layer	4.2	67.1	Not applicable as the fabric is very thin

11.6 Fabric Specifications and Interpreting Results

Table 11.7 presents the experimental values relating to fabric's physical characteristics, durability, aesthetics and fabric comfort, which are some of the vital prerequisites for a fabric to find its usage in sportswear and performance clothing. The woven material made of 100% polyester is a lightweight fabric intended for midlayer soft shell jackets that is water repellent, whilst the knitted fabric is medium weight fabric with a water-repellent finish intended to be used as tops for men and women. The woven fabric has a 1/1 plain weave structure in which both sides of the fabric appear similar,

TABLE 11.7

Example Fabric Specification

	Properties	01 Woven Fabric		02 Knitted Fabric	
Fabric physical characteristics	Area density (g/m²)	115		255	
	Bulk density (g/cm³)	0.64		0.39	
	Thickness (mm)	0.18		0.66	
	Fabric cover (K) factor/ stitch density	13.5 + 6.3		11.0	
	Fabric structure	1/1 plain weave		Single jersey	
	Fibre composition (%)	100% polyester		Polyester/Lycra	
	Yarn count (tex)	Warp 5.0 Weft 12.0		20.0	
Fabric density	Woven fabric (ends per inch × picks per inch)	100 × 72		NA	
	Knitted fabric (courses per cm and wales per cm)	NA		22 × 15	
Durability	Fabric pilling (grade)	5		4	
	Abrasion resistance @ 10,000 — Change in colour	Very minor change		Slight increase in shade	
	Rate of loss of mass	Nil		0.01	
	Threadbare	No		No	
Aesthetics	Fabric handle stiffness	Warp	Weft	Wales	Courses
	Flexural rigidity (μNm)	12.0	4.3	2.8	0.5
	Bending modulus (N/m²)	0.4	0.1	0.1	0
	Fabric drape coefficient (%)	55.3		17.41	
Comfort	Moisture transport (wicking)	Waterproof		Water-repellent fabric (see fabric fingerprint)	
	Stretch and recovery	Not applicable		Wale	Courses
	Mean extension			79%	146%
	Mean residual extension			2.66%	6.75%
	Water vapour resistance (Pa m²/W)	11.7		4.3	
	Absolute water vapour permeability (%)	45.5		65.3	
	Thermal resistance (m² K/W)	Not applicable due to fabric thickness <0.2 mm		53.0	

and the knitted fabric is a single jersey, which possesses a distinct technical face and back. It is made of polyester and Lycra. Fabric thickness of woven fabric is less than 0.20 mm whilst the knitted fabric is 0.66 mm. Fabric thickness plays an important role during joining of fabrics, especially in maintaining the pressure at the presser foot of a sewing machine. Fabric cover factor determines the extent to which a set of yarn covers the area of a fabric.

The warp cover factor 13.5 and weft cover factor 6.3 indicate that the area covered by warp yarns exceeds weft yarns. In the case of the knitted fabric, the tightness factor indicates the extent to which the area of the knitted fabric is covered by the yarn. Generally, yarn count affects the tightness factor of a knitted fabric. The tightness factor of 11.0 indicates the fabric has an optimal level of closeness. In the case of woven fabric, warp yarns are finer (5 tex) than the weft yarn (12 tex). In the case of knitted fabric, yarn count is of medium quality (20 tex). The fabric count for woven fabric of 100 × 72 indicates the fabric is an unbalanced weave, where warp yarns (100) exceed the weft yarns (72). In the case of knitted fabric, 24 courses per centimetre × 15 wales per centimetre indicates that the fabric has moderate closeness or compactness. This factor affects the ability of the fabric to transmit moisture between the skin and the environment.

Fabric pilling grade 5 reveals that the fabric performed well, with little or no surface fuzz; in the case of knitted, there was minor surface fuzz (pilling grade 4). Fabric abrasion resistance reveals that at 10,000 rubs there were minor colour changes with no change in the mass and threadbare. The abrasion resistance for knitted fabrics was good, with little change in mass and no threadbare and slight increase in shade. Fabric durability depends on fibre type as well as on fabric structure and yarn fineness. The drape coefficient of 55% indicates that the woven fabric possesses medium drape, such that it is neither a stiff nor a flexible fabric. However, in the case of knitted fabric, the drape coefficient of 17.4% indicates that the fabric is pliable. In the case of fabric stiffness, bending length is observed from which flexural rigidity is calculated. The flexural rigidity is a measure of stiffness that depends on fabric thickness, yarn count, fabric structure and finishes applied to the fabric which make the fabric compact. Flexural rigidity of 12.0 µNm in warp direction and 4.3 µNm in weft direction indicates that the fabric possesses stiffness in the lengthwise direction compared to the width-wise direction. In the case of knitted fabric, the flexural rigidity of 2.8 µNm indicates that the fabric possesses little stiffness; in other words, the fabric is pliable. The moisture assessment using MMT fabric print indicated that woven fabric is a waterproof fabric (as there was no wicking) and knitted fabric was water repellent due to its specific finish. Knitted fabric possessed good stretch in the width-wise direction (146%) compared to the lengthwise direction (7%), which is an important parameter, especially in designing base layer garments worn next to the skin that require stretch and recovery – for instance, in the seat area, knee flex and back arm flex. It is also important to note that fabric growth after extension is 2.6% in the wale direction and 7% in the course direction. This should be considered in designing products as close-fit garments such as tops for women that may become baggy after repeated usage that involves intense body stretch (yoga practice or aerobics). In the case of comfort assessment – water vapour permeability – the compact nature of the woven fabric resulted in average permeability (45%) compared to knitted fabric, which had a better permeability (65%). The water vapour resistance for woven fabric was 11.7 (Pa m²/W) and 4.3 (Pa m²/W) for knitted fabric. This meant that knitted fabric is comfortable to wear as the moisture permeability is marginally better than for the woven fabric. Thermal resistance of woven fabric could not be assessed as the fabric thickness was less than 0.20 mm.

Based on these findings, it could be inferred that the knitted fabric is durable and it is suitable for active wear applications as it offers good resistance to pilling and abrasion. In addition, the fabric is flexible, offers stretch in a width-wise direction and is water repellent, which can be used for outdoor wear. Fabric possesses good moisture permeability characteristics which enable the wearer to remain comfortable during an activity. In the case of the woven fabric, the fabric is suitable for soft shell jackets which offer resistance to wear and tear and have average moisture permeability. Both the fabrics can be recommended for performance applications targeted to a low to medium market where the garment usage is less frequent.

11.7 Summary

The sportswear and functional clothing sector drives innovation, particularly in the area of fabric and accessories. Textile testing has been instrumental in determining the performance of these innovative high-performance materials. In addition, different types of sporting activities require different performance and the choice of fabrics varies. For example, an outdoor cycling kit requires a fabric that is lightweight, possesses stretch and offers thermal balance next to the skin. However, ski wear requires good thermal insulation to protect the wearer from severe cold conditions. The chapter highlighted the importance of fabric evaluation in determining the fit for purpose using various textile parameters including physical characteristics, durability, aesthetics and comfort. Each test parameter was referred to in British standards and a brief description of the test method was presented along with the visual illustration of the test equipment. In addition, example results were also presented to enable the reader to understand the outcomes and their relevance in fabric assessment (grade). The chapter also emphasised the importance of physical characteristics of fabric on its performance. In the case of performance assessment, various test equipment, including a Taber abrasion tester, pilling box, stiffness tester, drape meter, moisture management tester, Fryma extensiometer and Permetest, was discussed. The final section, which outlined the fabric specification with example results between woven and knitted fabrics, will enable the reader to interpret test results and comprehend how performance is assessed using the outcomes. The test methods discussed in the chapter were presented in the context of evaluating fabrics used in performance clothing and will serve as an invaluable resource to professionals and novices alike.

References

Allsop, C. A. (2012). An evaluation of base layer compression garments for sportswear, MSc dissertation, Manchester Metropolitan University, Manchester, UK.

Anand, S. C. (2000). Technical fabric structures – Knitted fabrics. In *Handbook of technical textiles*, Anand, S. C. and Horrocks, A. R. (eds.). London: Woodhead Publishing.

Boguslawska-Baczek, M. and Hes, L. (2013). Effective water vapour permeability of wet wool and blended fabrics. *Fibres and Textiles in Eastern Europe* 21 (1): 67–71.

BS 3356. (1990). Method for determination of bending length and flexural rigidity of fabrics.

BS 4294. (1968). Methods of test for stretch and recovery properties of fabrics, BSI, London.

BS 5058. (1973). Method for the assessment of drape of fabrics.

BS EN ISO 5084. (1997). Determination of thickness of textiles and textile products.

BS EN ISO 5470-1. (1999). Determination of abrasion resistance of fabrics: Taber abrader.

BS EN ISO 12945-1. (2001). Determination of fabric propensity to surface fuzzing and to pilling – Pilling box method.

BS EN ISO 12947-2. (1998). Textiles. Determination of the abrasion resistance of fabrics by Martindale method.

Cohen, A.C., Johnson, I., Price, A. and Pizzuto, J.J. (2010). J.J. Pizzuto's Fabric Science, Fairchild, New York.

Collier, B. J. and Epps, H. H. (1999). *Textile testing and analysis*. Upper Saddle River, NJ: Pearson, Prentice Hall Inc.

Das B., Das, A., Kothari et al. (2009). Moisture flow through blended fabrics – Effect of hydrophilicity. *Journal of Engineered Fibres and Fabrics* 4 (4): 20–28.

Hes, L. (2014). Manual for Permetest, Sensora Instruments, Czech Republic, (unpublished).

Mandal, S. and Abraham, N. (2009). Effect of fabric weight on bending rigidity – A statistical analysis. *Man-Made Textiles in India* 52: 161–163.

McQueen, R. H., Batcheller, J. C., Mah, T. and Hooper, P. M. (2013). Development of a protocol to assess fabric suitability for testing liquid moisture transport properties. *Journal of the Textile Institute* 104 (8): 900–905.

Özdil, N., Kayseri, Ö. G. and Mengüç, G. S. (2012). Analysis of abrasion characteristics in textiles. In *Abrasion resistance of materials*, Adamiak, M. (ed.). Intech, http://www.intechopen.com.

Pocock, G. and Richards, C. D. (2009). The regulation of body core temperature. In *The human body: An introduction for the biomedical and health sciences*. Oxford, UK: Oxford University Press.

Roy, P. (2014). Making products that protect. *AATCC News*, US.

Saville, B. P. (1999). *Physical testing of textiles*. Oxford, UK: Woodhead Publishing and Textile Institute.

Voyce, J., Dafniotis, P. and Towlson, S. (2005). Elastic textiles. In *Textiles in sport*, Shishoo, R. (ed.). Cambridge, UK: Woodhead Publishing.

X-BIONIC. (2014). *X-BIONIC manual for automobili Lamborghini.*

Yao, B., Li, Y. and Kwok, Y. (2008). Precision of new test method for characterising dynamic liquid moisture transfer in textile fabrics, AATCC review, pp. 44–48.

Zhou, R., Yu, J. and Qin, J. (2014). Comfort properties of modified knitted polyester fabrics. *Melliand International* 20: 38–41.

12

Application of Pressure Sensors in Monitoring Pressure

David Tyler

CONTENTS

12.1 Introduction 289
    12.1.1 The Challenge of Measuring Pressure 289
    12.1.2 Units of Pressure 291
12.2 Pressure Sensors for Medical Applications 292
    12.2.1 Compression Hosiery: The Hatra Hose Pressure Tester 292
    12.2.2 Compression Hosiery: The Medical Stocking Tester 292
    12.2.3 The Kikuhime Tester 293
    12.2.4 Overview of Other Test Instruments 293
        12.2.4.1 The Oxford Pressure Monitor MkII 294
        12.2.4.2 The Talley Skin Pressure Evaluator 294
    12.2.5 Evaluation of Pressure Sensing Instruments 296
    12.2.6 The PicoPress Instrument 297
12.3 Pressure Sensors for Clothing Applications 298
    12.3.1 Use of Medical Instruments 298
    12.3.2 Tekscan Technologies (I-Scan® System) 299
    12.3.3 Tekscan Technologies (FlexiForce) 302
12.4 Discussion of Laplace's Law 303
12.5 Summary and Conclusions 306
Acknowledgements 307
References 307
Further Information 309

12.1 Introduction

12.1.1 The Challenge of Measuring Pressure

Monitoring pressure distribution using probes and sensors to ascertain the performance of a wide range of products in medical and clothing compression wear is important for understanding the efficacy of products.

The technology challenge is substantial, because surfaces are 3D contoured and deformable. Textiles can stretch and recover according to their construction and fibre type, and human bodies are covered in skin, below which are various permutations of fat and bone.

Pressure is a term that describes the force applied per unit area. The equation that allows quantitative measurement of pressure is as follows:

$$P = F/A \qquad (12.1)$$

where P = pressure, F = applied force and A = area affected by the applied force.

When an object (like a part of the human body) is in contact with a stretch fabric (a bandage or a compression garment), it experiences a compressive force. According to Equation 12.1, the average interface pressure is the total force divided by the interface area. However, the average pressure is only part of the story. The human body is not a smooth cylinder, but a complex surface of extensible skin under which are soft tissues and rigid bones. Furthermore, stretch fabrics are not simple materials to understand, as they have different stretch properties in different directions and exhibit the phenomenon of relaxation after extension. Consequently, localised interface pressure measurement is necessary to assess the distribution of pressure and to find concentrations of peak pressure.

Pressure measurement technologies are designed to map the location and magnitude of peak pressures and to gain information about pressure gradients across interfaces. To handle exponential increases in information gathered, computerised systems have been developed to analyse the data and provide visual representations of the interface being studied.

For medical products, there are numerous tools used for the measurement of compression. For compression hosiery, the Hatra Mk2A+ Hose Pressure Tester and the Salzmann MST Professional have been developed. For other applications, the Kikuhime® tester and the PicoPress® instruments are widely used. These are described in Section 12.2 (with brief mentions of other technologies).

At a research level, numerous additional sensors have been used for medical products as well as for clothing. The instruments are constantly changing, but emphasis is given in Section 12.3 to the use of Tekscan pressure sensors, including the FlexiForce® interface pressure sensors.

In medical contexts, where compression is applied frequently to limbs (which have cylindrical body forms), reference is often made to a variant of Equation 12.1, known as Laplace's law:

$$P \propto T/R \qquad (12.2)$$

where P (pressure) is directly proportional to T (tension) divided by R (radius).

This equation is the basis for data processing in the British Standard for compression hosiery (BS 6612, 1985). The medical background for compression bandages and stockings is summarised in Rotsch et al. (2011).

Laplace's law means that the smaller the radius (with constant tension), the higher is the compression pressure. Since the human leg is smaller in diameter nearer the ankle and larger nearer the knee, if bandages are wrapped at a constant tension, there will be a pressure gradient (known as graduated compression) with maximum pressure at the ankle and reduced pressure toward the knee. This graduated compression is considered to accelerate the venous flow rate, with medical benefits to the patient.

Equation 12.2 also suggests a potential problem when the radius is small. A pressure measuring device that has a thickness of a few millimetres has the potential of distorting the radius locally, thereby distorting the compression pressure locally. Questions have been raised about the accuracy of some instruments because of this effect.

12.1.2 Units of Pressure

Pressure is defined as force divided by area (with the Laplace law being a special case of this). The international system (SI units) recognised the pascal as the unit of pressure. Physicists have defined one pascal (Pa) as the pressure exerted by a force of one newton applied over an area of one square metre. The SI unit of pressure honours Blaise Pascal as a pioneering seventeenth century French scientist who made significant contributions relating to understanding pressure.

One pascal represents a low pressure, and there are many applications where other units are deemed more appropriate, sometimes for historical reasons. There are numerous metric and imperial units that were in common use before SI units were defined, and they continue to be employed. Examples of metric units are kilograms force per square metre (kgf/m^2) or grams force per square cm (gf/cm^2). An imperial unit of pressure is pounds per square inch (psi). Some important additional units of pressure in common use are the torr, millimetres of mercury (mmHg) and bar.

The torr is a unit honouring the seventeenth century Italian physicist Evangelista Torricelli, who invented the mercury barometer and was the first to explain the concept of atmospheric pressure. He found that the column of mercury in a barometer positioned at sea level measured 760 mm. One torr is the pressure needed to sustain 1 mm of mercury (Hg) in a barometer, so 1 torr is 1 mmHg. Most pressure-measuring medical instruments are calibrated in mmHg units. One torr is approximately 33 Pa.

One bar represents the mean atmospheric pressure at sea level. It is common to use this unit when referring to the pressure of water at depth (with reference to diving, for example). It is now defined as 100 kPa. Meteorological charts normally use hectopascals (hPa), where 1 hPa = 100 Pa and 1 bar = 1000 hPa.

12.2 Pressure Sensors for Medical Applications

12.2.1 Compression Hosiery: The Hatra Hose Pressure Tester

During the 1970s, a tool for measuring the properties of compression hosiery was developed by Derek Peat at the Hosiery & Allied Trades Research Association (Hatra, Nottingham, UK). The garment is stretched lengthways and widthways in a defined manner in a range of sizes. A measuring head utilises a strain gauge to record the compression provided at any position from the ankle upward. The head has a rectangular plate (25 mm wide) that is pushed onto the stretched hose and the resistive forces are recorded. The equipment provides reproducible test data and was incorporated into British Standard 6612 in 1985. The Hatra tester was also adopted by two other British Standards: BS 7672:1993 and BS 7563:1999. The MK2 Hatra was available before 1990, after which the Mk2A was released, allowing tailored leg profiles to be easily added. The current model is the Mk2A+, illustrated in Figure 12.1.

12.2.2 Compression Hosiery: The Medical Stocking Tester

In 1977, the first medical stocking tester (MST) was launched by Dr A. A. Bolliger in Switzerland. The concept is similar to the Hatra tester. The main difference is that the compression stocking is placed on a leg-shaped former and a flat measuring device (40 mm wide, 0.5 mm thick, linked to an air pump and a pressure transducer) is used to quantify the compression forces. A separate former is needed for each size to be tested. This tool has also been developed over time, and the current model is Mk V. Alongside this, the MST

FIGURE 12.1
The Hatra Mk2A+ hose pressure tester. (Courtesy Segar Technology.)

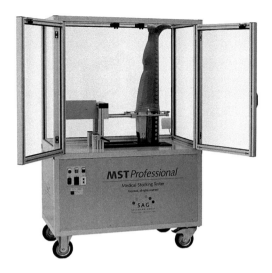

FIGURE 12.2
Salzmann MST Professional. (Courtesy Swisslastic AG.)

Professional has a variable leg form that is claimed to cover 95% of known leg sizes. The measuring probe has the capability of measuring the compression exerted by stockings worn by live subjects, which means it can be used additionally as a research tool. The instrument is illustrated in Figure 12.2 and an example of its use in research for both in vivo and in vitro measurements is provided by Liu, Lao and Wang (2013).

12.2.3 The Kikuhime Tester

The sensor used in the Kikuhime instrument is an oval polyurethane balloon containing a 3 mm thick foam sheet. This is connected to a syringe (for changing the air pressure within the balloon) and a measuring unit (with the pressure transducer). Testing starts by adjusting the syringe so that the balloon is at atmospheric pressure, and then it is placed in position (for example, between the leg and the compression garment). The transducer monitors the pressure experienced by the balloon and the output is a digital display (Figure 12.3). An assessment of the reproducibility and reliability of measurement was undertaken by Brophy-Williams et al. (2014), who concluded that the tester was suitable for use with sports compression garments.

12.2.4 Overview of Other Test Instruments

Numerous other instruments can be found in the literature and there are many new ideas coming to the fore each year. A survey of the field was undertaken at an international consensus meeting of medical experts and representatives from industry held in January 2005 in Vienna, Austria. In

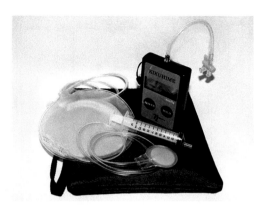

FIGURE 12.3
The Kikuhime tester. (Courtesy TT Meditrade and mediGroup Australia Pty Ltd.)

vivo measurement of interface pressure was considered highly desirable, and methods for measuring the interface pressure were considered with that in mind. Table 12.1 presents the different technologies considered.

It is necessary to point out that many of these instruments are no longer available. There are many design concepts that have been explored, but finding commercial viability has not been easy. The first two entries listed in Table 12.1 are described in the following two sections, as these have had a more significant impact in the literature. The Talley Group designed pressure sensors in the 1980s but production ceased around the year 2000. The company no longer offers instruments to measure pressure, but has concentrated on producing mattresses, cushions and other products providing pressure relief and compression therapy.

12.2.4.1 The Oxford Pressure Monitor MkII

This instrument was designed to monitor pressures between skin tissue and support media for chair- or bedbound individuals. The name originates because of collaboration with the Oxford Orthopaedic Centre. Multiple sensors were used to enable the simultaneous monitoring of pressures below a reclining patient. Each sensor was constructed from two thin plastic sheets that could be inflated by a pulse of air, allowing a measure of the compression forces. The MkII had 12 sensors, and a later device, the MkIII, was equipped with 96 sensors and also marketed as the Talley IPM (Interface Pressure Monitor).

12.2.4.2 The Talley Skin Pressure Evaluator

The Talley SD.500 Skin Pressure Evaluator was designed as a portable device for checking tissue pressures in medical wards, wheelchair pressures and

TABLE 12.1

Types of Interface Pressure Sensors

Pneumatic, Pneumatic-Electric or Pneumatic-Piezoelectric
Oxford pressure monitor (Talley Ltd, Ramsey, Hampshire, UK)
Talley pressure evaluator (Talley Ltd)
MST MKIII Salzmann (Salzmann Medico, St Gallen, Switzerland)
Digital interface pressure evaluator (Next Generation Co., Temecula, CA)
Scimedics pressure evaluator pad (Vista Medical, Winnipeg, MB, Canada)
Kikuhime (Meditrade, Soro, Denmark)
Juzo tester (Elcat, Wolfratshausen, Germany)
Sigat tester (Ganzoni-Sigvaris, St. Gallen, Switzerland)

Piezoelectric
MCDM-I (Mammendorfer Inst. Physik, Munich, Germany)

Fluid Filled, Fluid Filled Resistive
Strathclyde pressure monitor (University of Strathclyde, Scotland)
FlexiForce (Tekscan, South Boston, MA)
Skip air pack analyzer (AMI Co., Japan)

Resistive and Strain Gauge
FSR, FSA (Vista Medical, Winnipeg, MB, Canada)
Fscan, Iscan (Tekscan, South Boston, MA)
Rincoe SFS (Rincoe and Associates, Golden, CO)
MCDM (Mammendorfer Inst. Physik, Munich, Germany)
Fontanometer (Gaeltec Ltd, Dunvegan, Isle of Skye, Scotland)
Diastron (Diastron Ltd, Andover, Hampshire, UK)

Capacitive
Kulite XTM190 (Kulite Semiconductor Products, Leonia, NJ)
Precision (Precision Measurement Co., Ann Arbor, MI)
Xsensor (Crown Therapeutics, Belleville, IL)
Pliance (Novel, Munich, Germany)

Source: Partsch, H. et al. 2006. *Dermatologic Surgery* 32 (2): 227.

any scenario where tissue trauma is an issue. There were two main parts: a handheld control unit with a digital display of pressure and a balloon type sensor that could be inflated manually. The sensor contained platinum wires on both sides of its inner surface. To obtain a measurement, the sensor was placed between the skin and clothing before inflating using the pump bulb. As the sensor inflated, the electrical contact between the two sets of platinum wire was broken. As air was allowed to flow out of the sensor, the platinum wires touched and the circuit was reconnected. At this point, the pressure was recorded and displayed on the control unit.

12.2.5 Evaluation of Pressure Sensing Instruments

A detailed comparison of three instruments was undertaken by Flaud, Bassez and Counord (2010). They selected the Salzmann, Talley SD.500 and Kikuhime testers and compared their performance in terms of accuracy, repeatability and sensitivity to flexion on a curved surface. The first set of tests utilised a chamber that could be preset to defined pressures. The second set used a wooden leg model and inserted the sensors between the leg and compression stockings of known pressure. The results for the pressurised chamber tests are reproduced in Figure 12.4.

The equations of lines fitted to the data points are as follows:

$$\text{Salzmann:} \quad y = 0.99x + 3.32 \ (R^2 = 0.99)$$

$$\text{Talley:} \quad y = 0.97x + 0.47 \ (R^2 = 0.99)$$

$$\text{Kikuhime:} \quad y = 0.93x + 0.57 \ (R^2 = 0.99)$$

The researchers summarised their results in this way:

> In a pressurised chamber, the three systems gave linear responses and an overall error of 15.4%, 3.1%, and 4.3% for Salzmann, Talley, and Kikuhime, respectively. The repeatability error was less than 0.6 mmHg. On the leg model, the overall errors differ between the systems. Repeatability was comparable between the sensors. (Flaud et al., 2010, p. 1930)

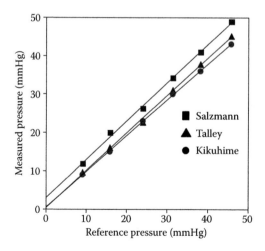

FIGURE 12.4

Measured pressures versus reference pressures for the three sensors in the pressurised chamber. (From Flaud, P. et al. 2010. *Dermatologic Surgery* 36 (12): 1930–1940.)

TABLE 12.2

Some Advantages and Disadvantages of Sensors

	Advantages	Limitations
Pneumatic transducers	Thin and flexible probes; cheap, easy, and handy	Dynamic measurement is only possible with additional special equipment; sensitive for temperature and hysteresis
Fluid filled	Flexible; dynamic measurements	Thick when filled; problems during motion
Resistance	Thin sensors; dynamic measurement	Sensitive to curvature; stiff and thick; not useful for long-term measurements

Source: Partsch, H. et al. 2006. *Dermatologic Surgery* 32 (2): 227.

This shows the sensors were capable of providing useful tools for medical practitioners. However, care must always be taken with stockings and bandages around legs and arms because material variability and user factors may introduce variability that is difficult to control.

Table 12.2 has some generalised comments on the advantages and limitations of different types of sensors.

12.2.6 The PicoPress Instrument

With the passing of time, the instruments listed in Table 12.1 require extensive editing. There are both additions and deletions. In particular, a new product designed for medical applications is worthy of note: the PicoPress produced by Microlab (Padua, Italy). The instrument has a manometer connected to a probe: a flexible circular plastic bladder (5 cm diameter). The bladder is placed in position in the deflated state and the bandage/compression garment is applied. To take measurements, the operator pushes on an embedded syringe to introduce 2 cm^3 of air to the bladder. The resultant expansion in thickness is constrained by the compression exerted by the bandage/garment and the manometer is used to record the pressure. Both static and dynamic measurements are possible, so the compression can be determined when resting, when walking and when standing. After collecting data, the probe is deflated and left in position until further readings are required. The PicoPress is shown in Figure 12.5. Partsch and Mosti (2010) evaluated this instrument by comparing its performance with two other commercial systems. They concluded: 'The results suggest that the PicoPress transducer, which also allows dynamic pressure tracing in connection with a software program and which may be left under a bandage for several days, is a reliable instrument for measuring the pressure under a compression device'.

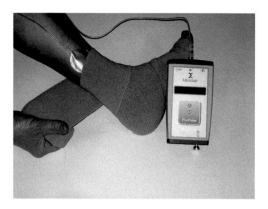

FIGURE 12.5
The PicoPress instrument. (Courtesy Microlab Elettronica.)

12.3 Pressure Sensors for Clothing Applications

12.3.1 Use of Medical Instruments

Several of the medical instruments discussed earlier have been used for measuring compression provided by clothing.

Chan and Fan (2002) used the Tally SD.500 skin pressure evaluator to clarify the relationship between subjective tightness sensation and the clothing pressure of girdles. This was followed by modelling work to predict the pressure of girdles on the human body (Fan and Chan, 2005).

Pressure garments are widely used in the treatment of skin damage caused by burns. Giele et al. (1997) expressed concern that devices like the Talley and Kikuhime testers were inadequate for three reasons:

1. Distortion of the garment, hence raising garment tension and increasing the pressure generated
2. Poor conformity of the device to the skin
3. Being unaware to what degree external pressure is transmitted to and through the skin

Consequently, they adapted a method of probing the subdermal cutaneous pressure using hypodermic needles connected to a continuous low-flow pressure transducer. They found that pressure garments have the effect of increasing pressure subdermally. They confirmed that the garment was responsible for controlling compression through and within the skin and were able to quantify the effects.

Another custom-built instrument to assess pressures obtained from garments worn for scar treatment was constructed by Teng and Chou (2006).

This was based on an 'air pack' sensor similar to those used in the Talley and Kikuhime testers.

However, despite this interest in customised instruments, research by Van den Kerckhove et al. (2007) into burned skin treatment went back to using the Kikuhime tester. They tracked reductions in pressure with time associated with different fabric constructions and concluded that 'the Kikuhime pressure sensor provides valid and reliable information and can be used in comparative clinical trials to evaluate pressure garments used in burn scar treatment'.

Another custom-made instrument has been called the 'Textilpress' (Maklewska et al., 2007). This has been designed to measure pressures exerted by compression bands, manufactured from knitted fabrics, on a cylinder surface of defined diameter. Its role is to test compression away from a human wearer, and it is not suitable for in vivo measurement. The device is based on tensometric sensors, to measure both the compression exerted by the fabric and the diameter of the cylinder.

12.3.2 Tekscan Technologies (I-Scan® System)

The I-Scan system provides ultrathin (0.15 mm) sensors of varying sizes for measuring compression forces. These sensors are formed from two sheets of thin polyester, each coated with linear electrical conductors and enclosing a pressure-sensitive material. The electrical conductivity of the sandwiched interlayer material changes linearly in response to applied pressures. The array of linear conductors on the upper sheet are at 90° to the array on the lower sheet. This creates a matrix of sensing locations (sensels) that is determined by the geometry of the sensor. Two examples of the many sensors available are shown in Figure 12.6.

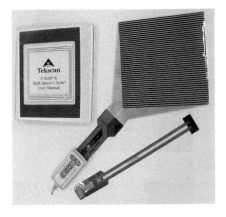

FIGURE 12.6
Two sensors supplied by Tekscan (model 5250 and model 4201). (Source: David Tyler.)

The smaller of these two sensors is model 4201, with a matrix width of 45.7 mm and a matrix height of 21.0 mm. The matrix itself is made up of 24 columns and 11 rows, making 264 sensels (see Figure 12.7). This gives a sensel density of 27.6 sensels/cm². The larger sensor is connected to a sensor handle (which carries the signals to a computer) and is model 5250. This is a square sensor with both matrix width and matrix height being 245.9 mm. The matrix itself is made up of 44 columns and 44 rows, making 1936 sensels.

An example of a screen plot of compression data using model 5250 is in Figure 12.8.

FIGURE 12.7
The matrix of model 4201. (Courtesy Sheena Tyler.)

FIGURE 12.8
Screen display showing sensels recording different pressures of a human hand against a flat surface. The sensels are colour coded, with black/deep blue representing the lowest pressures and, as higher pressures are recorded, the colours move through the spectrum to red (maximum pressure). In this black-and-white illustration, sensels with lighter shades of grey are recording higher pressures.

A review of technologies available in 2005 and suitable for measuring interface pressures relevant to the formation of pressure sores in patients was produced by Swain (2005). Six commercial systems were considered, including Tekscan and Tally. Tests with Tekscan found evidence for significant hysteresis and creep in the data output, but the author noted that clinicians preferred this system because of its real-time display capabilities, resolution and display options.

A detailed investigation of the calibration issues affecting one particular I-Scan sensor was undertaken by Macintyre (2011). The paper contains an overview of the measurement inaccuracies affecting many instruments, but there is a recognition that Tekscan sensors and the supporting software have become 'widely used in recent years'. The range of interface pressure of interest is between 6 and 50 mmHg, which is 'often at the lower end of the measurement range of commercial pressure sensors'. Using the Tekscan 9801 sensor, a very detailed evaluation was undertaken of standard calibration procedures to quantify the accuracy of test results. These are referred to as the '2-point power law calibration' and the 'Linear calibration' procedures. A difference of 3% between measured and applied loads was considered unsatisfactory:

> Despite considerable effort and many attempts to calibrate these sensors the results were disappointing and unsatisfactory. The 2-point power law calibration was most accurate in the middle of the calibrated range, while the linear calibration was most accurate towards the top of the calibrated range (and was completely inaccurate at low applied loads). [...] This level of variability was unacceptably high for precise product development work so another method of calibration was sought. (Macintyre, 2011, pp. 1176–1177)

The rest of this chapter is concerned with a revised calibration procedure and additional probing of sources of error. The findings are presented in a table of the mean differences between measured and applied pressures, where the largest difference is 2.1 mmHg. The results were shown not to be time or use dependent and the new method was adopted for 'the accurate measurement of pressures delivered by pressure garments (and compression bandages)'. This calibration method was used to evaluate design solutions for pressure garments used in the treatment of hypertrophic burn scars (Macintyre, 2007).

Brorson et al. (2012) used Tekscan technology (the I-Scan system) for in vitro measurements on compression garments for treating lymphedema. Their aim was to define a protocol for gaining quantified data relating to compression hosiery. Garments from three manufacturers were selected; wear and tear was simulated by washing the garments before putting them on plastic legs every day for 4 weeks. Whilst there were differences between garments from different manufacturers, no difference was found between garments from the same manufacturer. During the trial period, decreases of subgarment pressure were not observed. They concluded that 'Tekscan pressure measuring equipment could measure subgarment pressure in vitro'.

An example of Tekscan sensors being used in sports science is provided by Pain, Tsui and Cove (2008). The authors set out to measure in vivo impact intensities during enacted front-on tackling in order to assess the effectiveness of rugby shoulder padding for reducing peak forces experienced by players. The work reported limited benefits from using shoulder pads, but raised a number of issues about the selection and use of sensors:

> These potentially inaccurate force measurements exposed three issues with the Tekscan sensor. Firstly, the sensor area was too small as large forces were generated up to and beyond the sensor boundary. Secondly, despite performing a dynamic calibration, the sensor's method of multiplexing data within a long sampling window (4 ms) may have caused the total impact peak force to be missed. Thirdly, the sensor has a low dynamic response time. [...] These issues can be alleviated with the use of a larger sensor that employs a higher sampling frequency. A further limitation is the inability to measure shear force and the fact that the sensors will produce a response if creased or curved too acutely. (Pain et al., 2008, p. 862)

These reflective comments are noted here as they show that experiment design considerations have to be addressed carefully, so that the instruments used are capable of delivering useful results. Sometimes it is necessary to analyse activities in terms of several elements and then focus attention on those elements separately. This has been a way forward for the analysis of forces on rugby players. Usman, McIntosh and Fréchède (2011) looked specifically at the forces in tackling, using a tackle bag equipped with four Tekscan sensors. Participants were asked to tackle the bag in four different ways: (1) dominant side, (2) nondominant side, (3) dominant side with shoulder pads and (4) nondominant side with shoulder pads. With repeated tackling, an assessment of the variability of the forces experienced by participants was gained.

12.3.3 Tekscan Technologies (FlexiForce)

Another type of Tekscan sensor is a single element force sensor with the brand name FlexiForce. There are similarities between the construction of force sensors and pressure sensors but, instead of a matrix of sensels, the resistive layer uniformly covers the whole area of the sensor. Force sensors do not map pressure distributions but provide feedback about the aggregated force experienced. Inevitably, data acquisition and analysis are simplified. Applications for these sensors have been found in sportswear research (Lin et al., 2011, 2012), modelling the compression effects of high-performance sportswear. The FlexiForce sensors provided empirical data to compare with the simulation:

> Seven FlexiForce A201 force sensors (Tekscan, Inc., USA) were placed at seven important muscles that flex or extend when running, namely: (i) vastus lateralis (VL); (ii) vastus medialis (VM); (iii) rectus femoris (RF); (iv) tibialis anterior (TA); (v) semimembranosus (SE); (vi) gastrocnemius lateralis (GL); (vii) gastrocnemius medialis (GM). (Lin et al., 2011, p. 1472)

Undoubtedly, new interface pressure measurement systems will emerge, and there are numerous others that have been reported in the literature that do not appear in this review. There is a problem with custom-made instruments and with commercial products that have a short market life: There is simply not the time to assess the reliability and reproducibility of these instruments and to build a knowledge base to achieve 'best practice'. Consequently, the literature considered in this chapter has been selected to stimulate thoughts on experimental design and helping research aims to be achieved by an appropriate choice of sensors.

12.4 Discussion of Laplace's Law

Rotsch et al. (2011) point out that the compression pressure exerted by a bandage is dependent on four factors:

- The type of bandage, particularly its elasticity
- The prestretching applied during application
- The number of layers of bandage
- The state of the bandage as it is in use

Laplace's law means that the applied pressure is directly proportional to the tension in a bandage but inversely proportional to the radius of curvature of the limb to which it is applied. This has immediate relevance to the selection of the material to bandage the limb. When an inelastic bandage is applied, the tension in the material tends to be low and the contact pressure when resting tends to be relatively low. However, when moving about, the limb expands as muscles contract, and the tension in the bandage tends to increase rapidly so that the resultant applied pressure is high. By contrast, the use of elastic bandages will result in more even compression whether resting or exercising. This leads Rotsch et al. (2011) to refine the conceptual model by distinguishing between static pressure and operating pressure. The static pressure is effectively the pressure when the bandage is applied to relaxed tissue. The operating pressure results from changes in the volume of muscles during movement. Clearly, Laplace's law does not provide a comprehensive mathematical model of compression pressure.

It is also necessary to point out that the human leg has a complex shape and is not well represented either as a cylinder or a cone. There are solid bones covered by various types of soft tissue, and there are many permutations depending on the individuals being bandaged. Deformation of the skin may vary significantly when considering different parts of the leg. This raises further questions about the application of Laplace's law and its relevance to compression garments, whether for medical purposes or for sport.

Thomas (2003) referred to the widespread recognition of the Laplace equation, but pointed out that it has not been well understood. He made reference to a book he published in 1990 that set out a version of the equation that would be more useful to practitioners. He wrote: 'It is also necessary to consider two further factors: the width of the bandage and the number of layers applied. Although these variables may not appear initially to form part of the original Laplace formula, they are essential to obtain an accurate value of tension' (p. 22). The modified equation used units selected because they are familiar to practitioners and incorporated bandage parameters.

$$\text{Pressure(mmHg)} = \frac{\text{tension(kgf)} \times n \times 4260}{\text{Circumference(cm)} \times \text{bandage width(cm)}} \qquad (12.3)$$

where n = the number of layers applied.

The goal of bandaging is to achieve graduated compression, with the highest pressures close to the ankle and the lowest pressures closest to the knee. However, despite great care being taken to apply bandages correctly, there has been an ongoing problem of demonstrating graduated compression. Schuren and Mohr (2008) drew attention to the various explanations that have been proposed to explain the problem: poor operator technique, poor measurement technique and the difficulty of maintaining constant tension during application.

With growing scepticism, Schuren and Mohr (2008) reviewed three detailed studies of graduated compression by standardising on the leg shape. Study 1 involved 32 experts, four commercial compression bandage systems and an artificial leg fitted with three Kikuhime pressure sensors. The participants were asked to repeat the bandaging exercise three times for each commercial system. Study 2 was an evaluation of a commercial prototype compression bandage, using three experienced orthopaedic technicians, and an artificial leg with six strain-gauge force transducers. Each participant applied 40 bandages to the artificial leg. Study 3 used the same leg but selected a different compression bandage, and eight nurses comprised the expert practitioners. Altogether, these studies yielded a database of 744 sets of data relating to graduated compression.

Using the Laplace formula, theoretical compression pressures were calculated. These all showed the desired graduated compression, mostly in the range of 30–60 mmHg (Figure 12.9).

When comparing theoretical with measured values, there was a marked disparity. First, the experimental results typically showed that less than 10% of the bandages applied achieved graduated compression. Furthermore, the measured values were consistently lower than theoretical values. The aggregated test results are plotted in Figure 12.10.

Schuren and Mohr (2008) discuss their findings critically, acknowledging the problems of working with artificial legs and in a laboratory (rather than in a clinical) environment. However, they question the widespread belief in

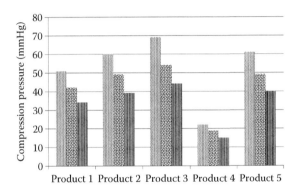

FIGURE 12.9
Theoretical compression according to Laplace's law. (Based on Schuren, J. and Mohr, K. 2008. *Wounds UK* 4 (2): 38–47.)

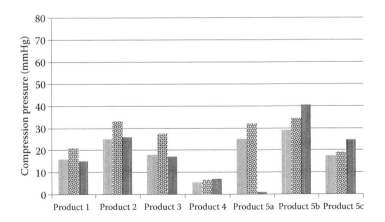

FIGURE 12.10
Measured mean pressure values from all studies. (Based on Schuren, J. and Mohr, K. 2008. *Wounds UK* 4 (2): 38–47.)

the usefulness of the Laplace equation and also the assumption that graduated compression is the norm:

> It is true to say that these studies should have produced data that presented Laplace's law in its best light as the environment, subject, and bandagers were well controlled. However, the pressure calculations made using the modified Laplace's law equation did not accurately predict the pressure values found in these three studies. In fact, true graduated compression was observed in only 53 of the 744 (7.1%) applications. (Schuren and Mohr, 2008, p. 46)

Of course, these findings do not mean that the Laplace equation should be abandoned, but merely that its limitations should be recognised along with the complexities of human anthropometrics. No one should assume graduated compression, but procedures are needed to check this experimentally. Even the goal of graduated compression should be questioned, as do Schuren and Mohr (2010). With sportswear, merely wearing a compression garment is no guide as to what effect it is having on the wearer. The very varied reports of no benefit/some benefit/measurable benefit coming from the sportswear research literature may be simply a pointer to the uncontrolled (and unmeasured) compression that these garments exert. For an example of research that has sought to model and measure compression forces in a more rigorous way, see Dias et al. (2003).

12.5 Summary and Conclusions

1. There are many different measurement systems for monitoring the compression pressures of garments. These use sensors based on a variety of technologies (see Table 12.1). Many of these have had a short lifetime and are of historic interest only.

2. To develop a standard test, many variables have to be excluded. This has been achieved with compression hosiery by eliminating the variability of the human leg and by establishing protocols for loading the garment to the instrument. The Hatra hose pressure tester is the test instrument for British Standards 6612:1985, 7672:1993 and 7563:1999.

3. Medical applications need simple test instruments that can be transported easily and where setting up and measurement times are short. Instruments need to be capable of making in vivo measurements. The most widely used test systems are the MST Professional (for compression hosiery), the Kikuhime tester and the PicoPress.

These have all withstood the test of time and have been developed over the years to incorporate enhancements.

4. Measurement of sportswear compression has made use of medical equipment, but there has been much interest in custom-made systems. There is a recent tendency to use Tekscan technologies. Researchers appreciate the paper-thin sensors, the variety of off-the-shelf sensors available, the sophisticated data-processing software and the visualisation tools. The main problem reported has been drift, and various approaches have been used to obtain reproducible outputs. With a combination of calibration and standardised measurement protocols, acceptable accuracies have been reported.

5. Medical practitioners and researchers appear to have underestimated the problems of getting a controlled and predictable compression. This is particularly apparent in the difficulties in producing graduated compression with leg bandages, but it is symptomatic of the variability associated with compression garments. There is an urgent need for sportswear compression research to be accompanied by detailed measurements of compression pressures. Without this, informed assessments of the value of compression garments cannot be made.

Acknowledgements

The author wishes to thank the following for supplying images, providing feedback on developments and granting permission to publish: Segar Technology (Hatra Mk2A+ hose pressure tester), TT Meditrade and medi-Group Australia Pty Ltd. (Kikuhime), Swisslastic Ag St. Gallen. (MST Professional), Microlab Elettronica (PicoPress), Adrian Smith of the Talley Group, and Sheena Tyler for the image used in Figure 12.7.

References

British Standards Institution (BS 6612). 1985. Graduated compression hosiery.

British Standards Institution (BS 7672). 1993. Specification for compression, stiffness and labelling of anti-embolism hosiery.

British Standards Institution (BS 7563). 1999. Specification for non-prescriptive graduated support hosiery.

Brophy-Williams, N., Driller, M. W., Halson, S. L., Fell, J. W. and Shing, C. M. 2014. Evaluating the Kikuhime pressure monitor for use with sports compression clothing. *Sports Engineering* 17: 55–60.

Brorson, H., Hansson, E., Jense, E. and Freccero, C. 2012. Development of a pressure-measuring device to optimize compression treatment of lymphedema and evaluation of change in garment pressure with simulated wear and tear. *Lymphatic Research and Biology* 10 (2): 74–80.

Chan, A. P. and Fan, J. 2002. Effect of clothing pressure on the tightness sensation of girdles. *International Journal of Clothing Science and Technology* 14 (2): 100–110.

Dias, T., Yahathugoda, D., Fernando, A. and Mukhopadhyay, S. K. 2003. Modelling the interface pressure applied by knitted structures designed for medical-textile applications. *Journal of The Textile Institute* 94 (3–4): 77–86.

Fan, J. and Chan, A. P. 2005. Prediction of girdle's pressure on human body from the pressure measurement on a dummy. *International Journal of Clothing Science and Technology* 17 (1): 6–12.

Flaud, P., Bassez, S. and Counord, J.-L. 2010. Comparative in vitro study of three interface pressure sensors used to evaluate medical compression hosiery. *Dermatologic Surgery* 36 (12): 1930–1940.

Giele, H. P., Liddiard, K., Currie, K. and Wood, F. M. 1997. Direct measurement of cutaneous pressures generated by pressure garments. *Burns* 23 (2): 137–141.

Lin, Y., Choi, K.-F., Luximon, A., Yao, L., Hu, J. and Li, Y. 2011. Finite element modeling of male leg and sportswear: Contact pressure and clothing deformation. *Textile Research Journal* 81 (14): 1470–1476.

Lin, Y., Choi, K.-F., Zhang, M., Li, Y., Luximon, A., Yao, L. and Hu, J. 2012. An optimized design of compression sportswear fabric using numerical simulation and the response surface method. *Textile Research Journal* 82 (2): 108–116.

Liu, R., Lao, T. T. and Wang, S.-X. 2013. Technical knitting and ergonomical design of 3D seamless compression hosiery and pressure performances in vivo and in vitro. *Fibers and Polymers* 14 (8): 1391–1399.

Macintyre, L. 2007. Designing pressure garments capable of exerting specific pressures on limbs. *Burns* 33 (5): 579–586.

Macintyre, L. 2011. New calibration method for I-Scan sensors to enable the precise measurement of pressures delivered by 'pressure garments'. *Burns* 37 (7): 1174–1181.

Maklewska, E., Nawrocki, A. and Kowalski, K. 2007. New measuring device for estimating the pressure under compression garments. *International Journal of Clothing Science and Technology* 19 (3/4): 215–221.

Pain, M. T. G., Tsui, F. and Cove, S. 2008. In vivo determination of the effect of shoulder pads on tackling forces in rugby. *Journal of Sports Sciences* 26 (8): 855–862.

Partsch, H. and Mosti, G. 2010. Comparison of three portable instruments to measure compression pressure. *International Angiology* 29 (5): 426–430.

Partsch, H., Clark, M., Bassez, S., Benigni, J.-P., Becker, F., Blazek, V., Caprini, J., Cornu-Thénard, A., Hafner, J., Flour, M., Jünger, M., Moffatt, C. and Neumann, M. 2006. Measurement of lower leg compression in vivo: Recommendations for the performance of measurements of interface pressure and stiffness: Consensus statement. *Dermatologic Surgery* 32 (2): 224–232.

Rotsch, C., Oschatz, H., Schwabe, D., Weiser, M. and Mohring, U. 2011. Medical bandages and stockings with enhanced patient acceptance, In *Handbook of medical textiles*, Bartels, V. (ed.). Cambridge, UK: Woodhead Publishing Ltd., 481–504.

Schuren, J. and Mohr, K. 2008. The efficacy of Laplace's equation in calculating bandage pressure in venous leg ulcers. *Wounds UK* 4 (2): 38–47.

Schuren, J. and Mohr, K. 2010. Pascal's law and the dynamics of compression therapy—A study of healthy volunteers. *International Angiology* 29 (5): 431–435.

Swain, I. 2005. The measurement of interface pressure. In *Pressure ulcer research, current and future perspectives*, Bader, D., Bouten, C., Colin, D. and Oomens, C. (eds.). Berlin: Springer-Verlag, 51–71.

Teng, T. and Chou, K. 2006. The measurement and analysis of the pressure generated by burn garments. *Journal of Medical and Biological Engineering* 26 (4): 155–159.

Thomas, S. 2003. The use of the Laplace equation in the calculation of sub-bandage pressure. *EWMA Journal* 3 (1): 21–23.

Usman, J., McIntosh, A. S. and Fréchède, B. 2011. An investigation of shoulder forces in active shoulder tackles in rugby union football. *Journal of Science and Medicine in Sport* 14 (6): 547–552.

Van den Kerckhove, E., Fieuws, S., Massagé, P., Hierner, R., Boeckx, W., Deleuze, J., Laperre, J. and Anthonissen, M. 2007. Reproducibility of repeated measurements with the Kikuhime pressure sensor under pressure garments in burn scar treatment. *Burns* 33 (5): 572–578.

Further Information

Hatra Mk2A+ hose pressure tester

Segar Technology

e-mail: segar@segartechnology.com

Web: www.segartechnology.com

I-Scan and FlexiForce

Tekscan, Inc.

e-mail: marketing@tekscan.com

Web: www.tekscan.com/

Kikuhime

mediGroup Australia Pty Ltd.

e-mail: sales@medigroup.com.au

Web: www.medigroup.com.au

Manufacturer: TT Meditrade, Soledet 15, DK 4180 Soro, Denmark

MST Professional

Swisslastic Ag St. Gallen (formerly Salzmann AG)

e-mail: info@swisslastic.ch

Web: www.swisslastic.ch/en/

PicoPress

Microlab Elettronica

e-mail: info@microlabitalia.it

Web: www.microlabitalia.it

13

Body Scanning and Its Influence on Garment Development

Simeon Gill

CONTENTS

13.1 Body Scanning: The Technology 311
13.2 Body Scanning and Its Benefits 312
13.3 Body Scanning and Classification of the Body 313
    13.3.1 Defining the Sporting Body 313
    13.3.2 Measurements Extracted from Body Scanning 315
    13.3.3 Creation of Avatars 316
    13.3.4 Measurements and Advances in Measurement 317
        13.3.4.1 Slices and Volumes 317
    13.3.5 Clothing Experiences of the Sporting Body 318
13.4 Preassessment of Garments in 3D Virtual Environments 319
13.5 Assessment of Functional Requirements and the Changing Body 319
    13.5.1 Postural Change and Sports 319
13.6 Custom Garment Provision 321
13.7 Future Developments and Requirements Using Body Scanning 322
References 323

13.1 Body Scanning: The Technology

Body scanning as a tool has been a commercially viable product with implications on clothing design and development since the start of the twenty-first century. Body scanning technology has been used extensively in national sizing surveys (SizeUK, Size USA, Size Mexico, Size Thailand, etc.) and has contributed to a number of developments in measurement, product development and assessment for fashion. Alongside the development of 3D body scanning technology there have been a number of developments in 3D software from companies like Gerber, Lectra, Optitex and V-Stitcher allowing for virtual modelling and enhancing the possibilities of body scanning in product development.

    Body scanners work by projecting structured light onto the body surface and then recording this through image capture technology. These data are then used to create a point cloud or between 300,000 and 1 million X, Y and Z coordinates which describe the body as a 3D object. The data can then be processed to create an avatar from which measurements can be extracted. The avatar enables the custom analysis of an individual in a fixed posture and the extraction of custom measurements, which can be used to classify the size and shape or to support product development.

13.2 Body Scanning and Its Benefits

Body scanning is recognised as a measurement tool that enables greater depth of analysis regarding the body and clothing than traditional methods (Simmons and Istook, 2003; Bye, LaBat and Delong, 2006). Body scanning can also be used to drive automation of product development (Voellinger Griffey and Ashdown, 2006; Yunchu and Weiyuan, 2007), though this is primarily for form-fitted garments. Whilst body scanning as a tool enables the collection of data, current interfaces for analysis require development to be more accessible to the clothing practitioners' skill sets (Brownbridge, Gill and Ashdown, 2013). Comparison has been made between body scan and manual measurements and the data have been found to be comparable (Bougourd et al., 2000). Variability in measurement definition is seen to exist between methods and often the reliance on manual measurement is not a suitable benchmark to support the benefits of body scanning (Tyler, Mitchell and Gill, 2012). These benefits can be seen in the speed of capture, limited requirement for contact with the subject, accessibility of the data and the ability to repeatedly analyse scans as techniques and understanding develop.

Scan data are generally limited to one fixed posture, often closely aligned to the postural definitions in the body scanner standards (BSI, 2010). Whilst the posture is generally fixed, having a copy of the body in a virtual environment greatly enhances the depth of analysis that is possible. As well as traditional 1D measurements, scans can be sliced, volumes determined, posture analysed and proportional relationships determined. This can be done often within the software (Sizestream.com; Tc2.com) or using other software tools. More recent developments with the technology and sensors allow for time-sequenced capture to provide the possibilities of analysing a moving scanned image (Sizestream.com); this provides the possibility of its integration with other systems such as motion capture. This ability to combine systems has been explored by Zong and Lee (2011), though further developments are required to fully realise the benefits. These developments further highlight the requirements for interdisciplinary skill sets to enable suitable analysis of scan data; this is also the case when employing third-party software.

13.3 Body Scanning and Classification of the Body

Body scanning provides a means to understand and classify the body in a manner that can inform population groupings (size charts, shape classifications) or more directly support innovations in product development (Istook, 2006; Brownbridge et al., 2013). Currently, the most accessible system for classification of shape is the female figure identification technique (FFIT) system (Devarajan and Istook, 2004; Simmons, Istook and Devarajan, 2004). This classifies the shape of the scanned body according to proportional relationships of key dimensions used in garment sizing and product development. Lee et al. (2007) used this system in their comparison of US and Taiwanese female body shapes and showed how body scanning can be used to understand proportional relationships and enable shape classification in a manner more suitable to product development. Having large volumes of data through scanning has also enabled a greater understanding of their classification for clothing product development (Hlaing, Krzywinski and Roedel, 2013). The use of proportional relationships derived from body scan data allows categorisation of the body by shape and 1D measurements that provide more objectively repeatable methods of classification than previous systems. However, current systems do not make distinctions between the morphology of sporting and nonsporting bodies.

13.3.1 Defining the Sporting Body

There have been a number of studies defining or comparing sporting bodies and their characteristics according to different sporting disciplines (Watts et al., 2003; Peeters et al., 2011; Aerenhouts et al., 2012; Zaccagni, 2012). These studies focus predominantly on collecting data like those documented by the International Society for the Advancement of Kinanthropometry (ISAK) (Stewart et al., 2011), which support the comparison of different sporting groups, but provide little data that can be used directly in product development. The body scanner cannot easily directly replicate the comparative data advised by ISAK, but it can provide suitable data in terms of measurements, shape and proportion that can be used to drive product development. However, without a clear understanding of how dimensions used in product development differ between athletes and a general population, it will always require iterative prototyping to create suitable garments. Having an understanding of the impact of sport on the body's morphology is important, as many pattern construction methods employ proportional rules based on the general population to generate block patterns (Gill and Chadwick, 2009). Body scanning allows the extraction of dimensional data individually, but only through accessible and suitably classified data sets can proportions of a sporting body be defined where they may differ from a general population.

Currently, the most accessible indication of variation within sporting populations is in the images of Howard Schatz (Schatz and Ornstein, 2002). However, these are limited to visual comparisons and whilst displaying specific sporting body morphologies, they do not provide the quantitative measurement data a body scanner can provide. Whilst body scanning offers the opportunity to collect sizing data of different populations quickly, there are limited accessible populations from which to make comparisons, in terms of both general population and specific sporting groups. There is also the difficulty of determining which measurements should be compared for clothing; key sizing dimensions or those used in product development (pattern construction) would be ideal as these would impact on clothing sizing and methods of product development.

A study conducted by Simeon Gill and Jane Ledbury at Manchester Metropolitan University during 2012 enabled body scans of 19 elite water polo players to be captured using a Tc2 NX16 body scanner. Existing sports-related studies suggest elite female water polo players have high muscularity due to the requirements of the sport (Marrin and Bampouras, 2007), though no indication is given as to how this may manifest itself in body morphology. Initial comparison between average measurements of the 19 water polo players and a comparable sample of 337 scans sharing some of the same markers (age, body shape) (see Figure 13.1) shows what may be considered key differences that need to be considered within product development.

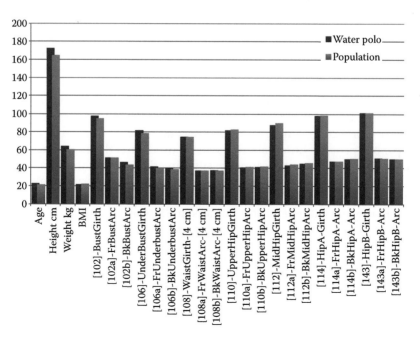

FIGURE 13.1
Comparison of elite water polo players to general population.

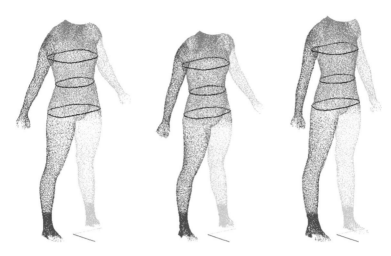

FIGURE 13.2
Examples of the female sporting body.

The water polo players appear to be taller and heavier, though with a slightly lower BMI. When looking at dimensions which would impact product development, their average bust and underbust girths are larger, though for the bust this is manifest in a larger back bust arc, suggesting increased muscularity on the back. Waist girths and hips are comparable, whilst the upper hip is smaller, suggesting differences in fat deposition around this area. These initial data show the potential of body scanning technology to provide data that can inform product development, especially regarding dimensional changes impacted by sports participation. Visual analysis is also possible using body scanning, with scans of the athletic figure showing a silhouette with good muscle tone and smooth curves and well defined muscles (Figure 13.2).

Visual assessment of the scanned water polo players indicates body morphology with good muscle tone and an upright posture. However, it is not until suitable comparative analysis is undertaken between measurements that the data can be used more objectively in product development. Developments in practice and methods of coding are required to ensure that captured data sets can be compared and suitable criteria identified which classify them into different groupings.

13.3.2 Measurements Extracted from Body Scanning

Capturing data using body scanners enables the opportunity to analyse the data; however, there is limited understanding as to existing definitions used by body scanners when defining measurements (Tyler et al., 2012). Variation will exist in measurement definitions; however, the current standards (ISO, 2010) are grounded in ergonomic rather than clothing anthropometric practices. Without a clear understanding of how measurements are defined using this technology, it is difficult to understand how these data may be comparable to previous manual data and applicable in existing product development practice.

13.3.3 Creation of Avatars

The body scanner as a tool offers the opportunity to capture a snapshot of an individual in time, in a fixed pose, and to then create a 3D computer representation or avatar. As current scan avatars have fixed postures, unlike the parametric avatars in CAD programs, it is not possible to transfer a scan of an individual into a virtual environment, which supports accurate assessment of movement and function. Recent developments with scanning technology support dynamic assessment, though currently only visually (Sizestream.com); whilst this is not applied in current research, it provides promise for future applications.

Scanning technology enables the integration of individual sporting bodies into the virtual environment, where software like Meshlab, Geomagic, Polyworks and others, often developed for engineering, can be used in analysis and classification to drive more informed product development (Figure 13.3). This third-party software requires development of operator skills outside those of the traditional product developer, though there must always be a clear appreciation of clothing measurement and its application in product development. The creation of a closed surface model is possible using the morphing features within Tc2. However; other scanners and the initial.obj

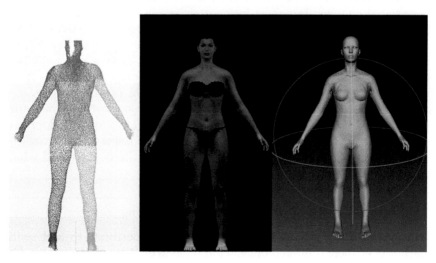

FIGURE 13.3
Basic scan to avatar in Tc2, created as .obj and opened in Meshlab.

files from Tc2 may require some manipulation before they can be employed in different software platforms.

13.3.4 Measurements and Advances in Measurement

Body scanning is recognised for providing greater depth of data than existing manual methods (Bye et al., 2006) and the following section provides examples of this.

13.3.4.1 Slices and Volumes

With the ability to collect data of participants' dimensions with a body scanner it is now possible to conduct detailed and specific analysis relative to particular sporting disciplines. The ability to take slices and volumes provides an opportunity to compare the unique morphologies of a sporting body with that of data taken from a general population.

Scan slices enable the visualisation of the body in a much more informative manner than surface analysis and can indicate how the body is balanced in certain circumferences. Whilst a sportsperson may have a large bust circumference, the scan slice can enable recognition of how much of this is influenced by musculature of the back, rather than increases in actual breast size (Figure 13.4). Scan slices are often used to visually represent changing size and circumferences (Hlaing et al., 2013) as well as a means to show distribution of dimensions around the body. Slice data can generally be transferred into other software platforms from Tc2; this is done using the .dxf format.

Having the closed avatar it is possible to determine volume data using body scanning or other software. For the Tc2 scanner this provides details of volumes for specific areas, whole body, torso, and the individual limbs, excepting extremities. This again allows for comparison of sporting and nonsporting populations as can be seen in Figure 13.5. Access to suitably classified scan data would enable determination of volumetric differences by key region, such as increased and focused muscularity of the legs or fat

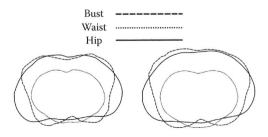

FIGURE 13.4
Scan slices of a sporting and a nonsporting body.

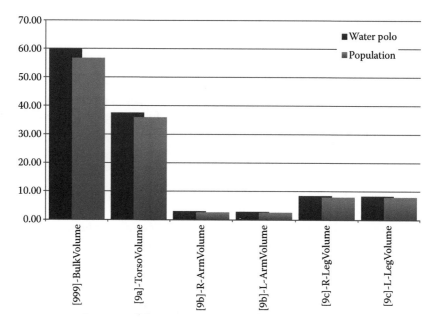

FIGURE 13.5
Examples of average volumes.

deposition in different regions as can clearly be observed between athletes in the images of Schatz and Ornstein (2002).

13.3.5 Clothing Experiences of the Sporting Body

Classification of the sporting body is undertaken currently using predominantly manual tools to determine the muscle, fat and fitness of an athlete (Stewart et al., 2011). Scanning offers a quick noninvasive method to capture the dimensions of the athlete (Stewart, 2012) and to monitor change in these dimensions over time. Whilst there is research into morphologies of athletes and existing research which sometimes contextualises them dimensionally, this is not suitable to drive product development. Furthermore, there is little understanding of how athletes' experiences of clothing may affect their body image. Clothing and body images have been found to be linked in general populations (LaBat and DeLong, 1990; Grogan et al., 2013) and it is not unusual for clothing to be used as a means to gauge satisfaction with body image (Grogan, 2008). These considerations are important regarding performance and ensuring that sporting participants have well fitting, aesthetic and functioning garments. Experiences of their body in relation to clothing can be negative and are rationalised by recognised changes to their morphology brought about by their participating in sports (Ledbury, 2009). Informal discussion with athletes often indicates they recognise the impact of sports participation on their bodies and this does not always lead to greater body image satisfaction. It is important that means are sought to recognise dimensional differences due to sport, which is clearly possible through body scanning. These data can be used both to develop products which satisfy the performance requirements of the athlete and to enable them to participate fully in the consumption of ready-to-wear clothing.

13.4 Preassessment of Garments in 3D Virtual Environments

Using 3D avatars in software like Optitex or Browzewear can allow for the virtual fitting of garments and evaluation of fit using tension maps. These tension maps, which relate to a dimensional difference between the body and garment, with consideration of body angles, allow some predictive capabilities regarding compression and for the virtual testing of garment fit on a larger population and in more depth than would be enabled manually. A limitation of this is the ability to relate this assessment of fit to methods of actual fit and wearer experiences and feedback to provide a more informed process.

Body scanning has been used as a tool to capture clothed participants and undertake fit assessment (Bye and McKinney, 2010; Song and Ashdown, 2010). Both studies compared 3D clothed scans with the same garments in a live-fit assessment. Whilst differences were determined between the results of live and 3D scan fit analysis, this research shows the opportunities for applying scanning technology as a tool in fit assessment. As the scan image can be rendered in a single tone, the ability to see stress folds and buckling is greater than during live fit; also, garment balance can be seen, though as has been indicated, seams can be difficult to locate on the scans' images (Song and Ashdown, 2010).

13.5 Assessment of Functional Requirements and the Changing Body

13.5.1 Postural Change and Sports

Assessment of functional change using scanning technology has been carried out by Choi and Ashdown (2011), who used a Vitus laser scanner and Polyworks software to determine changes to the lower body. A similar approach had been undertaken by Chi and Kennon (2006) with earlier scanning tools but with less success because the requirement to stabilise posture and resolution of the data was problematic. However, changes to surface dimensions can clearly be established and when correctly sited in relation to the pattern, can inform with ease requirements at key locations (Gill, 2009). Body scanning offers the most suitable tool for flexible analysis of postural change and the ability to define and refine measurement context within the software environment.

The ability to use scanning technology in combination with motion capture systems was explored by Zong and Lee (2011). Through largely exploratory work they show how the capabilities of body scanning and motion capture systems can be used to drive developments in the understanding of functional requirements in apparel. They also highlight the limitation of capturing and rendering reliable data in postures other than the standard scan posture (Zong and Lee, 2011). Further developments in scanning technology and the adoption of new IR depth-sensing technology promise opportunities for dynamic 4D capture with the scanner (SizeStream.com); however, the possibility to then extract usable measurement data is as yet unproven.

Using existing scanning technology it is possible to capture scans of participants in different sporting postures. These can range from what can be considered control (or pivot) postures (Figure 13.6) to dominant postures related to specific sports (Figure 13.7). However, current limitations of scanning technology with the necessity to define certain markers (crotch and armpits) before the point cloud can be rendered into a state where automated analysis can commence mean that postures which deviate too much from the standard scan posture cannot easily have measurements derived.

Manual manipulation of the automated landmarks on the scan shown in Figure 13.6 enabled the alignment of the waist to the angle of the torso; however, this was not possible with other measurements, which are fixed to a horizontal plane. As many automated measurements are aligned with the floor, the current ability within software to make comparisons between

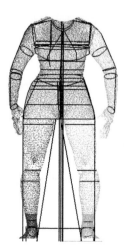

FIGURE 13.6
Control posture.

(a) (b) (c) (d)

FIGURE 13.7
Examples of dynamic scan images: (a) fencing, (b) golf, (c) speed skating and (d) tennis.

a standard and a changed posture is limited. There is a clear need for the development of automated measurements tied to features and planes of the body, rather than those of the scanning environment, that will allow the benefits of automated scanning to be fully utilised for sporting body analysis. As with the work of Choi and Ashdown (2011), it is often necessary to employ other 3D software and landmarking methods in the analysis of body scans to determine dimensions in postures other than the standard posture. Utilisation of specialist software also requires different skill sets from those that are often accessible to clothing practitioners.

Using third-party software like Geomagic or Polyworks requires a clear understanding of measurement positioning for clothing. Though these packages can be used to extract dimensions from nonstandard scan postures, a limitation of current scan software (Zong and Lee, 2011), they do not offer the automated processing capacity for large volumes of scan data. Using a Tc2 KX16 body scanner it was possible to capture body scans in different sporting positions (Figure 13.7) and whilst the software is better at dealing with nonstandard scans than earlier versions, the scans could not be rendered or automated measurements extracted. However, these files could be output for processing in other third-party software, though there are still issues with scan volumes capturing all of the participants and in the removal of unwanted data, like the floor or side curtain.

13.6 Custom Garment Provision

The body scan or avatars provide the perfect opportunity to model garments directly onto the body and flatten these 3D surfaces to 2D patterns. Whilst a number of publications exist outlining the principles of such an approach (Kang and Kim, 2000; Istook, 2006; Yunchu and Weiyuan, 2007) there is still a requirement for careful consideration of production and pattern practices, and these only really suit form-fitted garments with either knitted structures or using high levels of elastic fibres. However, for elite athletes who may exhibit large variations in body morphology dependent on discipline and may fall outside the norms imposed by current sizing and product development practices, the opportunity for bespoke clothing may help maintain performance by removing dissatisfaction with garments during sports participation. The need for greater consideration of fit is highlighted by Ledbury (2009) in her analysis of clothing for women participating in elite sports related activities. This is further supported through feedback from elite Olympic athletes who participated in a body scanning exercise during 2012. The development of automated garment provision from scalable models developed using scan data is a recent area of research development (Sayem, Kennon and Clarke, 2012; Hlaing et al., 2013; Tao and Bruniaux, 2013). Body scanning is key to enabling these developments, both in terms of understanding the body and in the translations of practice into an electronic environment.

13.7 Future Developments and Requirements Using Body Scanning

Body scanning capabilities continue to expand and data accuracy improves; the ability to capture time sequenced scans in real time (SizeStream.com) enables the possibilities of assessing function and fit using these tools. However, producing data that can inform the product developer not only on functional requirements but also on changes to the body by sports is required to fully realise the potential of body scanning in product development. In part this requires greater access to these facilities, but also to the scan data captured. Having suitably coded data sets of athletes from different sports will enable identification of differences unique to sporting disciplines and promote the development of methods to enable their morphologies to be considered within garment creation and product development practices. Existing methods of product development (pattern construction) are heavily influenced by the general population, and corrections currently applied are specific to practitioners.

The ability to utilise body scan data of individual sporting participants in creating form-fitted custom garments provides a clear area of development. Tailoring performance garments to the specific morphology of the athlete has the potential to improve performance, though this requires many developments in the practices of product development and careful considerations of how data are extracted from body scanners.

References

Aerenhouts, D., Delecluse, C., Hagman, F., Taeymans, J., Debaere, S., Gheluwe, B. V. and Clarys, P. (2012). Comparison of anthropometric characteristics and sprint start performance between elite adolescent and adult sprint athletes. *European Journal of Sport Science* 12 (1): 9–15.

Bougourd, J. P., Dekker, L., Grant Ross, P. and Ward, J. P. (2000). A compilation of women's sizing by 3D electronic scanning and traditional anthropometry. *Journal of The Textile Institute* 91 (2/2): 163–173.

Brownbridge, K., Gill, S. and Ashdown, S. (2013). Effectiveness of 3D scanning in establishing sideseam placement for pattern design. In *4th International Conference on 3D Body Scanning Technologies*, Long Beach, CA, 19–20 November 2013.

BSI (2010). BS EN ISO 20685:2010: 3-D scanning methodologies for internationally compatible anthropometric databases. London, British Standards Institute.

Bye, E. and McKinney, E. (2010). Fit analysis using live and 3D scan models. *International Journal of Clothing Science and Technology* 22 (2/3): 88–100.

Bye, E., LaBat, K. L. and Delong, M. R. (2006). Analysis of body measurement systems for apparel. *Clothing and Textiles Research Journal* 24 (2): 66–79.

Chi, L. and Kennon, R. (2006). Body scanning of dynamic posture. *International Journal of Clothing Science and Technology* 18 (3): 166–178.

Choi, S. and Ashdown, S. P. (2011). 3D body scan analysis of dimensional change in lower body measurements for active body positions. *Textile Research Journal* 81 (1): 81–93.

Devarajan, P. and Istook, C. L. (2004). Validation of female figure identification technique (FFIT) for apparel software. *Journal of Textile and Apparel, Technology and Management* 4 (1): 1–22.

Gill, S. (2009). Determination of functional ease allowances using anthropometric measurement for application in pattern construction (unpublished PhD thesis), Manchester Metropolitan University.

Gill, S. and Chadwick, N. (2009). Determination of ease allowances included in pattern construction methods. *International Journal of Fashion Design, Technology and Education* 2 (1): 23–31.

Grogan, S. (2008). *Body image: Understanding body dissatisfaction in men, women and children*, 2nd ed. London: Routledge.

Grogan, S., Gill, S., Brownbridge, K., Kilgariff, S. and Whalley, A. (2013). Dress fit and body image: A thematic analysis of women's accounts during and after trying on dresses. *Body Image* 10 (3): 380–388.

Hlaing, E., C., Krzywinski, S. and Roedel, H. (2013). Garment prototyping based on scalable virtual female bodies. *International Journal of Clothing Science and Technology* 25 (3): 184–197.

ISO (2010). 7250-1:2010: Basic human body measurements for technological design – Part 1: Body measurement definitions and landmarks. Brussels, International Standards Organisation.

Istook, C. L. (2006). 3D to 2D pattern generation and product design from 3D bodyscan data. *Proceedings of the Annual AATCC Conference*, Atlanta, GA, pp. 267–273.

Kang, T. J. and Kim, S. M. (2000). Optimized garment pattern generation based on three-dimensional anthropometric measurement. *International Journal of Clothing Science and Technology* 12 (4): 240–254.

LaBat, K. and DeLong, M. (1990). Body cathexis and satisfaction with fit of apparel. *Clothing and Textiles Research Journal* 8 (2): 43–48.

Ledbury, J. (2009). Beyond function: A design development case study. In *Fashion and Well-Being. IFFTI 11th Annual Conference*, London, UK, 31 March–3 April, 2009.

Lee, J. Y., Istook, C. L., Nam, Y. J. and Park, S. M. (2007). Comparison of body shape between USA and Korean women. *International Journal of Clothing Science and Technology* 19 (5): 374–391.

Marrin, K. and Bampouras, T. M. (2007). Anthropometric and physiological characteristics of elite female waterpolo players. In *Kinanthropometry X*, Marfell-Jones, M. and Olds, T. (eds.). Glasgow: Routledge, pp. 151–164.

Peeters, W. M., Goris, M., Keustermans, G., Plgrim, K. and Claessens, A. (2011). Body composition in athletes: A comparison of densitometric methods and tracking of individual differences. *European Journal of Sport Science* i: 1–8.

Sayem, A. S. M., Kennon, R. and Clarke, N. (2012). Resizable trouser template for virtual design and pattern flattening. *International Journal of Fashion Design, Technology and Education* 5 (1): 55–65.

Schatz, H. and Ornstein, B. J. (2002). *Athlete*. New York: Harper Collins.

Simmons, K. P. and Istook, C. L. (2003). Body measurement techniques: Comparing 3D body-scanning and anthropometric methods for apparel applications. *Journal of Fashion Marketing and Management* 7 (3): 306–332.

Simmons, K. P., Istook, C. L. and Devarajan, P. (2004). Female figure identification technique (FFIT) for Apparel. Part 2: Development of shape sorting software. *Journal of Textile and Apparel, Technology and Management* [online], vol. 1. http://www.tx.ncsu.edu/jtatm/volume4issue1/vo4_issue1_abstracts.htm (accessed pp. 1–15).

Song, H. K. and Ashdown, S. P. (2010). An exploratory study of the validity of visual assessment from three-dimensional scans. *Clothing and Textiles Research Journal* 28 (4): 263–278.

Stewart, A. D. (2012). Kinanthropometry and body composition: A natural home for three-dimensional photonic scanning. *Journal of Sports Sciences* 28 (5): 455–457.

Stewart, A., Marfell-Jones, M., Olds, T. and de Ridder, H. (2011). *International standards for anthropometric assessment*, New Zealand, The International Society for the Advancement of Kinanthropometry.

Tao, X. and Bruniaux, P. (2013). Toward advanced three-dimensional modelling of garment prototype from draping technique. *International Journal of Clothing Science and Technology* 25 (4): 266–283.

Tyler, D., Mitchell, A. and Gill, S. (2012). Recent advances in garment manufacturing technology; joining techniques, 3D body scanning and garment design. In *The global textile and clothing industry*, Shishoo, R. (ed.). Cambridge, UK: Woodhead Publishing, pp. 131–170.

Voellinger Griffey, J. and Ashdown, S. P. (2006). Development of an automated process for the creation of a basic skirt block pattern from 3D body scan data. *Clothing and Textiles Research Journal* 24 (2): 112–120.

Watts, P. B., Joubert, L. M., Lish, A. K., Mast, J. D. and Wilkins, B. (2003). Anthropometry of young competitive sport rock climbers. *British Journal of Sports Medicine* 37 (5): 420–424.

Yunchu, Y. and Weiyuan, Z. (2007). Prototype garment pattern flattening based on individual 3D virtual dummy. *International Journal of Clothing Science and Technology* 19 (5): 334–348.

Zaccagni, L. (2012). Anthropometric characteristics and body composition of Italian national wrestlers. *European Journal of Sport Science* 12 (2): 145–151.

Zong, Y. and Lee, Y.-A. (2011). An exploratory study of integrative approach between 3D body scanning technology and motion capture systems in the apparel industry. *International Journal of Fashion Design, Technology and Education* 4 (2): 91–101.

14

Eco-Considerations for Sportswear Design

Jennifer Prendergast and Lisa Trencher

CONTENTS

14.1 Introduction 327
14.2 Ethical Practices within Sportswear 328
14.3 Consumers and Sustainability 330
     14.3.1 Consumer Knowledge of Ethical/Sustainable Practices 330
14.4 The Sustainable Supply Chain 332
     14.4.1 Corporate Social Responsibility 332
14.5 Ethical Sportswear Design 334
14.6 New Fabric Developments 335
14.7 Technology in Sportswear Design 336
14.8 Conclusion 338
References 339

14.1 Introduction

The sportswear industry alone is worth approximately $145 billion (PWC, 2014) and is set to increase. It is an industry that caters to all sexes, ages and cultures, hence its popularity. The attraction is that it crosses casual wear, active sportswear and fashion. However, these consumer-driven factors have not always considered ethical and sustainable supply chain processes.

There have over the years been increases in sustainable practices, which have grown in popularity within many sportswear companies. However, the need to implement sustainable design practices can often be time consuming and costly – particularly as ethical supply chain processes can be more complex due to the nature of the product development such as fibre production, labour costs and sourcing (Stuart, 2011). This would suggest that there are many challenges for sportswear companies, which in the past have relied heavily on the fast-paced environment of the traditional supply chain to fulfil consumer demand (Kunz and Garner, 2007). Many of these challenges can be overcome by utilising sustainable and ethical programmes throughout the world, many supported by governments. For example, the US Department of Agriculture (USDA) regularly undertakes joint ventures with non-governmental agencies and academia to research the impact of sustainable initiatives on the consumer and environment, etc. Redress, founded in Hong Kong, promotes specific issues directly associated with the apparel industry with a collective of members worldwide (Redress .com.hk, 2014). In Australia, the Council of Textile and Fashion Industries of Australia Ltd (TFIA) focuses on the clothing and apparel industry and the impact on all resources from natural resources to the workforce (Tfia .com.au, 2014). Waste management is one of the major concerns for retailers and consumers alike and is undertaken by many agencies worldwide; one such non-governmental organisation, which is based in the UK, is WRAP (Waste and Resources Action Programme) (WRAP.org.uk, 2014). These agencies and many others, including those that consult with manufacturers in Bangladesh, India and Cambodia, deal with issues from human rights to health and safety. All of them have a wealth of information, with sustainability and ethics at the core of every initiative.

The benefits for sportswear companies are that they are in a fortunate position whereby the popularity in worldwide sporting events such as the International Cricket Councils (ICC) Twenty20 World Cup Cricket, the 2012 London Olympics and the 2014 World Cup in Brazil all provided a platform on which ethical sportswear design could have had a global impact. There have been some developments relating to ethical and sustainable design through the Nike 'Making' app (Nikemakers.com, 2014), which is a sustainability tool designed to assist in the design and development of the company's sporting goods. Nike's statement – 'Designers need to think "ethical" from the outset of their design, not just about fabric but 100% recyclable garments, e.g. polyester fabric, polyester buttons, fabric waste' – suggests that the ethical considerations are not only vast but also complex (Nikemaker.com, 2014).

The ethical clothing market value in 2011 was £50.76 billion in the UK, marking a notable growth since 2007. The market is predicted to reach almost £76.7 billion in 2016 (Clothing Retailing, 2012). Therefore, in relation to sportswear, many brands have the potential to establish a large market share within the sector, particularly with the shift in global economic wealth and the advent of new technological advances within the apparel industry. There are, however, obstacles to overcome prior to design conception which raise many ethical issues, particularly as the customer is at the core of any initiative. The issues range from CSR (corporate social responsibility) and consumer perception of ethical goods to quality, price and brand loyalty (Ellis, McCracken and Skuza, 2012).

14.2 Ethical Practices within Sportswear

Corporate social responsibility is underpinned by ethical and sustainable practices that provide long-term plans to minimise environmental impact through a transparent supply chain. This platform then provides consumers with products that have a unique point of difference from non-ethical goods (Carrigan and De Pelsmacker, 2009). However, the hierarchical nature of the supply chain and communication channels can often prove problematic, due to the complex nature of internal and external involvement; therefore, companies must invest in planning and implementing policies strategically (Stuart, 2011). It is therefore essential that those involved in the global supply chain are clear in their understanding of the product and consumer requirements, in order to influence consumer purchasing decisions that rely heavily on personal beliefs and values (Mirvis, 1994; Doran, 2009; Goworek, 2011). Investment in ethical and sustainable products is high, and in many cases the costs are passed on to the consumer through product purchase – particularly where the retailers have refined programmes of production to focus on environmental initiatives (Mirvis, Googins and Kinnicutt, 2010).

In order to demonstrate the impact of CSR on sportswear companies, Puma developed its PUMA environmental profit & loss ('E P&L') account (About .puma.com, 2014). This document details the measures Puma has undertaken to assess the product journey within the supply chain, 'from design conception to consumer interaction; in order to make the processes a transparent ethical, environmental audit programme' (Puma.com, 2014). This statement by Puma clearly defines how the company will implement its CSR policy through its environmental responsibilities within the supply chain.

As the population continues to grow and resources become ever more precious, international governmental strategies that support shared international wealth through ethical practices are key incentives to all parties involved (Gaskill-Fox, Hyllegard and Ogle, 2014). In terms of strategies, this has resulted in many ethical forums, each with its own unique initiatives. For example, the ETI (Ethical Trading Initiative) of which there are many prominent and popular retailers, is concerned with workers' rights: from education to fair working practices and fair wages. The membership does include several sportswear companies. Additionally, there is the Fairtrade Foundation, which develops sustainable trading practices with Third World workers in ensuring that everyone benefits throughout the supply chain. WRAP develops initiatives in waste reduction, including reuse, recycling and longevity practices. In relation to livestock, DEFRA (Department for Environment, Food and Rural Affairs) is a UK-focused governmental department involved in animal welfare and environmental issues that impact the UK. Finally, the OECD (Organisation for Economic Co-operation and Development) is an international organisation whereby governments can discuss and implement ways to develop and improve sustainable and ethical practices throughout the world, together. These organisations represent some of the initiatives available that promote and encourage retailers to act responsibly. However, as stated previously, there are many initiatives; therefore, it could be argued that many retailers will find the implementation of their own CSR policies difficult if trying to cover all of the different areas of ethical practice. This has led many sportswear companies to undertake their own ethical/sustainable initiatives to raise consumer awareness whilst implementing their own CSR policies. Nike has undertaken initiatives on the production of organic cotton in conjunction with the London Fashion Week's Sustainable Fashion Show (Nike.com, 2009). Through the 'Common Threads Initiative' Patagonia has undertaken work to promote fair trade practices, sustainable production and longevity of its products through reuse, recycling and repairing (Patagonia.com, 2014).

CSR for many sportswear companies is a complex practice. The need to be socially and environmentally aware and place the issues at the forefront of traditional business poses many challenges to retailers. It is difficult to control all aspects of the implementation as there is an understanding that everyone involved shares the same beliefs. However, as seen with the development of new products within sportswear, such as the Nike running 'singlet' made of recycled plastic bottles (Nike, Inc., 2014) and Puma's 'Incycle' products that adhere to the 'cradle to cradle' recycling theory (Leader, 2013; Environmentalleader.com, 2014), these sustainable and ethical products have been developed to enhance the lifestyle of the producers as well as those of the consumer. We have yet to see these penetrate the mass market. The implementation of CSR policies will take some time, but with the assistance of government and NGOs (nongovernmental organisations), consumers can be educated in how sportswear production affects them as well as the wider international community.

14.3 Consumers and Sustainability

The issues associated with consumer understanding and participation in sustainable initiatives have yet to be explored in great depth. However, further investigation and insight into consumer requirements will provide an overview of the consumers knowledge, ethical practices, ethical consumption, sustainable literature and marketing evidenced by academic literature and press reports. It is important to add that the eco-considerations for sportswear design are often complex processes, which the consumer may not understand or simply choose to ignore, in return for a product that is aesthetically pleasing and/or performs well. However, many academics have argued that the consumer is the key to the success of any sports products; sportswear brands and press journalists who regularly review sustainable and ethical innovations implemented by sportswear companies support this.

14.3.1 Consumer Knowledge of Ethical/Sustainable Practices

Ethically, reuse practices are widely acknowledged and accepted in contemporary society (Doran, 2009). Particularly in the disposal of clothing, a garment's identity (i.e. garment construction, country of origin, fabric composition) is clearly presented to the consumer and the important factors and considerations are the reuse purpose (e.g. disassembly, remake, etc.) (Dickson, 2000). Doran (2009) suggested that the ethical global supply chain is often too complex for consumers to understand and those who do understand the processes involved are often most likely to purchase ethical clothing. Ethical consumers often assess such practices according to their personal values and align them according to their understanding of sustainability. The consumer who does not purchase sustainable clothing may do this because of the cost, a misunderstanding of ethical issues or simply because the alternative options are cheaper (Doran, 2009). Despite this, purchasing behaviours are clearly embedded in consumers' psyches; the decision to purchase must fulfil their individual requirements but the need to satisfy the physical and psychological demands is high, whether they are ethical consumers or not. Cross-generational studies suggested that consumers are united in supporting fair and ethical trade but their individual needs and societal beliefs are determinants against the value and cost of the goods. However, consumers who purchase ethical goods on a regular basis have an understanding of how this benefits them as well as global supply chain (Cheah and Phau, 2011). Hustvedt and Dickson (2009) highlighted a significant link between product understanding and purchase intentions, suggesting that consumers who purchase non-apparel organic products are likely to be aware of popular types of organic fibres in clothing and are able to purchase such clothing with a high level of confidence in the product. Understanding the consumers' requirements is essential when using organic fabrics in clothing; of equal importance are the aesthetic qualities and product serviceability such as function and durability. Most retailers offering this type of product are either niche or specialist clothing companies, providing garments for loyal customers who are willing to invest not only in the quality of the garment but also in the companies' brand values (Delong, 2009; George, 2009). With regard to product serviceability, consumers often link the perceived value of ethical garments to the price and expect them to perform to standards the same as/ or higher than their man-made counterparts (Hustvedt and Dickson, 2009). Function, aesthetics, performance and the aftercare of ethical garments are important to consumers, as they pay more for such garments. In addition to these qualities, they also want exclusive aesthetic qualities and expect the cost of the garment to represent the ease of aftercare (Ha-Brookshire and Norum, 2011).

According to the Ethical and Green Retailing Report compiled by Mintel (2009), in the UK consumer awareness of issues such as sustainability and ethical production is higher than ever. In the United States, aspects such as environmental issues and a lack of sustainable literature have forced many consumers to question political and business motives in sustainable practices. However, other barriers to the purchase of ethical clothing are that prices are high and there is a lack of availability in terms of product ranges. In addition, consumers are not clear about various complex terms used, such as 'ethical clothing', 'fair trade', 'organic fabric', 'sustainable garment', etc. There is also lack of trust in labelling. In addition, consumers choose quality and longevity economy against quantity when making a sustainable choice.

14.4 The Sustainable Supply Chain

Kunz and Garner define the supply chain as 'a total sequence of business processes involving single or multiple businesses and countries that enables demand for products or services to be satisfied' (2011:436). Within the sportswear market, there are some examples of 'business processes' in a traditional supply chain model.

In order to operate sustainably within the supply chain there are many key issues to understand; some include the use of materials. This encompasses everything from fabric waste to packaging waste. There are also the recycling elements, which can involve fabric, including biodegradable fabrics and, of course, packaging. However, one of the major issues that affects all our daily lives is pollution. Minimising pollution is an essential element within the sustainable supply chain process. There is potential from the pollution of dyestuff, transportation and pesticide use – all of which, along with many others, contribute to water pollution.

14.4.1 Corporate Social Responsibility

Some of these factors highlight that there are many important factors to consider regarding the production of basic sportswear garments. Therein, this highlights the complexities of implementing a sustainable supply chain. In addition, the sustainable supply chain, as with the traditional supply chain, will also encounter communication issues associated with sustainable meanings and practices as the supply chain links are worldwide.

Published within the Keynote Report (2012) are some of the key barriers to implementing a sustainable supply chain such as cost, consumer awareness, cultural barriers and health and safety. According to the report, the costs are in relation to 'the ongoing uncertain economy [which] remains the largest threat to the market as consumers often reduce spending in the industry if faced with an economic downfall' (Keynote Report, 2013). Therefore, brands may carefully consider their sustainability strategies given the costs, some of which will inevitably be passed on to the consumer. In relation to the consumer, there is a lack of knowledge or understanding of the wider effects of sustainable and ethical practices in relation to the environment, the producer and the economies of less developed countries (Koszewska, 2011). In addition to this, as brands source in existing and new regions, there is a need for collaboration to share best practice and educate new stakeholders in the sustainable supply chain mechanisms (De Brito, Carbone and Blanquart, 2008). For example, in April 2013 the Rana Plaza Garment Factory in Dhaka, Bangladesh, collapsed, killing over 1,000 people (BBC News, 2012; Taplin, 2014). Whilst not directly linked to traditional sustainability issues, questions were asked in relation to whether stakeholders, local governments, factory owners and large retailers can allow these practices to continue.

Directly in relation to sportswear and the support of workers' rights worldwide, governmental and non-governmental agencies collaborated on a project to influence major sports companies prior to the 2012 Olympics, in considering the direct personal impact of the increase in the production of sports products prior to the event (*Safety and Health Practitioner*, 2010). However, these issues had been raised prior to both the 2004 and 2008 Olympic Games (Casabona, 2008). This resulted in the major sportswear companies addressing some aspects of the transparent supply chain; however, workers' rights were not addressed, suggesting that their position within the supply chain was only as the receiver of goods not the producer. However, public exposure and reports of alleged malpractice within the workplace forced many to reconsider their responsibilities within the process (Doorey, 2011). Since this and other unfortunate incidents, the Alliance for Bangladesh Worker Safety was established by a group of retailers in an attempt to provide a transparent platform on which to present their supply chain safety processes and procedures in an accessible online report (Radhadkrishnan, 2014).

There are, however, positive examples of good practice within the sportswear brands. The outdoor wear company Patagonia is dedicated to minimising environmental impact through the initiative '1% for the Planet', which is an alliance of businesses that actively implement programmes that necessitate protecting the natural environment. Sportswear retailers are under pressure to deliver and produce at greater speed than ever due to the fast-paced nature of the industry, putting the supply chain under greater pressure. New Balance is also a company that has for some time developed environmental initiatives relating to production costs, many of which have been developed in conjunction with its suppliers, suggesting that the responsibility for ethical and sustainable practices is shared (Newbalance.com.au, 2014). Therefore, in relation to the sportswear market and the level of competition within this area to produce low-cost products for a high consumer demand, sportswear retailers are under pressure to deliver at a greater speed than ever due to the fast-paced nature of the industry that puts the supply chain under greater pressure.

14.5 Ethical Sportswear Design

Ethical or sustainable design is often more complex than the traditional forms of fashion design, which in many cases rely heavily on fashion trend analysis and catwalk collections. Most ethical design principles focus on developing products that reduce the impact on the environment and improve quality of life (Orzada and Cobb, 2011). The considerations for ethical design range from consumer requirements to fibre choice, product disposability and product development processes. With these and many other ethical considerations that have practical, environmental and social implications, sportswear designers could consider a more lateral thinking approach to creativity and product development as discussed by Taura and Nagai (2013), De Bono (1992) and De Brito et al. (2008). These authors highlight the many benefits to designing within the ethical supply chain with the advent of new technological advancements, which can lead to cleaner and leaner processes throughout the chain. The authors proceed to explain that many of the new fabric developments using recycled materials, which are often discarded in bulk, are now used within products sold by some of Europe's largest retailers. However, the sustainable supply chain needs clear channels of communication due to its multifaceted business relationships (De Brito et al., 2008; Caniato et al., 2012).

Berchicci and Bodewes (2005) investigated several different theories regarding new product development in sustainable design. Their investigations highlighted that product placement and marketing and strategic positioning of the product were at the forefront of product design well before design conception. However, the marketing of green products needs careful consideration in order to ensure that these products can be sold universally to a wide range of consumers (Beard, 2008).

In the marketing of green products, sportswear companies often present promotional literature regarding their sustainable practices via different new media platforms, and it is questionable as to whether many consumers are aware of them. These platforms are used to promote and encourage sustainable practice within sportswear, such as the app, 'Making of Making Powered by Nike MSI' (Nike.com, 2014). There are also social networking sites and dedicated web pages that specifically focus on companies' sustainability reports, corporate social responsibilities and associated initiatives, such as those produced by Puma, Patagonia and the Adidas Group, who all have an abundance of information. However, the question remains as to whether these enhance the consumer's desire to buy the goods or not (Joergens, 2006). It is also questionable as to whether the ethical products meet the performance requirements of the companies' standard products.

In order to develop products that meet the sustainable and ethical requirements, the team, consisting of all of the individuals involved in the process from design to production, need to expand upon their current knowledge and access information which will allow for innovation to be implemented into consumer-driven products (Esslinger, 2011). There are, however, many complexities relating to the consumer that often include the emotional attachment to the garment as identified by Niinimaki and Koskinen (2011), whereby consumers often use different aspects of the garments' construction and general make up to satisfy their ethical practices; this can lead to longevity of the garment through multi-use and function. In fact, the argument for consumer input into garment design is supported by the work of Gam and Banning (2011), whose hypothesis is based on the sustainable design process. This involved trends being established from the consumers, their requirements of garments and emotional needs. Emotional needs which lead to emotional attachment to a garment are complex and multi-faceted (Lee and Soo, 2012; Muehling, Sprott and Sultan, 2014). In fact, many aspects (such as the neural system, which controls our sensory perceptions such as smells, sound and feelings) play a role in our product preferences (Muehling et al., 2014). In addition to this, there are personality traits that we project onto products in order to align ourselves within certain categories, such as social groups or how we wish others to perceive us (Malar et al., 2011; Patwardhan and Balasubramanian, 2011).

These factors form the foundations for a more conceptual design development process and highlight elements for the garment requirements, such as longevity, function and individuality. Important findings from the research of Patwardhan and Balasubramanian (2011) and Malar et al. (2011) were that natural resources are still limited, leading to less creativity. It can therefore be argued that in order to develop a high level of creativity, sustainable design should be innovative, similar to new products designed to reduce environmental impact, which many consumers use on a daily basis, such as low-energy appliances that benefit both themselves and the environment (Nidumolu, Prahalad and Rangaswami, 2009). However, using the most suitable resources from a diverse range of stakeholders – from the farmer to the product developer, who have the expertise with and understanding of the product requirements – is essential (Margolin, 2007).

14.6 New Fabric Developments

Fabrics are the bulk of any apparel product, with cotton as one of the most popular, due to its qualities associated with comfort, ease of dyeing and availability (Cotton.org, 2014). However, the resources used to propagate and grow this fibre are damaging not only to the environment but also to society. The fibre is grown over several hundreds of acres of land and, along with the use of pesticides, causes harm not only to the environment but also to the health of the workers (Ellis et al., 2012). As with all crops, organic cotton also needs water; however, the impact of pollution is significantly lower than for conventional cotton fibre growth. However, the costs of organic cotton are higher, as there is a time in which it takes to propagate and grow due to the limited use of pesticides and other harmful pollutants. There are, however, other options to replace cotton, such as cellulose blends, which are plant-based fibres that can then be woven. These blends have similar qualities to cotton, such as drape and strength, but the process of producing them is more complex. Silk is another fibre that is grown organically and is slowly being introduced into mainstream fashion. It has an abundance of qualities such as insulative and breathable, comfortable, and lightweight and takes dyes well; in addition, it is also a very strong fibre (Post and Orth, 1997). However, non-organic silk, which is mainly produced by a silkworm, can be extracted in an inhumane fashion. In recent years there has been the advent of organic silk; this process allows the moth to emerge from the cocoon before the fibre is extracted, making this a more humane process. In order to mimic some of the qualities found in natural fibres there are silk alternatives such as Lyocell, which is a sustainable fabric regenerated from wood pulp cellulose; it has good draping qualities, is crease resistant and can be blended with other natural fibres, which is why it is used frequently in yoga wear (Copeland, 2014). Developments that are more recent include fabrics that combine man-made and plant fibres such as 'plant-based spandex', which is more environmentally friendly due to the content and production process. As with all fabric developments they are experimental; however, many of them are viable alternatives that are available for use in mainstream sportswear product development (Ecouterre.com, 2014).

14.7 Technology in Sportswear Design

Technology to many designers is at the forefront of sustainable and ethical design. Some are very innovative and others are based on traditional skills, but at the heart of the process is the ability to produce products that consumers require.

Technology facilitates the design process (Park et al., 2014). For example, replacing the traditional manual skills, such as component cutting, sample development and hand embroidery, with 3D CAD visualisation, laser cutting and computerised embroidery machines makes the eco-design process more efficient.

Three-dimensional printing, once used only by architects to visualise their projects, is being utilised by sportswear companies to develop more innovative sportswear products. This has been visible in athletic footwear whereby many of the major sportswear retailers have started to produce customised footwear for professional athletes (Nike.com, 2000). The 3D process involves using different flexible materials, inks or even fibres and then printing the product using a specially designed printer; this process has many features that ethical consumers would relish. These include the bespoke element: The consumer is able to input his or her own measurements in the system. This enables the garment to be printed without fabric wastage; it also means that customers can customise the garments to their individual tastes and requirements. Some 3D printers also produce interconnecting pieces, which enable the wearer to adjust or change the garment function for their particular needs, such as a bespoke fit and comfort factors. This in turn extends the product's life cycle, whereby the consumer will see the ethical values relating to, for example, waste disposal, multi-use benefits and product versatility (Continuumfashion.com, 2012). These and other benefits significantly put the retailer in a favourable position as this will attract more consciously ecologically minded consumers and untapped markets and, amongst other things, positively contribute to the development of its CSR policies.

With the advancement of technology, the laser cutting once used for decorative purposes, which is prevalent in many fashion design collections in the form of engraving a fabric and the cutting out of elaborate, decorative designs, has more recently been used within technical product design. Having similar benefits to 3D printing, it offers the opportunity to be used directly in conjunction with organic fabrics. The benefits, for example, include the sealing of raw edges, eliminating the need for the neatening process; the cut pieces are very accurate and precise, which is not always possible when the garment is cut by hand. In addition, the computer can be set to ensure that the patterns are positioned in such a way that fabric wastage is kept to a minimum. In relation to the consumer, garments can be cut based upon the specific requirements (Ondogan et al., 2005; Qiu and Hu, 2014). Within these processes, innovation is addressed through the sustainable product development process. For example, the design innovation can be derived from utilising the whole fabric yardage and using moulding techniques, such as draping, where the fabric is manipulated to define a shape such as a sleeve or armhole rather than separate pieces being constructed to form these shapes. This suggests that the designer may have more creative freedom within the design process by starting at the point of the fabric rather than the sketch, which is how it is traditionally done. This will then lead to more individual and innovative designs rather than those that follow trends. Additionally, fabric wastage may be eliminated because the retailer is using all of it. Again, this will appeal to the consumer on the same level as 3D printing, through efficiency practices.

Body scanners are used in sportswear to develop custom-made garments for professional athletes. They are more commonly known as 3D body scanners. The technology takes several measurements of the body, which are then used to produce garments for the specific wearer. There are several examples of use within the professional sports arena. The benefits of this software are that individual measurements are recorded, making the product unique to the wearer (Hollings.mmu.ac.uk, 2013). This also means that performance enhancements were developed, such as those by Speedo for the 2008 Beijing Olympics for the Australian swimming team – the Fastskin suits that allowed the water to travel over the wearer with more efficiency, effectively allowing the wearer to navigate through the water at a faster speed (Speedo, 2008). More recently, the Brazilian football team for the World Cup 2014 had its kit designed by Nike; this enabled the company to scan each individual team member and design garments that were unique to his physiology. These football kits allowed the wearer to play football in clothing that was specifically designed to cater to each of his movements and his physiology (Nike.com, 2012).

Technology within the apparel industry is developing at a fast rate: For example, circular knitting allows a garment to be constructed using fewer seams, which provides comfort and reduced friction to the wearer, particularly if the garment is worn for long periods, such as in mountaineering and trekking (Rodie, 2009). There are spray-on fabrics, which can be used to provide layers in specific areas, such as those that require regular movement or need more thermal protection. Many of these spray-on fabrics can also be easily applied and removed, eliminating the need of assistance (Fabricanltd.com, 2010). There are also technologies for garment decorations that can include the addition of thermally regulated technologies, which can change colour according to the wearers' temperature to monitor their health during an activity (Dunn, 2003). Therefore, the design and technology process will forever be evolving and, with the application of sustainable materials, has the ability to provide consumers with clothing that will be of use to them over a longer period.

14.8 Conclusion

To summarise, eco-considerations for sportswear design fall into four main areas:

- CSR regulations
- The consumer
- The design process
- Technology

Each of these areas encompasses sustainable and ethical processes that influence the design and production of sportswear. There are many international governing bodies and non-governmental organisations, such as WRAP, The Fairtrade Foundation, TFIA, etc. that encourage and support policies that focus on the environment, workers' welfare, animal welfare and many others that sportswear retail companies have access to. Positively, many sportswear companies also actively promote their own CSR programmes and policies, but there is still a long way to go. The consumer plays the most important role within the entire design and development process. Many ethical consumers want to understand what their product is made of. How was it made? Who produced it? Where was it produced? How was it shipped? However, supply chain issues are just the beginning; there are also questions relating to the longevity of the product and its uniqueness to the consumer. Therefore, it is a very complex and lengthy process, which for a sportswear company to undertake could be quite daunting. But as evidenced, many sportswear companies are doing this through large sporting events such as football, skiing, cycling, etc. If they can achieve a balance, there is an opportunity to reach untapped markets, such as those consumers who would not normally purchase ethical goods and are fast fashion purchasers. In order to create interest from these consumers, the designers must create garments that not only are appealing but also have a point of difference from mainstream goods. This will demand a more lateral approach to design, one where the traditional design process of trend first is replaced with consumers' requirements, such as uniqueness relating to aesthetics, performance, function and the psychological needs informed by emotional attachment to the product. Again, this will mean that designers are able to experiment with more criteria in order to produce ethical sportswear. There are, however, benefits to this. If technology is introduced and combined within the design process, there is the ability to attract consumers that would not normally be involved in ethical and sustainable causes. The ability to produce more individual and aesthetically pleasing sports garments for the consumer is key to heightening ethical practices.

References

About.puma.com. (2014). PUMA®—Environmental Profit and Loss Account. [online]. Available at http://about.puma.com/en/sustainability/environment/environ mental-profit-and-loss-account (accessed 21 August 2014).

BBC News. (2012). Sweatshop campaign targets Adidas. [online]. Available at http://www.bbc.co.uk/news/uk-18348247 (accessed 6 November 2013).

Beard, N. (2008). The branding of ethical fashion and the consumer: A luxury niche or mass-market reality? *Fashion Theory: The Journal of Dress, Body & Culture* 12 (4): 447–468.

Berchicci, L. and Bodewes, W. (2005). Bridging environmental issues with new product development. *Business Strategy and the Environment* 14 (5): 272–285.

Caniato, F., Caridi, M., Crippa, L. and Moretto, A. (2012). Environmental sustainability in fashion supply chains: An exploratory case based research. *International Journal of Production Economics* 135 (2): 659–670.

Carrigan, M. and De Pelsmacker, P. (2009). Will ethical consumers sustain their values in the global credit crunch? *International Marketing Review* 26 (6): 674–687.

Casabona, L. (2008). Report looks to draw attention to sportswear working conditions. *WWD* 86:12.

Cheah, I. and Phau, I. (2011). Attitudes towards environmentally friendly products: The influence of ecoliteracy, interpersonal influence and value orientation. *Marketing Intelligence and Planning* 29 (5): 452–472.

Clothing Retailing. (2012). 5th ed. Keynote Ltd.

Clothing Retailing. (2013). 10th ed. Keynote Ltd, p. 53.

Continuumfashion.com. (2012). Continuum 3D printed fashion. [online]. Available at http://www.continuumfashion.com (accessed 7 November 2013).

Copeland, T. (2014). How eco-friendly is bamboo fabric, really? [online]. Ecouterre. com. Available at http://www.ecouterre.com/how-eco-friendly-is-bamboo-fab ric-really/ (accessed 12 February 2014).

Cotton.org. (2014). National Cotton Council of America. [online]. Available at http:// www.cotton.org (accessed 17 April 2014).

De Bono, E. (1992). *Serious creativity*. New York: HarperBusiness.

De Brito, M., Carbone, V. and Blanquart, C. (2008). Towards a sustainable fashion retail supply chain in Europe: Organisation and performance. *International Journal of Production Economics* 114 (2): 534–553.

Delong, M. (2009). Innovation and sustainability at Nike. *Fashion Practice* 1 (1): 109–114.

Dickson, M. (2000). Personal values, beliefs, knowledge, and attitudes relating to intentions to purchase apparel from socially responsible businesses. *Clothing and Textiles Research Journal* 18 (1): 19–30.

Doorey, D. (2011). The transparent supply chain: From resistance to implementation at Nike and Levi-Strauss. *Journal of Business Ethics* 103 (4): 587–603.

Doran, C. (2009). The roles of personal values in fair trade consumption. *Journal of Business Ethics* 84 (4): 549–563.

Dunn, R. (2003). Looking good, feeling great. *Textile World* 153 (4): 39.

Ecouterre.com. (2014). Plant based spandex. Ecouterre. [online]. Available at http:// www.ecouterre.com/index.php?s=plant+based+spandex (accessed 15 July 2014).

Ellis, J., McCracken, V. and Skuza, N. (2012). Insights into willingness to pay for organic cotton apparel. *Journal of Fashion Marketing and Management* 16 (3): 290–305.

Environmentalleader.org. (2014). [online] Available at http://environmentalleader .org (accessed 3 August 2014).

Esslinger, H. (2011). Sustainable design: Beyond the innovation-driven business model. *Journal of Product Innovation Management* 28 (3): 401–404.

Ethical and Green Retailing. (2009). [online] Available at http://academic.mintel .com.ezproxy.mmu.ac.uk/display/395927/# (accessed 12 September 2014).

Ethicaltrade.org. (2014). Ethical trading initiative. Respect for workers worldwide. [online]. Available at http://www.ethicaltrade.org/ (accessed 10 March 2014).

Fabricanltd.com. (2010). Fabrican home: Spray-on fabric non-woven coating aerosol technology cotton fibres. [online]. Available at http://www.fabricanltd.com (accessed 1 June 2014).

Gam, H. and Banning, J. (2011). Addressing sustainable apparel design challenges with problem-based learning. *Clothing and Textiles Research Journal* 29 (3): 202–215.

Gaskill-Fox, J., Hyllegard, K. and Ogle, J. (2014). CSR reporting on apparel companies' websites: Framing good deeds and clarifying missteps. *Fashion and Textiles* 1 (1): 1–22.

George, M. (2009). Trending upward. [online]. Wearablesmag.com. Available at http://Wearablesmag.com (accessed 8 February 2014).

Goworek, H. (2011). Social and environmental sustainability in the clothing industry: A case study of a fair trade retailer. *Social Responsibility Journal* 7 (1): 74–86.

Ha-Brookshire, J. and Norum, P. (2011). Willingness to pay for socially responsible products: Case of cotton apparel. *Journal of Consumer Marketing* 28 (5): 344–353.

Hollings.mmu.ac.uk. (2013). Body scanner MMU. [online]. Available at http://www.hollings.mmu.ac.uk/bodyscanner (accessed 13 November 2013).

Hustvedt, G. and Dickson, M. (2009). Consumer likelihood of purchasing organic cotton apparel: Influence of attitudes and self-identity. *Journal of Fashion Marketing and Management* 13 (1): 49–65.

Joergens, C. (2006). Ethical fashion: Myth or future trend? *Journal of Fashion Marketing and Management* 10 (3): 360–371.

Koszewska, M. (2011). Social and eco-labelling of textile and clothing goods as means of communication and product differentiation. *Fibres & Textiles in Eastern Europe* 19 (4): 87.

Kunz, G. and Garner, M. (2007). *Going global*. New York: Fairchild Publications.

Leader, E. (2013). Puma launches cradle-to-cradle sportswear line. [online]. *Environmental Management & Sustainable Development News*. Available at http://www.environmentalleader.com/2013/02/13/puma-launches-cradle-to-cradle-sportswear-line/ (accessed 9 June 2014).

Lee, H. and Soo, M. (2012). The effect of brand experience on brand relationship quality. *Academy of Marketing Studies Journal* 16 (1): 87.

Malar, L., Krohmer, H., Hoyer, W. and Nyffenegger, B. (2011). Emotional brand attachment and brand personality: The relative importance of the actual and the ideal self. *Journal of Marketing* 75 (4): 35–52.

Margolin, V. (2007). Design, the future and the human spirit. *Design Issues* 23 (3): 4–15.

Mirvis, P. (1994). Environmentalism in progressive business. *Journal of Organizational Change Management* 7 (4): 82–100.

Mirvis, P., Googins, B. and Kinnicutt, S. (2010). Vision, mission, values: Guideposts to sustainability. *Organizational Dynamic* 39 (4): 316–324.

Muehling, D., Sprott, D. and Sultan, A. (2014). Exploring the boundaries of nostalgic advertising effects: A consideration of childhood brand exposure and attachment on consumers' responses to nostalgia-themed advertisements. *Journal of Advertising* 43 (1): 73–84.

Newbalance.com.au. (2012). Made to move: New Balance responsible leadership report. [online]. Available at http://www.newbalance.com.au/Responsible-Leadership-Flipbook/responsible_leadership_flipbook,en_AU,pg.html (accessed 17 April 2014).

Nidumolu, R., Prahalad, C. and Rangaswami, M. (2009). Why sustainability is now the key driver of innovation. *Harvard Business Review* 87 (9): 56–64.

Niinimaki, K. and Koskinen, I. (2011). I love this dress: It makes me feel beautiful! Emphatic knowledge in sustainable design. *Design Journal* 14 (2): 165–186.

Nike, Inc. (2014). Kenyan marathon champion to wear Nike uniform of innovative sustainable materials. [online]. Available at http://news.nike.com/news/kenyan-marathon-champion-to-wear-nike-uniform-of-innovative-sustainable-materials (accessed 11 March 2014).

Nikemakers.com. (2014). Nike makers. [online]. Available at http://nikemakers.com/ (accessed 12 June 2014).

Ondogan, Z., Pamuk, O., Ondogan, E. and Ozguney, A. (2005). Improving the appearance of all textile products from clothing to home textile using laser technology. *Optics & Laser Technology* 37 (8): 631–637.

Orzada, B. and Cobb, K. (2011). Ethical fashion project: Partnering with industry. *International Journal of Fashion Design, Technology and Education* 4 (3): 173–185.

Park, J., Morris, K., Stannard, C. and Hamilton, W. (2014). Design for many, design for me: Universal design for apparel products. *Design Journal* 17 (2): 267–290.

Patagonia.com. (2014). *Patagonia—Environmentalism: Becoming a responsible company.* [online]. Available at http://www.patagonia.com/eu/enGB/patagonia .go?assetid=2329 (accessed 23 August 2014).

Patwardhan, H. and Balasubramanian, S. (2011). Brand romance: A complementary approach to explain emotional attachment toward brands. *Journal of Product & Brand Management* 20 (4): 297–308.

Post, E. R. and Orth, M. (1997). Smart fabric, or 'wearable clothing'. In: *Wearable Computers, 1997. Digest of Papers., First International Symposium on.* IEEE, pp. 167–168.

PWC. (2014). Changing the game: Outlook for the global sports market to 2015. [online]. Available at http://www.pwc.co.uk/consulting/issues/changing -the-game-outlook-for-the-global-sports-market-to-2015.jhtml (accessed 12 June 2014).

Qiu, C. and Hu, Y. (2014). The academic thinking based on the perspective of fashion design. *International Journal* 2 (1): 17–26.

Radhakrishnan, S. (2014). Application of Biotechnology in the Processing of Textile Fabrics. *Roadmap to Sustainable Textiles and Clothing.* S. S. Muthu, Springer Singapore, 277–325.

Redress.com.hk. (2014). Redress. [online]. Available at http://redress.com.hk/ (accessed 12 June 2014).

Rodie, J. B. (2009). *Innovations in knitting.* Textile World 159 (5): 42–43.

Safety & Health Practitioner Magazine. (2010). Worker abuse rife among Olympics supply chains. [online]. Available at http://go.galegroup.com/ps/i.do?id=GALE %7CA232896896&v=2.1&u=mmucal5&it=r&p=AONE&sw=w&asid=1ed6a84e a0db1b09a47f38daacd784af (accessed 16 June 2014).

Speedo. (2008). Fastskin: Feel faster. [online]. Available at http://www.speedo.co.uk /technology/fastskin (accessed 2 March 2014).

Stuart, H. (2011). An identity-based approach to the sustainable corporate brand. *Corporate Communications: An International Journal* 16 (2): 139–149.

Taplin, I. M. (2014). Who is to blame? *Critical Perspectives on International Business* 10 (1/2): 72–83.

Taura, T. and Nagai, Y. (2013). *Concept generation for design creativity.* London: Springer.

Tfia.com.au. (2014). TFIA—Home. [online]. Available at http://www.tfia.com.au /home (accessed 12 June 2014).

USDA.gov. (2014). US Department of Agriculture. [online]. Available at http://www .usda.gov/wps/portal/usda/usdahome (accessed 8 May 2014).

WRAP.org.uk. (2014). WRAP—Circular Economy & Resource Efficiency Experts. [online]. Available at http://www.wrap.org.uk/ (accessed 12 June 2014).

Index

Page numbers followed by f and t indicate figures and tables, respectively.

A

Abrasion resistance, 79
  of fabrics, 80, 269
  good/poor, 66
  Martindale abrasion tester, 269
  percentage change in mass, 271
Absorption, 26, 27
Absorption–desorption processes, 67
AC, *see* Acromioclavicular (AC)
AC/DC supply, 156
Acromioclavicular (AC), 156, 216, 221
Activewear, *see* Sportswear
Adidas, 9, 20, 41, 154
  brand architecture portfolio strategy,
    10
  expansion, 13
  market position of, 4
  retail sales, 13
  team sponsorships, 19
Admiration needs, 6
Adsorption, 27
ADVANSA Thermo°Cool®, 36, 42, 78
Affiliation needs, 6
Air permeability, 77, 78, 165
  Coolmax fabric, 69
  need for, 137
  pine cone, 46
  test fabrics, 79
American Association for Textile
    Chemists and Colorists
    (AATCC), 279
Antibacterial fabrics, 97, 188, 189t–190t
Anti-G trousers, 134
Apparel; *see also* specific entries
  characteristics, 54
  fabric types, 55
Apple, 167
Arm/leg sleeves, 188
Asian economies, 6, 7

Athletes
  Caucasian, 71
  perceive comfort, 277
  stretch/recovery, 48
Avatar, in Tc2 scanner, 316

B

Baggy fit, 251
Ballistic protective vests, 107
Bandaging, 304
Base layer clothing, 71
Bespoke clothing, 322
B-guard, 213
Bicomponent fibres, 33, 35, 36
Bicomponent filament extrusion, 36
Biomimetic fabrics, 115
BISFA (International Bureau for the
    Standardization of Man-Made
    Fibers), 31
'Black Diamond' cycle, 107
Block patterns, 313
Blood-borne pathogens, 107
Body perspiration, mechanism of, 30
Body scanners, 315, 337
Body scanning, in garment
    development, 311, 313, 319
  benefits, 312
  capabilities, 322
  classification of body, 313
    avatars, creation, 316–317
    measurements extracted, from
      body scanning, 315–316
    sporting body, defining, 313–315
  custom garment provision, 321–322
  functional requirements, assessment
    of, 319
    postural change/sports, 319–321
  future developments/requirements,
    322

measurements/advances
in measurement, 317
slices/volumes, 317–318
national sizing surveys, 311
sporting body, clothing experiences,
318–319
Body shaping, 240
Body stretch
movements, 32
types, 179
Bonded zippers, 133
Brain redirects blood flow,
hypothalamus region, 108
Branding strategies, 9–10
Brand loyalty, 328
Brands, 5; *see also* specific brands
building tool, 17
personality/consumer, three-way
relationship, 17
Brazil, economy-based issues, 15
Breathability
fabrics, 111
resistance to evaporative heat
transfer (RET), 44
BRIC countries (Brazil, Russia, India
and China), 12
British Standard (BS 7209:1990), 113, 114
Browzewear, 319
Bulkiness, 31, 48, 223, 235, 266
Burberry, 14
Burlington's smart fabric, temperature
management, 54

C

CAD
3D visualisation, 336
programs, 316
Caffeine, 181
Cannibalisation, 9
Canterbury, 218
rugby tops, 219
Carbohydrate consumption, 183
Caucasian athletes, 71
CBRNE hazards, *see* Chemical,
biological, radiological,
nuclear, and explosive
(CBRNE) hazards
C-change™, 115

Celebrity endorsement, 16–19, 17
Celliant, 36, 38, 39
Centers for Disease Control and
Prevention (CDC), 12
CEP, compression wearables, 179
'C-form' structure, 34
Channels to market
factory outlets, 13
flagship stores, 13
in-store formats, 13
online, 13–14
wholesale, 12
Charge whilst you wear, 167
Chemical, biological, radiological,
nuclear, and explosive
(CBRNE) hazards, 106
China
economy-based issues, 15
market, sportswear brands, 12
Chromatic function, category of, 163
Circular knitting, 231
garment, 232, 239
Circular machinery, 232
Circular weft knitting machinery, 232
Clever garment engineering techniques,
94
Climbing, 250
Close-fit garments, 54, 284
Clothing applications, pressure sensors
athletes' experiences of, 318
medical instruments, use of, 298–299
Tekscan technologies, 299–302
FlexiForce, 302–303
I-Scan® System, 299–302
Clothing comfort, 166, 252
Clothing fit
ease levels, within garment, 254
fabrics, 246–247
factors affecting, 245–246
function of, 247–249
and pattern, 246
performance expectations, 252–254
sensorial comfort, 249
to gender, 251–252
pressure comfort, 250–251
wearer perceptions, psychological
considerations of, 249–250
Clothing insulation, 146
Clothing layers, 68

Clothing physiology, 71
Cloverbrook Fabrics, 99
Coatings
    microporous, 105
    yarns, 157
Coats Epic thread, 132
Cohesiveness, 26
Cold water bathing, 183
Cold weather clothing system, 134, 137, 138
Cold weather survival, 136
Cole Haan brands, 9
Comfort, in clothing, 250
Complete garment technology, 233, 236, 240
Composite fabrics, for functional clothing, 104
    aircrew challenges, designing cold weather clothing systems, 135
        cold weather, human factors, 137–138
        environmental considerations, 135–137
        thermal balance/layered clothing systems, 138–139
    application of, 106
        chemical/biological protection, 106–107
        flame/heat protection, 107
        high visibility, 108
        outdoor clothing/sportswear, 108
        physical/mechanical protection, 107
        protection/survival, 106
    characteristics, 104
    cold weather survival clothing materials, 141
        comfort layer, 144
        construction, 146–147
        fit, 146
        insulation/ventilation, 144–145
        mobility, 146
        protection layer, 143–144
        selection of fabrics, 141–143
    design process, 139–140
    functional outerwear, 109–113
        biomimetic fabrics, 115
        drop liners, 113
        ePTFE, 114–115

        European Outdoor Group, 109
        four-layer laminates, 113
        functionality considerations, 111–113
        hill walking jackets, in-depth analysis of, 110
        2.5-layer laminates, 112
        soft shells, 113
        sustainability, 115–116
        technologies, 110–111
        thermally adaptive fabrics, 115
        three-layer laminates, 112–113
        two-layer laminates, 111–112
        UV absorbing/reflecting fabrics, 115
    garment design/development, 130
        assembly techniques, 132–133
        features, 131–132
    hill walking jacket, case study, 116
        consumer requirements, 117–121
        evaluation of prototype, 121–130
        PTFE laminates, 116
        rates and water vapour transfer, 117
        ventilation design feature, 121
    human physiological response/ functional requirements, 108–110
    laminates/coatings, 105
    measurement techniques/ performance comparison
        performance comparisons, 114
        water resistance, 113
        water vapour permeability, 114
    microencapsulation, 105–106
    military flight crew, case study
        current clothing analysis, 133–134
        design brief, 134–135
        survival clothing, 133
    personal protective equipment (PPE), 104
    staple for outerwear performance, 109
Composite fabrics, for outdoor clothing, 108
Compression bandages, 301
Compression garments, 173, 180
    athlete wearing, 174
    average retail selling price, 191

base layer, 189–190
breakdown of sales, 191
close-fitting, 195
compression wear sales, 192
cyclist, 183, 184
designing, 193
layer clothing, 175
meta-analysis, 182
physiological, benefits of, 175
rugby, 185, 186, 187
skiing, 185
    base layer, 185, 186
for sportswear and leisure
        applications, 178
systematic review of evidence
        relating, 199
tapes, use of, 187
venous blood return, 175
Compression hosiery, 292
Hatra hose pressure tester, 292
medical stocking tester (MST),
        292–293
Compression modalities, 172
Compression sportswear,
        applications of
background/rationale, 172–173
compression garments applications
    cycling, 183–184
    rugby sport, 185–188
    skiing, 184–185
compression therapy treatment
        for lower limbs, 173–175
contextual factors affecting
        compression garment
        performance, 192
    body shapes, 194–195
    fabric panels, 195–196
    garment sizing, 193–194
    sizing/designing with stretch
        fabrics, 195
elastic compression bandages, 172
evaluation for sportswear, 178–183
    compression garments, using
        effects, 181–183
market trends, 188–192
for medical uses, 175–178
physical movement of body, 171
Compression sportswear market, 199
Compression stockings, 173

Compression therapy, 173, 178
Compression wearables, 179
Compression-x, compression wearables,
        179
Conductive yarns, 156
Consumer apparel market, 9
Consumer requirements, to fibre choice,
        334
Control posture, 320
Conventional woven fabrics, 89
CoolCore LLC, 47
CoolCore Technologies, 47
Coolmax®, 47
    fabric, 26, 69, 96
    thermoregulation of, 70
Cordura 1100, 75
Corporate social responsibility (CSR),
        328
    on sportswear companies, 329, 330
Cotton, 27
    fibres swell, 90
    garments, conventional, 31
    production, 25
Courses per centimetre (cpc), 265
Cradle to cradle, 330
Crimp, 26
Criteria for Extreme Cold Weather
        Survival System for Military
        Aircrew, 141
Cross-generational studies, 331
CSR, *see* Corporate social responsibility
        (CSR)
Custom-built equipment, 217, 298, 299
Custom garment provision, 321–322
CW-X, compression wearables, 179
Cyclist, cross-country, 63

D

Defender M9180, 142
Deflexion, 214
DER, *see* Dynamic elastic recovery
        (DER)
Designer
    challenges, 136
    criteria, 146
    ethical/sustainable, 334
    process, 140
    user-centred approach, 139

Design process, structured user-centred approach, 139–140
Diaphragm pumps, 121
Domestic brands, 12
Drape coefficient, 275, 276
Draped specimen, shadow of, 275
Drop liners, 113
Dry heat transfer, 30
Dynamic elastic recovery (DER) of elastic knitted fabrics, 33
Dynamic scan images, 321
Dyneema, 79
    abrasion-resistant fabric, 80
    high thermal conductivity, 81
    hydrophobic, 80
    tear resistance of fabric, 79

E

Eco-considerations, for sportswear design
    consumers/sustainability, 330
        ethical/sustainable practices, consumer knowledge of, 330–332
    ethical practices, 328–330
    ethical sportswear design, 334–335
    International Cricket Councils (ICC), 328
    new fabric developments, 335–336
    sustainable practices, 327
    sustainable supply chain, 332
        corporate social responsibility, 332–333
    technology, 336–338
    US Department of Agriculture (USDA), 327
Economic downfall, 332
Ectomorph shapes, 194
Elastane yarns, 33, 42, 49
Elastic compression bandages, 172
Elastic fabrics, 33
Elastomeric fibres, 25
Elastomeric filaments, 195
Electrocardiogram (ECG), 158
Electrochromic, 163
Electronic illumination, 166
Electrostatic discharge (ESD), 141
Endomorph shapes, 194, 195

Endorsement, celebrity, 16, 17
Environmental impact, 167
EPTFE, *see* Expanded polytetra-fluoroethylene (ePTFE)
Ergonomic comfort, 238
Ethical/sustainable design, 334
Ethylene vinyl acetate (EVA) foams, 214, 218, 219
ETI (Ethical Trading Initiative), 329
Evaporative polyester fabric, 31
Excellent, grade, 279
Expanded polytetrafluoroethylene (ePTFE), 111
    laminated membanes, 111

F

Fabric behaviour, 64, 81
    characteristics, confident understanding, 53
    external factors influencing fabric behaviour, 67–70
        clothing layers, requirements, 68
        dry state measurements, 69
        outer wear, commercial examples, 70
        wet state measurements, 69–70
    factors, influence, 64
    internal factors influencing fabric performance, 64–67
    specific fabric property, 70
Fabric bulk density, 266
Fabric characteristics, 56
Fabric coating, 68
    with Outlast phase change material, 161
Fabric composition, 82, 97
    importance of, 77–78
    natural fibres, 78–79
    on sportswear performance, 76
    synthetic/smart fibres, 79–80
Fabric count, 268
Fabric density, 66
Fabric design construction, 267
Fabric engineering, to enhance performance, 247
Fabric formation, 74
Fabric/garment performance, 25
Fabric gsm cutter, 264

Fabric handle, 274
Fabric–human body interaction, 72
Fabric panels, for muscle groups, 198
Fabric parameters, 59
Fabric performance, internal factors, 65
Fabric pilling, 273, 284
Fabric properties, 57, 60–62
 essential/desirable properties, 57–64
 product performance, influence on,
  54–57
Fabric selection, two-layer system, 143
Fabric specification, 283
Fabric structure/characteristics, 73
 discussion of, 80–83
 phase change materials, 74
 special multilayer fabrics,
   for protection, 74–76
Fabric thickness tester, 266
Fabric weight, 264
 affects, 265
 fabrics used in sportswear, 265
Fabric wicks moisture, 280
Fabrics/garments, properties of, 58
Fabrics' physical parameters, 263
Fabrics, with specific properties, 70
Factory outlets, 13
Fashion influences, 19
Fast jet (FJ) aircraft, 136
Fastskin FSII, 180
Fastskin FS-PRO, 180
Fastskin suits, 338
Female sporting body, 315
Fiber resiliency, 26, 34, 78
Fibre absorption, 28
Fibre blend ratio, 67
Fibre composition, 47
Fibre elasticity, 28
Fibre flexibility, 28
Fibre resiliency, 26, 34, 78
Fibres, for sportswear, 24, 33
 absorption, 28
  fibre elasticity, 28
  fibre flexibility, 28
  hygroscopic, 28
  skin discomfort, 28
  static buildup, 28
  thermophysiological, 28
  water repellence, 28
  wrinkle recovery, 28

bicomponent, 35–36
body perspiration/temperature
 regulation, mechanism of,
 30–31
discussion, 47–49
hollow fibres, 34–35
innovation, 25
market trend/overview, 41
 market drivers, 41
microfibres, 34
modifications, 26
moisture management, 43–47, 45–46
 breathability, 44
 maintaining body temperature, 47
 moisture vapour transfer rate
  (MVTR), 44
 resistance to evaporative heat
  transfer (RET), 44
 wicking, in activewear products,
  46–47
new developments, in industry,
 40–41
performance assessment, 26–28
 absorbency, 26
 absorption, 27
 adsorption, 27
 hydrophilic, 27
 moisture regain, 27
 wicking, 27
performance of garments, 42–43
physiological parameters, 29
properties, 26
stretch/recovery, 31–33
thermoregulation, 36–40
 cellulose blends, 40
widely used, 41–42
 elastane, 42
 merino wool, 42
 polyester, 41
Fine-diameter fibres, 34
Fingerprint, for moisture management
  properties, 280
Flagship stores, 13
Flame resistance, 54
Flankers, 187
Flat bed knitting machinery, 231
FlexiForce®, 302
 interface pressure sensors, 290
 sensors, 303

Flexural rigidity, 275
Flight clothing, 134
  pilot's, 134
Force-deformation behaviour, 209
FR viscose blend yarns, 107
Fryma extensiometer, 277
Functional clothing, performance-
    driven, 2
Functional outerwear, 130
  design feature, 132
Functional surface changes, 249

G

Game on World, 6
Garment design, contextual factors,
    172
Garment pressing process, 238
Garment silhouettes, 252
Garments, fabric properties, 55
Garments preassessment, in 3D virtual
    environments, 319
Geomagic, 316, 321
Gerber, 311
German-based fibre company, 45
GH, *see* Glenohumeral (GH)
Gilbert, 218
Glenohumeral (GH), 221
  joints, 216
Glove, outlast phase change materials,
    162
  hand before test, 162
GORE-TEX fabric, 76
  outer fabric, 73
GORE-TEX film, 105
  laminated to nylon, 107
GPhlex, 214, 219
Graduated compression, goal of, 306
Grapheme-based devices, 166
Gsm (grams per square metre) cutter,
    264

H

Hatra hose pressure tester, 292
Hatra Mk2A+ hose pressure tester, 292
Hazards
  categorisation, 106
  loose-fitting garments, 146

Head protectors, for cricketers, 216
  compression therapy, 175
  elastic compression bandages, 172
  monitoring of soldiers, 154
  smart textiles, development of, 154
Health monitoring, 158
Healthcare
HealthVest, 164
Heat flux sensing, 281
Heat thermal protective clothing, 107
Hematoma, 208
Hierarchy of Needs, 6
Highly flexible batteries, 166
High-performance sportswear, 302
High thermal conductivity, 81
Hill walking jacket, 131
  anatomy of outdoor, 131
  performance of composite fabrics,
      116–117
  prototype, evaluation of, 121–130
  schematic diagram, 122
H&M's fitness tights, 6
Hohenstein Institute, 47
Hollow fibres, 33, 34
Hollow polypropylene fibres, 34
Hologenix, 39
Hood designs, fixed/removable, 131
Hookers, 187
Human musculature, 198
  anatomy of, 198
  fabric panels, 197
Hydrophilic fabrics, 27, 111
  property, 76
  proportion, 78
Hydrophobic fibers, 28, 42, 46, 72, 78,
    80, 82
Hydrostatic head test, 113
Hygroscopic fibers, 28, 113
Hypothermia, 71, 138

I

ICD+ jacket, 155, 167
Impact attenuation test, 217–221
  energy absorption characteristics,
      206
  findings from, 218
  impact protection, regulations for,
      221–222

material properties, 211
pressure sensors, 217
rig, 217
Impact forces, reduction/broadening
     of, 220
Impact injuries, 206
Impact-resistant materials, 206, 221
     benchmarking of, 222
     commercially available, 213–216
     garment design/new product
          development, 222–226
     high-contact sports, 205
     injuries sustained, during sporting
          activities, 206–213
     insertion, 215
     methodology, rationale for using, 216
India, economy-based issues, 15
Individual sports, different types, 263
Indoor cycling, 252
Industrial bump caps, 216
Infrared camera, 162
Infrared (IR) light, 38
Injury, sport/types, 207
Injury sustained, proportion of, 210
Inner jacket construction, fabric
          mapping of, 145
Inspiring, mainstream fashion, 2
In-store formats, 13
Insulation, clothing, 109, 146
Integral shaping, 237
Integrated sensors, 155
Intelligent textile technologies, 154
Intense body stretch, 285
Interface pressure sensors, 295
Interlock knitted fabrics, 213, 278
International Cricket Councils (ICC), 328
International Events Group (IEG), 18
International Rugby Board (IRB), 218, 221
     regulations, 208, 216
          shoulder protection in rugby
               tops recommended, 222
International Society for the
          Advancement of
          Kinanthropometry
          (ISAK), 313
International Textile Machinery
          Association (ITMA), 233
Internet, shoppers, 13
INVISTA Coolmax fabric, 27

IR depth-sensing technology, 320
I-Scan system, 299, 301

J

Jackets
     air temperature, 123, 124
     humidity measured, 124, 125
     skin temperature, 125, 126
JJB Sports, 14, 41
Jockey®, 37
Just do it, 6

K

Keynote Report, 332
Kikuhime instrument, 293, 294
Kikuhime pressure sensors, 304
Kikuhime® tester, 290, 299
Knit structures, 91, 234, 235
     classification of, 90–91
     properties of, 234
     on sleeve, 240
     warp, 277
     weft, 91
Knitted fabrics, 178
     air permeability, 82
     elastic, 33
     fabric characteristics, 56
     fabric count, 268
     microdenier polyester, 43
     polyester microdenier, 65
     properties of, 91
     structure, 268
     tightness factor for, 268
     wool-blend, 81
Knitted ribs, 234
Knitting techniques, 243
     on flat bed machinery, 235
     manufactured using, 214
     seamless, 243
     synthetic filaments, 88
Knitwear industry, 242
Kooga, 218
     market brands, 190
     PVA foam, 209
     rugby tops, 219, 224
     shoulder pads, 209
KX16 body scanner, 321

L

Lamination
    composite fabrics, 112
    Defender M9180, 147
    four-layer, 113
    2.5-layer, 112
    PTFE laminate, 122
    three-layer, 112
    two-layer, 111
Laplace equation, 173
Laplace's law, 290, 291, 303, 304
    measured mean pressure values,
        305
    theoretical compression, 305
Latent heat storage, 48
Layering, clothing system, 138
Lectra, 311
LED detector, 158
Leggings, compression garments,
    181
Leisure Trends Group's Running
    RetailTRAK™, 188
Leisure wear, 33
Life cycle model
    innovation and imitation, 11
    product, 10–11
Limiting oxygen index (LOI) test, 142
LincSpun™ technology, 41
LincSpun yarns, 40
Linear calibration procedures, 301
Liquid transport properties, 46
Lithium ion batteries, 156
LOI test, *see* Limiting oxygen index
    (LOI) test
Lycra®, 2, 283
    containing knitted fabrics, 88
    revolutionised woven fabrics, 88
    sport beauty fabric, 33
    sport energy fabric, 33
    SPORT fabric, 31
Lyocell, 336
LZR Pulse fabric, 94
LZR Racer Suit, 180

M

Magnifying glass, 268
Mainstream fashion, inspiring, 2

Making of Making Powered by Nike
    MSI, 334
Man-made fibres, 25
Manual manipulation, 320
Market drivers/emerging trends
    celebrity endorsement, 16–19
    economy-based issues, 14
        aging population, 14–15
        growth of, 15
        major sporting events, 15–16
        sports participation, 16
    fashion *vs.* function, 19–20
    mass customisation, 20
    sports sponsorship, 16–19
    technology, 19
Marketing, of green products, 334
Martindale abrasion tester, 269
Maslow's hierarchy, of needs, 7
MECS, *see* Medical elastic compression
    stockings (MECS)
Medical elastic compression stockings
    (MECS), 177
    bandages, 177
    made-to-measure, 177
Medical stocking tester (MST),
    292–293
    Salzmann MST professional, 293
Merino fibre, 42
Merino wool, 40, 42, 43, 78, 82, 99
    advantages of, 78
    aerobic sports, 78
    garment, 42
    hygroscopic properties and soft
        handle, 113
    natural fibres, 82
    in sportswear, 42
    types of fibres, 40
Mesh fabric, 48, 53, 256
Meshlab, 316
Mesomorphs shapes, 194
Metal conductive fibres, 157
Meteorological charts, 291
MiAdidas, 14
Micro-denier fibres, 64, 65
Micro-encapsulated materials, in
        sportswear applications, 106
Microfibres, 33, 34
    fine-diameter fibres, 34
    woven fabrics, 96

Microporous fabrics, 111
    composite, 105
    polymers manufacture, 111
Microporous membrane, GORE-TEX
        fabric, 76
Military aircrew, extreme cold weather
        survival system, 135
Military clothing, 143
Military operations, 135
MMT, *see* Moisture management tester
        (MMT)
Mobile phone, 155
Model 4201 matrix, 300
Mogliano Veneto, 212
Moisture accumulation, 73
Moisture control fibres, 46
Moisture diffusion, 68
Moisture management, 30, 43–47,
        45–46
    breathability, 44
    maintaining body temperature, 47
    moisture vapour transfer rate
        (MVTR), 44
    properties, 278
        of fabrics, 67
    reactive/proactive materials, 44
    resistance to evaporative heat
        transfer (RET), 44
    wicking, in activewear products,
        46–47
Moisture management tester (MMT),
        278, 279
    fabric print, 284
    grades fabric, 279
    water location *versus* time, 280
Moisture regain, 27
Moisture regulation, 97
Moisture-repellent polyester fibres, 46
Moisture transmission, 54
Moisture vapour permeable (MVP), 141
Moisture vapour transfer rate (MVTR),
        44
Moisture vapour transmission, 43
Moisture wicking, 165
Monitoring pressure distribution, 289
Mosaic from Experian, 8
Motorcycle clothing, 142
MP3 player, 155
Multilayer fabrics, 77

MVP, *see* Moisture vapour permeable
        (MVP)
MVTR, *see* Moisture vapour transfer
        rate (MVTR)

N

Nanofibres, 24
NanoSphere, 105
Nanotechnology, 97
Nano-TiO_2, UV protection, 68
National Electronic Injury Surveillance
        System (NEISS), 206
National Sports Medicine Institute, 207
Natural fibres, 24
    Merino wool, 82
Next-to-skin knitted fabrics, 71
NFL league, 18
NGOs (nongovernmental
        organisations), 330
Nike, 9, 20, 41, 154, 188
Nike flagship store, 13
NikeID, 14
Nike 'Making' app, 328
Nike, market position of, 4
Nike's brand, 18
    architecture, 9
Nike's revenue, 12
Nike's sponsorship, Indian cricket, 15
Niketown, 13
Nike, website, 14
Nilit Aquarius, 41
Nomex fibre, 144
Nonfast jet (NFJ) flight, 136
Non-governmental agencies, 327
Non-Newtonian fluid, 213
Non-phase-change material, 163
    glove, 162
Nonsporting body, scan slices, 317
Nonwoven fabrics, 280
Normal body movement, 277
NuMetrex, 164
Nylon, 24, 195
Nylon 6.6 yarns, 31, 40

O

Olefin, 27
Olympic effect, 15

Olympic team kit sponsorship, 16
Olympic village, 5
OMMC, *see* Overall moisture
    management capacity (OMMC)
Online, 13–14
Optitex, 311, 319
Organic cotton, 335
Outerwear garments, 109
Outlast fabric, 69
Outlast phase change materials
    glove, 162
Outlast Technologies, 30, 37
    acrylic fibre, 40
    acrylic filament with thermocules, 39
    adaptive comfort, 37
    phase change materials (PCMs), 38
    polyester fibre, 39
Overall moisture management capacity
    (OMMC), 279
Oxford Orthopaedic Centre, 294

P

Padding, rugby tops, 224
Parkwood Topshop Athletic Ltd, 5
Pascal, 291
Pattern development, 246
PCM, *see* Phase change material (PCM)
Pentland Group, 41
Performance Apparel Markets, 29, 39
Performance clothing, fabrics
    application in functional clothing, 88
    basic structures/influence, on
        sportswear performance, 89
        conventional woven structures,
        89–90
    conventional knitted structures
        classification of, 90–91
        warp knitting, 92
        warp/weft-knitted structures,
        92–93
        weft knitting, 91
    fabric structures, application of, 95–96
    overview of, 87
    woven/knitted fabrics, future
        developments, 96
        fabric developments, 97–99
        nanotechnology, 97
        wovens *vs.* knits, 93–95

Performance evaluating, of fabrics
    comfort measurement, 277
        functional clothing, moisture
            transfer influence, 280–282
        moisture management tester
            (MMT), 278–280
        wicking, 278
    design and development
        of sportswear, 262
    durability, evaluating, 269
        abrasion resistance, 269–270
        fabric pilling, 271–273
    fabric specifications/interpreting
        results, 282–285
    handle, in sportswear, 274
        drape, 275–276
        stiffness, 274–275
        stretch/recovery, 276–277
    individual sports, 263
    woven/knitted fabrics, physical
        measurements, 263
        fabric area density, 264–265
        fabric bulk density, 266
        fabric construction, 267
        fabric count, 268
        fabric cover factor, 267
        fabric thickness, 265–266
Performance swimwear garments, 90
Permetest, 281, 282
Personal protective equipment (PPE),
    104, 106
    textile coatings, applications of, 108
Personal protective systems (PPS), 105
pH, 72
Phase change material (PCM), 38, 70, 74,
    105, 161, 162, 164
    absorb energy, 74
    professional garments, 82
    thermoregulation of performance,
        74
Philips ICD+ jacket, 153
Photochromatic, 163
Photovoltaic (PV) cells, 156
PicoPress® instruments, 290, 297, 298
Piezochromic materials, 163
Piezoelectric fibres, 158
Pilling, 271, 273
    assessment photographs, 273
    test, sample preparation, 273

Pilling box, measure pilling resistance of fabrics, 272
Plain weave, 267
Plain-woven fabric, 53
Plant-based spandex, 336
Plastic films, 106
    membranes, coated, bonded/ laminated, 106
Pocket bags, 133
Polartec Thermal FR R2206 flame-resistant fleece fabric, 144
Pollution
    impact of, 336
    minimising, 332
Polyaniline (PANI) salts, 157
Polyester fibre, 24, 25, 41, 43, 49, 53, 132, 195, 282
    filament fabrics, 42, 65
    outlast phase change material, 160
    plain-woven fabrics, 66
Polyethylene terephthalate (PET) fibre, 158, 278
Polymer-based materials, 213
Polyproplylene (PP), 158
Polypyrrole (Ppyr), 157
Polystretch micropile fabric, 76, 77
Polytetrafluoroethylene (PTFE) films, 107
Polyurethane (PU), 111
Polyvinylidene fluoride (PVDF), 158
Polyworks, 316, 321
    software, 319
Poron XRD, 213, 219
    impact forces, 218
    material, 213
PPE, *see* Personal protective equipment (PPE)
Pressure comfort, 251
Pressure garments, 298
Pressure measurement technologies, 290
Pressure sensors application, in monitoring pressure
    clothing applications, pressure sensors, 298–303
    compression hosiery, 292
        hatra hose pressure tester, 292
        medical stocking tester (MST), 292–293

Kikuhime instrument, 293
Laplace's law, discussion of, 303–306
measuring pressure, challenge of, 289–291
medical applications, 292
PicoPress instrument, 297–298
pressure sensing instruments, evaluation of, 296–297
test instruments, overview, 293–295
units of pressure, 291
Product life cycle, 10–11
Proskins, day-to-day clothing, 181
Protection, clothing engineered cold, 139
Protective clothing, 106
Psychological comfort, 239
PTFE composite fabrics, 117
Puckering, 83
Puma, 6, 18, 36
PUMA environmental profit & loss ('E P&L') account, 329
Puma UK–protective layers, 255–256
Pump operation, schematic diagram, 122
PVA foam, 209

Q

Quick-dry ability, polyester, 53
Quicksilver, 180

R

Ready-to-wear clothing, 319
Recovery, 29
    athletes, 48
    compression garments, 183
    stretch fabrics, 31, 276
    wrinkle, 28
Reflective materials, 108, 165
Regional sweat rate (RSR)
    female participants, 29
Relative water vapour permeability (RWVP), 281
Relaxed fit, 251
Researchers at Nottingham Trent University, 157
Resiliency of fabrics, 26, 34, 78
Resistance to evaporative heat transfer (RET), 44

Respiratory rate, 158
Revolutionary change, 19
Rib knit fabrics, 73
Rubber
    membranes, coated, bonded/
        laminated, 106
Rugby
    player tackling, 210, 212
    shoulder injury, 224
    tackle types, 208
Rugby tops, 222
    design issues, 223
    padding, 224
    peak forces acting, 220
    protective gear for sports, 223
Russia, economy-based issues, 15

S

Safety and Health Practitioner, 333
Sales revenue, 5
Salzmann MST professional, 293
Samsung, 167
Santoni, temperature control by, 240
Scan data, 312
Scan slices, 317
Scanning technology, 316
SCCS yarn, *see* Spandex core cotton
        spun (SCCS) yarn
Scotchlite®, 108
    c-change materials, 159
Seam distortion, 83
Seamless draped top, 237
Seamless garment fit, 236–238
Seamless knitted garments, 238
Seamless knitting, application, 231
    application of body measurement, 235
    circular knitting, 231
        seamless, 232
    complete garment knitting, 233
    current limitations within industry,
        242–243
    flat bed knitting machinery, 231
    garment aesthetics, 240–241
    garment comfort, 238–240
    garment fit/knit structures,
        implications of, 233–234
    knitted fabrics, implications of,
        233–234

specific fit-related opportunities,
        241–242
    stitch manipulation, creating
        garment shape, 235–236
Seamless machinery, 232
Seamless technology, protect
        intellectual property, 242
Sebum, 72
Sensels recording, 300
Sensorial comfort, 238
Sensors
    advantages and disadvantages, 297
    measured pressures *vs.* reference
        pressures, 296
Shape memory effects, schematic
        diagram, 160
Shape memory polymers, 165
Shape memory textiles, 159
Sharkskin-based design fabric, 93
Sharkskin swimsuit design, 94
Shirley stiffness tester, 274
Shoulder
    anatomy of, 225
    impact, 211
    padding, 221
        for rugby, 215
    protection, in rugby tops, 222
Ski clothing, 164, 250
Skin
    blood flow, 30
    discomfort, 28
    fabric, psychological interactions, 72
    model, 281
    sensational wear comfort, 29
    stretches, 31
SKINS, compression wearables, 179
Skin temperature, 125, 126
    at abdomen, 127, 128
    at chest, 126, 127
    at lower back, 128, 129
    at upper back, 125, 126
SmartLife, 158, 164
Smart materials, for sportswear
    conductive yarns, 156–157
    definition of, 155
    discussion of, 166–167
    future developments, 166
    integrated sensors, 157–158
    'intelligent' textiles, 153

leisure activity, 154
Philips ICD+ jacket, 153
power supplies, 155–156
specific applications, 163
  athletics, 163–164
  ski wear, 164–165
sportswear performance, influence
  of, 158
  automatic adjustments, 159–163
  health monitoring, 158–159
Smart textiles, 155, 164, 192
  development of, 154
  market, 167
Smock-style jackets, 118
Snug-fitting garments, 95
Socks, compression garments, 181
Soft shell, 113
Soft-tissue injury, 171
Solar energy, 156
Sol-gel technique, 68
Solvachromic materials, 163
Sorbothane, 214
Spacer fabric, 214
Spandex back-plated cotton (SBPC) yarn, 33
Spandex core cotton spun (SCCS) yarn, 33
Speedo, 179
  LZR Pulse fabric, 97
  swim apparel manufacturer, 93
Sponsorship, 16–19
Sport England, 57
Sports
  adults taking part, 56, 57
  bra, NuMetrex, 164
  clothing, 30, 58
    performance expectations, 253
  development encompasses, 90
Sports Direct, 41
Sports injury, cramps, 176
Sports-inspired clothing, 3
Sports jackets, 57
Sports participation influences
  sportswear, 16
Sports performance, factors affecting, 193
Sports sponsorship, 16–19
Sportswear
  applications, 159
  function of, 8
  industry, 327
  market, *see* Sportswear market

Sportswear market, 3
  aerobics/indoor fitness, 3–4
  branding strategies, 9–10
  channels to market
    factory outlets, 13
    flagship stores, 13
    in-store formats, 13
    online, 13–14
    wholesale, 12
  competitive position, 4–6
  and consumers, 6–9
    fashion segments, using, 8
    Maslow's hierarchy, 6, 7
    women, 8–9
  definition of, 2–3
  golf, 4
  key markets, 12
  leisure wear excluding outdoor wear,
    3
  outdoor pursuits, 3
  product, life cycle, 10–11
  racket sports, 4
  sportswear, 3–4
  swimwear, 4
  team sportswear, 3
Static buildup, 28
Sternoclavicular (SC), 221
Stiff fabrics, 276
Stitch density, 265
Stockings, 33
Stretch, 29
  athletes, 48
  fabrics, 31, 290
  key areas of, 178
  properties, of fabric, 276
Summer Olympics, 4
Surface fuzz, knit fabric, 272
Sweat accumulated, in clothing system,
  130
Sweat patterns, 29, 30, 48, 71
Swedish Olympians, 4
Swimsuit, 87
Swimwear, 33, 88
Synthetic fibres, 83, 99, 272

T

Taber abrasion test, 271
Taber abrasion tester, 269, 270

Tackling, rugby player, 187
Talley IPM (Interface Pressure Monitor), 294
Talley SD.500 Skin Pressure Evaluator, 294
Tally SD.500 skin pressure, 298
Tc2 NX16 body scanner, 314
Tear resistance, of fabric, 79
Teflon coating, 77
Teijin Limited, 46
Tekscan model, 299
Tekscan pressure sensors, 196
Tekscan sensors, 209, 302
Tekscan technology, 301
Temperature regulation, mechanism of, 30
TenCate, 141
TenCate Defender M9180, 144
TENCEL, 40
Textile and Fashion Industries of Australia Ltd (TFIA), 328
Textile clothing application, 161
    personal protective equipment (PPE), 108
Textile fibres, 25
Textile materials, 53
Textile testing, 144, 262
Textiles, stretch, 289
TEXTRENDS 2015/2016 categorises fabrics, 263
Texturing process, 26
Thermal absorptivity, 73
Thermal burden, 133, 136
Thermal comfort, 138
    properties, of elastic knitted fabrics, 69
Thermal heating patterns, 29
Thermal insulating layer, 107
Thermally adaptive fabrics, 115
Thermal mannequin, 73
Thermal resistance, 54
    conducting heat, 109
    dry, 281
    physiological comfort, 69
    values, of fabrics, 74
    woven fabric, 285
Thermal underwear, 26
Thermochromic textiles, 163
Thermocules™, 37, 38f–40f

Thermophysiological comfort, 28, 239
Thermoregulation, 25, 28, 48
    challenges, 116
    of Coolmax, 70
Thermoregulatory properties, 160
Thick fibres, 26
Third-party software, 316
3D avatars, 319
Three-dimensional printing, 336
Three-dimensional technology, 257
Time-consuming postproduction processes, 233
TiO_2 nanocoating, protection from UV rays, 82
Tissue oxygen levels, 39
Toothbrush bristles, nylon, 24
Torr, 291
Trevira Perform Moisture Control, 45, 46
Trevira profile, drying time, 46
Trinomax AQ®, 40
Triplet torque, 183
Trousers, 26
T-shirts, compression, 180
2XU, compression wearables, 179

U

UK consumer awareness, 331
Ultraviolet protection factor (UPF) values, 68
Umbro, 9
Unbroken patterns, 240
Upper body pressure, 252
Upper body surface change, control/ movement postures, 248
US-based Invista, 26
US Department of Agriculture (USDA), 327
US Sports and Fitness Industry Association, 16
UV protection, 68
UV rays, TiO_2 nanocoating, 82

V

Variability in measurement definition, 312
Various sports, requirements, 262
Vascular system, compression pressure gradient, 174

Velcro, 99
Venous blood return, 175
Venous diseases, 175
Very good, grade, 279
Viscose fibres, 43
    outlast phase change material, 160
Vitamin E, 181
Vitus laser scanner, 319
V-Stitcher, 311

W

Wales per centimetre (wpc), 265
Warmth, 26
Warp knitting, 92
    fabric, 222
    military mesh, 98
    spacer fabric, 98
    structures, 93
Waste management, 328
Water adsorption, mechanism of, 27
Water polo
    population, comparison, 314
    Tc2 NX16 body scanner, 314
    visual assessment, 315
Waterproof breathable fabrics, 75
Waterproof fabrics, 111
Waterproof tape, 133
Waterproof zippers, 131
Waterproofing, 54
Water repellence, 28
Water-resistant hard shell, 117, 119
Water-resistant jacket, 121
Water vapour absorbency, 43
Water vapour permeability, 43, 67, 73,
    77, 281
Water vapour resistance, 281
Water vapour transfer rates, 117
Wearable motherboard garment, 154
Weave density, 66
Weft knit structures, 92
Weft-knitted fabric, 233, 268
Weft knitting, 91

Wet-cling behaviour, 40
Whilst fit analysis, 246
Whilst garment fit assessment, 246
Wholesale, 12, 20
Wicking, 27, 46, 54, 278
Wireless protocols ANT™, 159
Wiring system, 156
Women's clothing, designing, 251
Wool, 25
Wool blends, 49
Wool Comfort Metre (WCM), 64
Wool fibre, 78
World Cup advertising campaign, 18
Woven basket-weave construction, 89
Woven fabrics, 88
    structure, 89, 268
    thickness, 282
Woven structures, twill and satin, 90
WRAP (Waste and Resources Action
    Programme), 328
Wrinkle recovery, 28

X

X-BIONIC®, 95
X-BIONIC® bib tight (X-BIONIC), 262

Y

Yamamoto, Yohji, 20
Yarn count, 66
Yarn density, 82
Yarn fineness, 284
Yarn formation, 31
Yarn linear density, on thermal comfort
    properties, 67
Yarn twist, 66
Yoga pants, 20
Yoryu fabrics, 66

Z

Zensah, compression wearables, 179, 188